COMING CLIMATE CRISIS?

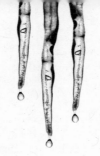

COMING CLIMATE CRISIS?

CONSIDER THE PAST,
BEWARE THE BIG FIX

CLAIRE L. PARKINSON

ROWMAN & LITTLEFIELD PUBLISHERS, INC.

Lanham • Boulder • New York • Toronto • Plymouth, UK

Published by Rowman & Littlefield Publishers, Inc.
A wholly owned subsidiary of
The Rowman & Littlefield Publishing Group, Inc.
4501 Forbes Boulevard, Suite 200, Lanham, Maryland 20706
http://www.rowman.com

10 Thornbury Road, Plymouth PL6 7PP, United Kingdom

British Library Cataloguing in Publication Information Available

The hardback edition of this book was previously cataloged by the
Library of Congress as follows:

Parkinson, Claire L.
Coming climate crisis? : consider the past,
beware the big fix / Claire L. Parkinson.
p. cm. — (Why of where)
Includes bibliographical references and index.
1. Climatic changes. 2. Climatic changes—Environmental aspects. I. Title.
QC981.8.C5P377 2010
551.6—dc22

2010005874

ISBN 978-0-7425-5615-7 (cloth : alk. paper)
ISBN 978-1-4422-1326-5 (pbk. : alk. paper)
ISBN 978-0-7425-6830-3 (electronic)

*To my sister Jean L. Harris, for her friendship,
and my uncle Thomas I. Parkinson, Jr.,
for his much appreciated
kindness and encouragement*

CONTENTS

PART V

Avoiding Paralysis despite Uncertainty

FOREWORD

THIS book discusses the potential pitfalls associated with geoengineered solutions to global climate change. Dr. Claire Parkinson is an accomplished, respected, and widely published scientist whose opinions on climate change and its solutions are well worth our attention. Dr. Parkinson shows great command of the scientific literature and presents her arguments in a thoughtful and respectful manner. Her research into the background of climate change science and the controversy surrounding it is very thorough. Although I do not share all of Dr. Parkinson's views, particularly regarding the scholarly contributions and motives of some climate contrarians, having known her since our time as graduate students at The Ohio State University, I applaud her for having the courage to express her opinion and concerns. Moreover, I am confident that she embarked on this book with no hidden agenda and that the views expressed therein reflect her genuine concern for humanity's collective future.

The primary message that Dr. Parkinson conveys in this book is that "global warming" is indeed under way, that human activities are an important contributor, and that undertaking large-scale geoengineering projects without better understanding the complexity of the climate system and without proper consideration for unintended consequences is an unwise course of action. Instead, she posits that all the world's populations must alter their relationships with carbon-based energy sources to ameliorate the effects of increasing atmospheric concentrations of radiatively active gases. Although she notes and applauds the recent rise of society's consciousness concerning the climate change issue, she takes the mainstream media and some scientists to task for oversimplifying and even sensationalizing new findings.

The book includes an excellent overview of 4.6 billion years of global change and in more recent times its interactions with human activities. She provides a primer on the record of atmospheric gases and aerosols, both natural and anthropogenic, and their effects on climate, the environment, and the biosphere. Dr. Parkinson presents an overview of the Earth system and its ever-changing nature to highlight that over its history Earth has undergone natural large-scale climate variations in response to changes in the Sun and the amount of volcanic dust in the atmosphere as well as internal variability of the coupled atmosphere-ocean system (e.g., the El Niño/Southern Oscillation) and the role of sea ice, on which she is an expert. She follows this with a discussion of the human forcing mechanisms, such as emissions of greenhouse gases and aerosols and land use changes that alter Earth's reflectivity, that are almost certainly inadvertently changing the Earth's climate. In fact, this is an example of a global-scale geoengineering experiment with unintended consequences. Dr. Parkinson discusses abrupt climate changes and highlights that paleoclimate records confirm that these occurred long before humans influenced the Earth system. Her message is that a prudent person would view this evidence as a wakeup call. The fact that such climate disruptions are naturally inherent in the system suggests the possibility that the system may also be vulnerable to abrupt changes in response to anthropogenic forcing(s).

Dr. Parkinson provides a thorough review of a number of smaller geoengineering projects that had negative or no overall lasting effects. She includes historical examples of instances where small-scale projects have been successful and alternatively where good intentions have gone awry and where solutions have backfired, resulting in different but in many cases worse problems. I share her concerns regarding many of the proposed solutions that she discusses, such as injection of sulfate aerosols into the stratosphere, oceanic iron fertilization, and placing reflectors in the atmosphere or in space. Although at present most of these should be considered unworkable, unrealistic, or even dangerous, it is prudent for scientists and engineers to identify and examine options that might be available in a crisis situation and to assess their relative risks.

The book includes cautionary considerations regarding our incomplete understanding of climate, but Dr. Parkinson speaks of avoiding paralysis despite uncertainty. Unlike the certainty about the shape of the Earth and its movement around the Sun, we will never attain 100% confidence about our projections of climate change because the climate system is so complex. Thus, we must look at the balance of the evidence and make our judgments accordingly. The best solutions will be those that offer win-win outcomes. For example, the future demand for energy will be enormous, and how we meet that demand will have huge implications for our national security as well as the security of the world. This poses a tremendous opportunity, as we must eventually transfer to alternative energy and the countries who lead in this technology will be the economies that thrive in the future. This offers a win-win situation by combating global warming and air pollution in a positive way. Should global warming prove to be a smaller threat than most of the climate science community anticipates (based on our best science), then the world will still be a better and, it is hoped, more peaceful place for all Earth's inhabitants.

Dr. Parkinson reminds us that experts are not always right. She includes examples from history demonstrating that societies suffer when scientific ideas are unfairly attacked or even suppressed because of religious dogma or political expediency (often the two are intertwined). There are examples of skepticism being carried to the point of absurdity. When ancient Greek scientists and philosophers formulated their theories about the spherical shape of our planet, skeptics asserted that the Earth was flat. This is understandable since this was a radical concept based on calculations and relatively limited observations. Similar objections were raised against the scientists who pioneered the idea of anthropogenic global warming. Skepticism of new theories and ideas is absolutely essential for the development of science and the accrual of observations, and other data should eventually point the way. However, today, as in times past, the motivations for skepticism may vary and not always be science driven. In fact, I prefer to call those whose motivations appear to extend well beyond scientific inquiry "contrarians." Dr. Parkinson does not introduce this term, and, in fact, the major issue I have with the book is that she ascribes nearly equivalent validity to the contributions of those in the climate change community who rely on the peer-review system to disseminate ideas and the smaller group of "climate skeptics" or contrarians. Many in the latter group are not climate scientists, and their ideas and work are often disseminated in white papers, editorials, privately funded foundation documents, blogs, and other attention-getting media outlets. Unfortunately, Dr. Parkinson does not alert the reader to this situation, although she does make it quite clear that they are in the minority. A closer look at the background and history of some skeptics would call their motivations into question.

Dr. Parkinson lists the efforts that have been taken by governments and industries to reverse the effects of climate change and other environmental hazards for which humans are partly or wholly responsible. She then lists the actions that large institutions and individuals can take to conserve materials and resources and reduce emissions. Most of her suggestions involve voluntary participation at both the institutional and individual level, that is, living by example, and I agree with her on the need to make important changes and advances in education. People must understand from an early age how their actions affect their environment and the climate and that if each person practices restraint in his or her consumption and use of resources, we can collectively make solid progress in dealing with our environmental problems. This must be done even if climate change falls short of the worst-case projections of the Intergovernmental Panel on Climate Change, simply because Earth's resources are finite.

Dr. Parkinson provides an excellent overview of Earth history, the factors affecting Earth's climate and environment, and how those have changed over time, both naturally and anthropogenically. The book discusses geoengineerng from an historical perspective and provides examples from which Dr. Parkinson concludes that extreme caution is essential. She highlights that uncertainty should not lead to paralysis and inaction but should propel governments and individuals to take personal responsibility for their interactions with the environment and for their energy consumption. Numerous examples for individual responses are provided in the book. In summary, I recommend this book to scientific colleagues, students, and general readers interested in issues associated with global climate change and potential actions to mitigate anticipated disruptions. In short, Dr. Parkinson recommends that humanity exhibit extreme caution when considering geoengineering projects, as she considers unintended negative consequences very probable. At this juncture in Earth history, the stakes could not be higher as 6.8 billion people currently rely on the Earth system for survival.

Lonnie G. Thompson, Distinguished University Professor
and Senior Research Scientist, The Ohio State University

PREFACE

SCIENCE has yielded many awe-inspiring results over the past few centuries, from topics as diverse as the structure of the universe, the existence of atomic particles, and the origin of life on Earth. In its totality, it is an incredibly exciting field of endeavor, aimed at deepening and improving our understandings of all aspects of the universe. However, science is continually evolving, and its results should never be viewed as final answers. Although exciting and impressive, the development of science is far from a monotonically upward progression. Time and time again, one or another major piece of the evolving edifice crumbles. On a whole, this has worked out well, as the crumbling leads scientists to construct a stronger edifice, continuing the circuitous journey to new advances.

Our understanding of Earth history provides an impressive example of the success that scientists have had in unraveling from very limited clues a basically coherent picture of a very nonobvious sequence, in this case 4.6 billion years of climate and biological evolution. At the same time, it also provides a poignant example that our understandings frequently change. Over the past several decades, dramatic revisions have come to our understandings of such factors as the movement of the continents, when and where life first arose, and how rapidly climate change can occur. Every year, stories appear in the news of instances in which new evidence revises our understanding of something significant in Earth history, for instance regarding whether the Earth was ever entirely ice and snow covered or regarding which earlier hominid species were direct ancestors of modern humans.

In most cases, the imperfections and changes in our understandings do not have major implications beyond the individual scientific fields. However, broader implications can arise when actions need to be taken based on the necessarily incomplete understandings at hand. We are now at such a point regarding climate change, as fears of an uncertain but anticipated upcoming climate crisis are leading scientists, politicians, and others to serious consideration of a wide range of policy options that will be expensive and might or might not accomplish the desired results.

Some of these options center around cutbacks in our emissions to the atmosphere and oceans in order to lessen the contributions that humans are inadvertently having on climate through these emissions. Fortunately, the consciousness of society at large has been raised concerning the issue of human impacts on the environment, and many people, both as individuals and within local, regional, national, and even international governmental and other organizations, are changing behaviors to help lessen these impacts. These actions can have multiple benefits both for our lifestyles and for the environment and hence should definitely be continued and enhanced.

However, many people claim that the types of actions currently being undertaken do not go nearly far enough to prevent a serious climate crisis that they anticipate will be upon us within the next few decades unless we make much more dramatic changes. Consequently, aggressive options are being put forward that go far beyond the actions already under way. To satisfy those who are most concerned, either we need to severely cut back on emissions, forcing us to forgo some of the tremendous and by now expected conveniences that the modern industrialized economy has provided, or we need to implement schemes to counteract the impacts of the emissions. It is the counteracting schemes that have become the most ominous. Some individuals fearful of the presumed upcoming climate crisis are advocating massive schemes to geoengineer the climate, with the express purpose of cooling the Earth and thereby avoiding an out-of-control global warming. These schemes involve such actions as placing reflective coverings over vast areas of deserts and pouring huge volumes of particulate matter into the upper atmosphere. Not only would these efforts address only part of the problem of human-induced changes, but they could create quite significant additional problems.

Since the 1970s, I have worked in the field of climate research, with a personal emphasis on the Arctic and Antarctic sea ice covers and their connections with the rest of the climate system. As a working climatologist, I have played a role, with many colleagues, in the evolving mainstream view of the seriousness of ongoing climate change, and I share the concern that humans may be causing major damage to our climate system. I am not, however, convinced that the mainstream view is necessarily correct in its many details; and I fear that accelerating apprehension regarding potential upcoming climate change could lead to decisions that might increase rather than decrease the human-induced problems. That is one of the primary reasons why I have written this book. Another is my dismay that polarization within the scientific community and elsewhere on the topic of future climate change and the role of humans in it is hindering the reasoned discussion that we should be having.

Among the personal factors helping to shape my views, the following have been particularly important: 1) prior to graduate school, my central focus was on theoretical mathematics, and the grounding in mathematics has made me acutely aware of the absence of proof regarding climate predictions, and 2) since my early twenties, I have read extensively in the history of science, and these readings have left me well aware that consensus views in science have on many occasions been overturned. Both of these factors make me less confident than many of my colleagues in the totality of the mainstream position on climate change. As a result, even though I share many of the mainstream positions, including a strong advocacy for restricting our carbon emissions, I have doubts about the extent of our ability to predict future climate, and these doubts have kept me from making the definitive, assertive statements about the future (e.g., statements that we are about to reach an irreversible tipping point or that we have less than 10 years to prevent a major catastrophe) that others have made.

Despite their drawbacks, my doubts also have some advantages. In addition to holding me back from making dogmatic statements that might not be justified, they help me to see and explain different sides of a complicated issue and to understand why many people are confused by the very contradictory statements made and widely broadcast about climate change. Further, because of the doubts, I can fully empathize with scientists and others labeled "skeptics" or "contrarians" or "climate change deniers" and can recognize the value of the points of view that they bring to the table.

In this book, I attempt to be fair to all sides as I discuss the issues of climate change and what we should or should not do regarding them. I accept that both the "alarmists" and the "skeptics" truly believe the basic positions they espouse; that is, I accept that the alarmists believe that humans are causing severe damage to the global climate and that this damage needs to be reversed or stemmed, and I accept that the skeptics believe that humans are not causing severe damage to the climate and that if we proceed to implement some of the proposals to stem our impacts, we could instead severely damage the global economy and the well-being of millions or even billions of people. I try to give the alarmists and the skeptics and all those in between a reasonable level of respect. Because of the severe potential consequences of some of the suggestions now being offered to alter the course of climate change, it has become more important than ever to have a reasoned discussion, not denigrating anyone and not allowing passions to lead to policy implementations that have not been adequately thought through.

In this book, I try to explain or at least provide a reasonable context for many of the confusing items regarding climate change. This includes confusion stemming from the fact that climate was changing well before humans had any impact, it includes confusion stemming from the large size of the Earth and hence the contrasting changes in different regions, it includes confusion stemming from incomplete data sets and imperfect models, and it includes confusion stemming from strident controversy and exaggerated sound bites.

Part I of the book deals largely with the nonhuman aspects of the Earth system. It is centered on a summary of Earth's long-term evolution that shows that the Earth has experienced climate change throughout its 4.6 billion years, with almost all those years coming prior to the emergence of human beings. During this long history, there were periods when the Earth was much warmer than today, had much more atmospheric carbon dioxide than today, and had much less ice and higher sea levels than today (and other periods when it was colder, with less atmospheric carbon dioxide and more ice). Part I also shows that the Earth has experienced widespread climate changes that were far more abrupt than anything yet attributed to humankind.

However, the fact that climate changed significantly prior to humans does not alter the fact that humans could now be causing major undesirable changes that we should attempt to reduce. Never before has the Earth had to accommodate a human population of 6.8 billion people. It is certainly within the realm of possibility that humans could cause serious climate change and that we could even trigger a disastrously (for humans) abrupt climate change. Hence, part II centers on the human factor: what impacts we have already had, what impacts are likely in the future, and why so many people are so concerned while others explicitly reject those concerns. It is undeniable that humans have had some impacts, both favorable and unfavorable, with substantial evidence that the unfavorable outweigh the favorable, but other impacts are much less certain.

Part III gets to the crux of why it has become important that nonscientists be made aware of the ins and outs of various climate issues. Because of the fears some people have regarding projected future climate (see part II) and the sense that global warming is at the core of the anticipated upcoming climate crisis, well-meaning individuals are advocating massive geoengineering to cool the planet. Some of the geoengineering schemes have the potential to backfire horrendously; they are particularly troublesome given all the uncertainties in our understandings of climate and climate change. Part III begins with a chapter warning, through various historical examples, that good intentions do not necessarily lead to good results (a fact that most people can readily attest to through examples in their personal lives). It then summarizes several of the most-discussed geoengineering schemes, with comments about the dangers involved and the potential for serious unintended consequences. Finally, it looks at earlier, smaller attempts to modify the environment and how those attempts have fared. The limited success achieved so far in small-scale attempted modifications should be cautionary.

Because of the importance of the issues at hand, the current inadequacies of our scientific understandings need to be made clear. Legislators and the general public are often awed by scientists to an unwarranted extent. Scientists are not all-knowing, and even a strong scientific consensus does not guarantee that a position is correct. Part IV provides cautions against being overly confident about what we think we know, starting with a historically based chapter illustrating that strong consensuses are sometimes proven wrong. This is followed by a chapter describing limitations of the computer models used to generate climate predictions. Although hugely valuable, computer climate modeling is still a very young science, having started only in the 1950s, and is by no means sufficiently mature that it can reliably give accurate predictions of climate changes decades into the future. The simulated predictions are probably our best guesses of what would happen under the assumptions inserted into the models, and as such they should be taken seriously; even so, the predictions remain only educated guesses, and it is vitally important to keep that in mind. The final chapter in part IV turns from the technicalities of computer modeling to relevant social considerations, describing some of the social pressures that contribute to making the current consensus view on climate change sometimes appear stronger than perhaps it is.

By the end of part IV, it should be clear that I am not entirely certain whether the fearful predictions of a coming climate crisis are correct; they could be seriously overestimating the coming difficulties, or they could alternatively be seriously underestimating them. However, despite all the uncertainties, the global community is left with the unnerving reality that sophisticated computer models and many climate scientists are predicting dire consequences if we fail to make major adjustments, either severely limiting our emissions or counteracting them. It would be preferable if we could act based on perfect knowledge of the consequences of each alternative, but we cannot wait until we have the means for perfect predictions, as that means will never be available, at least not in the foreseeable future. Furthermore, as indicated in part II, the predicted climate crisis is not the only problem being produced by human activities (or even the only problem being produced by our voluminous emissions to the atmosphere and oceans). This brings up another primary difficulty with many of the massive geoengineering schemes to cool the planet: not only are they fraught with the potential of increasing rather than decreasing human-induced damage (see part III), but they address only a portion of the problems caused by human emissions and not other difficulties that might be just as important, such as increased ocean acidification, one of the many major concerns for life in the world's oceans (see part II). Hence, for us to deal appropriately with human impacts, we need to go to the core of the problem. This includes reducing emissions, which is a central topic of part V. This last of the book's five parts makes it clear that despite my doubts about the soundness of the science and the predictions, I am decidedly not advocating a continuation of the status quo. Even though I am not certain that we are headed toward a climate crisis as severe as some people predict, I am certain that our emissions are having some negative impacts, for instance on human health and ocean acidification, and that they could conceivably also be leading to a climate crisis. For me, this is quite enough to advocate reducing those emissions where reasonable and, even more strongly, to warn against implementing any geoengineering scheme that purposefully adds further emissions to the atmosphere or oceans, with the frightening potential of considerably compounding the problems.

The first chapter of part V describes ways in which individuals, corporate groupings, and government entities have lessened and can further lessen their damaging environmental and climate impacts. It highlights plane travel as a primary culprit in human-caused emissions to the atmosphere, with a clear implication that the people who are most vocal in sounding the alarm about a coming climate crisis should conscientiously limit their plane-based travels, something that most of them have not yet done. The chapter also goes into the difficulties of cutting emissions back as much as some people suggest is needed while offering some encouragement that not all reductions will be onerous and some hope that viable energy solutions will be found. This is followed by a concluding chapter summarizing the main thrust of the book and pleading that we not compound our problems by undertaking massive geoengineering with uncertain but potentially disastrous consequences.

As a climate scientist generally aligned with the consensus view on climate change but with many serious doubts, I can appreciate the confusion that exists in the minds of thoughtful nonscientists who are trying to make sense of the countering views they encounter in media reports. I hope that by articulating some of the key contentious issues in a nonadversarial way, this book can help lessen the confusion and improve the tone of the discourse on climate change and related issues.

A second hope is that the book will be effective in convincing its readers that we do not yet know enough to warrant undertaking massive geoengineering that could further damage the global environment. I do not oppose all geoengineering possibilities, but some of the proposed schemes have the potential for appalling damage, and our ability to predict the consequences of these schemes is overwhelmingly too primitive to risk their implementation. We must be careful not to compound the damage we have already done to the environment by doing even greater damage in the attempt to correct our past and ongoing impacts. Our scientific understanding is much too incomplete to risk major further damage to the Earth system by implementing massive geoengineering based on imperfect computer projections. If we do not exercise caution, good intentions could backfire in horrendous and unanticipated ways. It would be far better for us to limit our emissions, and in doing so, we just might find that cutting back on our excesses improves rather than reduces the quality of our lives as well as the quality of the environment.

ACKNOWLEDGMENTS

First and foremost, my thanks go to the series editor Professor George Demko of Dartmouth College for his enthusiastic support of the book from its inception and for oral and written comments on several versions of the manuscript. Next, I wholeheartedly thank John Walsh of the University of Illinois and the University of Alaska, Fairbanks; Lonnie Thompson of the Ohio State University; and Robert Gurney of the University of Reading for each carefully going through the entire draft manuscript and providing written review comments that helped me in revising the text into an improved final product. I also thank Thorsten Markus and Franco Einaudi of NASA Goddard Space Flight Center, Warren Washington of the National Center for Atmospheric Research, and my mother, Virginia Parkinson, for helpful conversations regarding the book. Although my father died in 1992, his influence remains as important to me as anyone's, and I thank him for the ideals that he instilled. If he were alive, he more than anyone else would likely support my willingness to voice the concerns I express in this book, despite how difficult it is for me to do so. Finally, I thank Susan McEachern from Rowman & Littlefield Publishers for her support and enthusiasm for the book and for facilitating the passage of the book through the numerous stages of the publication process; and I thank Jehanne Schweitzer, also of Rowman & Littlefield Publishers, for her much-appreciated assistance once the book reached the production stage.

PART I

The Earth System and
Its Ever-Changing Nature

CHAPTER 1

Introduction

The issue of what (if anything) to do about climate change has become a topic of considerable debate. The issue is filled with confusing aspects, but the stakes are too high to ignore. This book aims to explain why there is so much confusion and to warn against well-intentioned but potentially disastrous extreme actions currently being proposed specifically to adjust future climate. First, however, there is a need to lay out some of the basics regarding the relevant aspects of the Earth system.

CLIMATE change has been much in the news in recent decades and for good reason, as the climate changes that are occurring and are projected to occur are impacting both human societies and whole ecosystems and are affected—although to an unknown extent—by human activities.

Some scientists, politicians, and members of the media have predicted unmitigated climate disasters within decades if humans continue "business as usual," especially regarding emissions to the atmosphere that are contributing to a global warming with numerous undesirable propagating impacts. Others equally adamant claim that the predictions of upcoming climate disasters are fundamentally flawed and that the real disaster will occur if we make the adjustments called for by those most concerned about climate change and thereby destroy the underpinnings of the global economy and add considerably to human poverty and misery. Both sides can be passionate and at times convincing in their arguments.

By now the topic has gone far beyond lively debate and into the realm of determined actions, for instance by individuals and organizations voluntarily limiting their emissions and more carefully disposing their waste products and by local, state, and national governments through legislation. Much of this activity can be of value regardless of whether the climate predictions are accurate, as it can lessen pollution of our air and waterways and encourage conservation. These activities most definitely should be continued and enhanced. In recent years, however, the fear of further warming and its consequences has grown so great that people are seriously considering a very different level of activity, termed "geoengineering," that proposes massive efforts to alter the Earth's future climate purposely in a way to avoid the predicted climate disasters. These geoengineering schemes include pouring particulate matter into the upper atmosphere, covering large areas of oceans or deserts with reflective surfaces, and other extremely expensive projects with potentially serious unintended consequences (described further in chapter 7).

Yet despite the research and activity, most people remain at least somewhat confused by all the discussion and vociferous disagreement about climate change. Given what is at stake, especially with geoengineering schemes that have the potential of backfiring horrendously, this book is an attempt to explain and perhaps lessen the confusion regarding the climate change debate and to warn of the danger that in the attempt to "solve" the perceived climate change problem, humans could unwittingly precipitate even greater disasters. I try throughout the book to respect both those in the majority and those in the minority on climate change issues, in the firm belief that each side has important points to make.

REASONS FOR THE CONFUSION

Six primary reasons go a long ways toward explaining why there is so much confusion regarding recent and future climate change and the role of humans in it.

First, the Earth system is very large, and changes occurring in it vary greatly from location to location and from time to time. Hence, even if a dangerously serious problem with global warming exists, there are bound to be solid facts showing that one location or another is experiencing cooling rather than warming and, relatedly, that one region or another is experiencing ice expansion rather than ice retreat.

Second, the Earth has undergone a very long evolution, with vastly different conditions at different periods during that evolution. There were indeed long periods when the Earth system was much hotter than today and long periods when the Earth was much colder than today. There were even periods with climate changing much more rapidly than today. None of this, however, affects how serious the damage might be from today's climate changes.

Third, our data sets are painfully incomplete regarding both the past and the ongoing climate changes. Eventual more complete data sets might force major revisions in our current understandings.

Fourth, the Earth system has many intertwined elements, changes in each of which can affect the others, and scientists are still learning about many of these interconnections. Additional major revisions could come as advances to our understandings continue.

Fifth, the models used to predict the future have numerous limitations despite the tremendous progress made over the past several decades, since the invention of electronic computers and the advent of computer modeling.

Sixth, the issues have become mired in strident controversy, with both sides at times presenting only one side of the issues and not giving the reader or listener even close to a balanced picture. Many pressures on scientists and reporters add to the difficulty of getting a balanced picture out to the public.

No book can be devoid of the author's individual perspective and influences, but this book at least attempts to present a balanced picture and to explain and lessen the confusion in each of the six arenas just mentioned. However, before embarking on a tour of the Earth's past 4.6 billion years that will set the scene for exploring the current situation, it will be useful to go over some basics. Hence, most of the rest of this introductory chapter provides a short primer on relevant aspects of the Earth system.

THE EARTH SYSTEM

The Earth system consists of the solid Earth and everything within it, the oceans and other water features, the atmosphere, and all the life forms living in and on the Earth and in its atmosphere. This is a highly interconnected system, with each component influencing and being influenced by the others.

The solid Earth is approximately spherical, with a radius of about 6,400 kilometers (about 4,000 miles). The highest point on its surface is the peak of Mount Everest, at 8.85 kilometers above sea level, and the lowest point is in the Pacific Ocean's Marianas Trench, at 10.9 kilometers below sea level. Oceans cover approximately 71% of the Earth's surface area, the remaining 29% being land, with some of the ocean and some of the land at any time being covered by ice. Snugly surrounding the solid Earth and its partial covering of water and ice is the slender halo consisting of the Earth's atmosphere. The atmosphere, maintained in place by the Earth's gravity, extends outward to about 550 kilometers, getting considerably less dense the higher one goes, with no sharp outer boundary but instead just fewer and fewer air molecules in the higher reaches. By the altitude of the top of Mount Everest, the air has thinned sufficiently that breathing becomes extremely difficult. That, plus the lack of an atmosphere on the Moon, made the first astronaut photographs showing the slender halo surrounding the Earth in the 1960s all the more powerful in instilling concern about the finite atmosphere that is so essential to our lives and to the global ecosystem.

The Earth's life forms not only depend on the Earth's atmosphere and water but also impact them in significant ways, as will become clear in chapter 2 for the nonhuman life forms and in chapter 4 for humans. All of Earth's species together constitute the Earth's "biosphere," whereas all the ice, snow, and permafrost (permanently frozen ground) constitute the "cryosphere," all the water constitutes the "hydrosphere," and the solid Earth is the "geosphere."

The Earth resides 150 million kilometers (93 million miles) away from its major energy source, the Sun, and it orbits around the Sun once every year, with a tilt to its axis that creates the opposing seasons in the Northern and Southern hemispheres. The Sun's energy powers the atmosphere and oceans as well as providing the energy for photosynthesis, essential to the world's ecosystems.

THE EARTH'S RADIATION BUDGET

Warming and cooling of the Earth system are tied closely to imbalances between how much energy the system receives and how much it gives off. Hence, to understand either global warming or global cooling requires understanding something about the energy flows into and out of the Earth system.

Energy coming into the Earth system is overwhelmingly radiation from the Sun, termed "solar radiation," whereas energy leaving the Earth system includes both reflected solar radiation and radiation emitted outward by the Earth system, the latter termed "terrestrial radiation." Solar and terrestrial radiation are distinctly different because of the contrasting temperatures between the two sources. Basically, the hotter a body is, the more radiation it has the potential of emitting and the shorter the wavelength of its peak emission. At the very hot temperatures of its emitting surface (about 6,000°C), the Sun's emission peaks at visible wavelengths of the electromagnetic spectrum, whereas at the much cooler temperatures of the Earth, the Earth emits predominantly at infrared wavelengths.

Natural mechanisms exist to keep the amount of radiation leaving the Earth system roughly in balance with the amount of radiation it receives. If, for instance, the Earth is sending out (by reflection and emission) more radiation than it is receiving, it will have a tendency to cool down, thereby causing it to emit less because of being cooler; by emitting less, the Earth will return to a rough equilibrium in its radiation budget, although at a cooler temperature than when the sequence began.

Similarly, if the Earth is sending out less radiation than it is receiving, it will have a tendency to warm up because of the radiation gain and hence will emit more because of being hotter and will thereby again return to a rough equilibrium although this time at a higher temperature. Hence, concern about global warming becomes concern about not enough radiation leaving the Earth system to balance the radiation received by the Earth system from the Sun.

The amount of radiation leaving the Earth system can change either because the amount of reflected solar radiation changes or because the amount of radiation emitted outward by the Earth system changes. Humans can influence—wittingly or unwittingly—either or both of these outgoing radiation amounts; both of these amounts also have varied significantly over time because of natural changes in the Earth system.

Turning first to the reflected solar radiation, currently about 31% of the radiation the Earth system receives from the Sun is reflected away by clouds and other matter in the atmosphere and by the Earth's various surfaces (ocean, ice, land, and whatever is on the ocean, ice, and land). Different surfaces vary substantially in what percentage of the radiation incident on them they reflect, and hence the relative expanses of different surface types can impact the reflectivity of the Earth system as a whole. For any surface, the percentage of incident solar radiation reflected is called the "albedo"; and for the Earth system as a whole, the percentage reflected (currently about 31%) is called the "planetary albedo." Approximate percentage ranges for the albedos of some common natural surfaces include the following: 2% to 15% for ice-free ocean and other liquid water bodies (close to 2% if the Sun is high in the sky but perhaps well over 15% if the Sun is low in the sky), approximately 10% for tundra, approximately 17% for bare soil, 9% to 20% for forests, approximately 25% for green grass, 30% to 40% for sand-based deserts, 35% to 70% for old and dirty snow, 50% to 80% for ice, and 80% to 97% for a fresh snow cover. The albedos for two common manmade surfaces are 4% to 12% for asphalt (close to 4% for fresh asphalt) and approximately 55% for concrete.

Consider the consequences that the different surface albedos can have for the Earth's radiation budget and hence for warming and cooling of the Earth. If a large region with a high albedo surface is replaced by a low-albedo surface (e.g., if an ice-covered ocean region loses its ice cover, resulting in the replacement of a high-albedo ice surface by a low-albedo ice-free ocean surface), then less radiation will get reflected back to space, tending to warm the system. Pouring of fresh asphalt is particularly troublesome in this regard because fresh asphalt reflects so little solar radiation back to space and typically is replacing a more reflective surface. Under some circumstances, a simple paint job could solve the reflectivity problem, as a layer of white paint can convert a highly absorptive black surface into a highly reflective white surface, something well demonstrated by walking barefoot on an asphalt parking lot on a hot summer day, when the feet can burn on the black asphalt but be comfortable on the white parking stripes. What to paint, however, needs to be considered carefully, for instance to avoid blinding drivers by a reflective white road. Knowledge of the albedo effect can and should be used effectively in various construction projects (see chapter 12), although several proposed geoengineering schemes go to extremes by proposing installation of gigantic and highly disruptive reflective surfaces (chapter 7).

Other changes in the radiation budget arise through changes in the amount of radiation emitted outward by the Earth system. One key reason why changes occur in the outward-going radiation involves changes in the composition of the atmosphere, as certain gases tend to hinder terrestrial radiation from passing through and leaving the Earth system. In fact, the human-induced changes in atmospheric composition are at the core of current concerns over human-induced (or "anthropogenic") climate change, so we turn now to the atmosphere and the reasons why changes in the atmosphere matter to the radiation budget and the Earth's temperature.

THE ATMOSPHERE

Atmospheric Structure

In the lower portion of the atmosphere, not only does the density decrease with height, as already mentioned, but the temperature does also, as is generally quite obvious to anyone climbing a sizable mountain. This general tendency for decreasing temperature with increasing height tends to be true throughout the lower atmosphere, up to an altitude ranging from about 8 kilometers to about 17 kilometers, depending on the specific location and specific time. This lower portion of the atmosphere, with temperatures generally decreasing with height, is called the "troposphere," and the top of the troposphere is called the "tropopause." Almost all the weather activity we normally think of takes place in the troposphere, and even the topmost point of the highest mountain on Earth, Mount Everest at 8.85 kilometers, is generally well within the troposphere. Jet aircraft typically fly in the upper troposphere, with the air within the plane controlled for passenger well-being to approximate the air density at ground level.

Immediately above the troposphere, temperatures no longer decrease with height, instead tending to stay constant or increase with height up to an altitude of about 50 kilometers. This region from the top of the troposphere up to about 50 kilometers is called the "stratosphere" and has received particular attention in recent decades because of containing much of the atmosphere's ozone. Above the stratosphere are much sparser regions of the atmosphere called, in order, the "mesosphere" (immediately above the stratosphere), the "ionosphere," and the "thermosphere," but those do not enter into the discussion in this book and so are not mentioned further other than to say that the International Space Station orbits the Earth at an altitude of about 380 kilometers, which places it within the thermosphere, in the outer reaches of the Earth's atmosphere.

Overall Gas Composition of the Atmosphere

The Earth's atmosphere is dominated by gases but also contains liquid water (e.g., rain), solid water (e.g., in clouds and precipitating as snow or hail), and particulate matter. In terms of gas composition, the atmosphere is composed predominantly of nitrogen and oxygen. The exact percentages are confused somewhat by the presence of water vapor (H_2O), which is distributed very unevenly around the globe and constitutes up to 4% of the atmosphere's volume. Hence, the percentages for the rest of the gases in the atmosphere are generally expressed as percentages of the "dry" atmosphere, meaning the atmosphere exclusive of its water vapor content. Following this convention, nitrogen (N_2) constitutes 78% of the dry atmosphere, and oxygen (O_2) constitutes 21%. The remaining 1% of the dry atmosphere consists of numerous so-called trace gases, meaning gases that occur only in trace amounts. Water vapor is also a trace gas because it too constitutes very little of the total atmosphere.

Of the trace gases other than water vapor, by far the most abundant is argon (A), constituting about 0.93% of the dry atmosphere. Next in abundance is carbon dioxide, at about 0.04%, then, in order of abundance, neon (Ne), helium (He), methane (CH_4), krypton (Kr), and hydrogen (H_2), all at far lesser percentages than carbon dioxide. There are also a host of additional trace gases, at even smaller abundances, and these include carbon monoxide (CO), sulfur dioxide (SO_2), nitrogen dioxide (NO_2), nitrous oxide (N_2O), sulfur hexafluoride (SF_6), and ozone (O_3).

Even though nitrogen and oxygen dominate the atmosphere, they rarely enter into discussions of climate change for the very fortunate reason that they are not exhibiting any significant changes. If oxygen were to change significantly, this could be disastrous, as too much oxygen could fuel hard-to-stop fires around the world and too little oxygen could be deadly to humans and other animals. Fortunately, nitrogen and oxygen are holding steady. This is not true, however, of several of the trace gases.

Even though the trace gases constitute a very small percentage of the atmosphere, their presence and abundance matter a great deal. A key reason is that several trace gases allow radiation from the Sun to pass readily through to the Earth's surface but absorb a substantial portion of the upwelling Earth radiation that otherwise would escape to outer space. This combination helps to retain radiation within the Earth/atmosphere system and contributes to warming the atmosphere. The trace gases with these properties are called "greenhouse gases," and the result of keeping the atmosphere warmer than it would otherwise be is called the "greenhouse effect." Today, the Earth's major greenhouse gas is water vapor. The next most abundant greenhouse gas is carbon dioxide, and other greenhouse gases of importance include methane, nitrous oxide, ozone, and sulfur hexafluoride.

Over the course of Earth's history, the greenhouse effect has been a major reason why the Earth's average temperature has generally been comfortably well above freezing (see chapter 2). Calculations suggest that without the greenhouse effect, the Earth's average surface temperature would be −18°C (about 0°F) versus today's 15°C (about 60°F).[1] Although many people now fear that the greenhouse effect is causing current global temperatures to rise too high and too fast, no reasonable person would want the average surface temperature to be down at 18°C below freezing. Quite counter to the negative connotations often given in discussions of the greenhouse effect, the natural greenhouse effect has been not only beneficial but also a key factor in making the Earth a habitable planet for humans and other life forms.

The reason for the current concerns regarding the greenhouse effect is that the beneficial natural greenhouse effect is now contending with unnatural additions as a result of human activities. Specifically, because of manufacturing and construction, electricity, use of such vehicles as aircraft and cars, artificial heating and cooling, agriculture, land use change, and additional fuel combustion, humans are inserting large quantities of several important greenhouse gases into the atmosphere, described in greater detail in chapter 4. These additions are increasing the Earth's natural greenhouse effect, with rising concern that the resulting warming might have devastating consequences (chapter 5).

Ozone

Although the trace gas ozone constitutes an extremely small percentage of the Earth's atmosphere, it plays a vital role for life at the Earth's surface. Specifically, ozone in the stratosphere absorbs much of the ultraviolet radiation that reaches the Earth/atmosphere system in the stream of ultraviolet, visible, and infrared radiation coming to us from the Sun. The ultraviolet absorption by the stratospheric ozone considerably heats the upper atmosphere and, more important, protects humans and other life forms at the Earth's surface from damaging overexposure to ultraviolet radiation. In the case of humans, we are protected from what would otherwise be much more frequent occurrences of serious cases of sunburn, skin cancer, immune deficiencies, and various eye problems. Hence, ozone in the upper atmosphere is highly desirable.

Because of its importance, the region in the stratosphere with the most ozone has been given its own name, alternatively the "stratospheric ozone layer" or simply the "ozone layer." This so-called ozone layer is not a layer of pure ozone but instead a layer of the atmosphere that has higher ozone concentrations than most of the rest of the atmosphere.

The reason the ozone layer has garnered considerable attention and interest is the justified concern that this protective layer could be endangered. More specifically, measurements as of the 1980s have revealed damage to the ozone layer, especially over the Antarctic region. This damage is tied directly to human activities, and hence discussion of the ozone layer reappears in this book, in much greater detail, in connection with human impacts.

In sharp contrast to the situation in the upper atmosphere, in the lower atmosphere ozone is not desirable, as there it has unfavorable impacts as a greenhouse gas and as a pollutant. Although the ozone molecule (O_3) remains the same wherever it is, the contrast in its desirability has led to occasional use of the terms "good ozone" for ozone in the upper atmosphere and "bad ozone" for ozone in the lower atmosphere.

Aerosols

In addition to gases, the atmosphere also contains airborne particles called "aerosols." Like the greenhouse gases, the airborne particulate matter arises from both natural and anthropogenic (i.e., human) sources. Natural sources include dust, sea salt, organic particles, sulfate particles, and volcanic emissions. Humans affect some of the natural sources, for instance through land use changes or stirring dust into the atmosphere when plowing the land or driving along a dusty road. Humans also insert additional particulate matter into the atmosphere, for instance through fossil fuel burning, biomass burning, and use of aerosol sprays.

Aerosols are important for several reasons. At ground level, they contribute to pollution of the air breathed in by humans and other life forms. Throughout the troposphere, they can serve as "cloud condensation nuclei," meaning that aerosol particles can facilitate the condensation of water vapor into liquid water droplets, a process critical to cloud formation and precipitation. They also impact atmospheric temperatures, although their temperature impact is more complicated than the relatively straightforward warming impact of the greenhouse gases.

Like the greenhouse gases, some aerosols provide a warming effect, doing so in their case by absorbing solar radiation, but the aerosols also reflect solar radiation, which provides a cooling effect. Different aerosols behave differently, with soot—a black substance composed largely of carbon and formed from the incomplete combustion of fuels such as coal, oil, and wood—in particular being highly absorptive and fostering warming, while many other aerosols contribute a cooling effect through the reflection of solar radiation back to space. It is widely felt that, on a whole, the net effect of aerosols on atmospheric temperatures is a cooling. This net cooling effect becomes important later in this book, in several contexts, regarding both unintentional human impacts on climate and proposed intentional geoengineering efforts.

In addition to impacting temperatures by reducing the amount of solar radiation that reaches the Earth's surface, aerosols also reduce the amount of sunlight visible to our eyes. The term used for the decreased light caused by more aerosols and clouds in the atmosphere is "solar dimming," and the corresponding term for increased light under conditions of decreasing aerosols and clouds is "solar brightening."

Aerosols have numerous other effects tying them to the rest of the Earth system. For instance, natural mineral dust from the Sahara Desert is routinely transported westward across the Atlantic Ocean in the lower atmosphere, providing a source of iron that fertilizes phytoplankton growth in the Atlantic and a source of nutrients for the soils of the South American rain forests.

Of all the sources of particulate matter to the atmosphere, none has been more important to the evolution of the Earth system than volcanic eruptions (see chapter 2). However, because volcanic emissions contain a vast mixture of gaseous, liquid, and solid materials, they warrant a separate section.

Volcanic Emissions and Their Impact on Temperature

Today it is common to label violent volcanic eruptions as natural disasters. They are certainly natural and can be quite disastrous in terms of the death tolls and destruction they cause in the short term. However, the landscape damage suffered immediately after an eruption is often followed by a very favorable boon to agriculture, as soils fertilized by the nutrients provided by volcanic ash can be unusually nourishing and productive. Similarly, the unpleasant pollutant aspects of volcanic emissions to the atmosphere are countered by the fact that volcanic emissions have been a key factor responsible for the robust composition of the Earth's atmosphere and its evolution (see chapter 2). Without volcanoes, our atmosphere would have evolved very differently from the way it did, perhaps never becoming amenable to the major life forms we now have.

In studies of weather and climate, volcanic emissions are particularly noteworthy because of the cooling effect they have on tropospheric temperatures. This effect was recognized as early as 1784 by the American statesman and scientist Benjamin Franklin when, while ambassador to France, he explicitly mentioned the 1783 eruption of Iceland's Laki volcano as a possible cause for the unusually cold winter that he and others endured in Europe in 1783–1784. Although this case was regional, three decades later the volcano/weather connection gained many more adherents and a recognition of much greater geographic impact when unusually cold weather was experienced in much of the Northern Hemisphere in the two years following the gigantic April 10–11, 1815, eruption of Mount Tambora in Indonesia. In addition to killing tens of thousands of people, the Tambora eruption injected an estimated 150 cubic kilometers of material into the atmosphere. This material blocked and reflected solar radiation, limiting the amount of radiation that penetrated through to the Earth's surface. The resulting cooling, at an estimated 0.4°C to 0.7°C for the Northern Hemisphere as a whole, was enough that 1816 was declared "the year without a summer" in both Europe and North America, far distant from the point of the eruption itself.[2]

The 1815 Tambora eruption was much larger than any eruption that has occurred since that time,* but even much smaller eruptions can have noticeable temperature impacts (as well as cause major death tolls and other destruction). When the Philippines' Mount Pinatubo erupted on June 15–16, 1991, it injected an estimated 3 cubic kilometers of material into the atmosphere,[3] much of it entering the stratosphere, and climate scientists quickly forecast a cooling impact on global tropospheric temperatures.[4] Indeed, the Mount Pinatubo eruption is widely considered the cause for cooler global tropospheric temperatures in 1992 and 1993 than had been forecast prior to the eruption or would be expected from the other factors known to influence global temperatures.[5] At the same time that tropospheric temperatures decreased, stratospheric temperatures increased[6] as a result of the

---

* In fact, the 1815 Tambora eruption is the largest known eruption since the beginning of written history. Still, it was minor compared to some prehistoric eruptions, such as the eruption of the island of Toba in Sumatra approximately 71,000 years ago, estimated to have been almost 100 times larger (Mayewski and White 2002).

absorption of radiation by the massive amount of volcanic material injected into the stratosphere. Thus, the stratosphere warmed because of increased absorption of radiation in the stratosphere, while the troposphere cooled because of less solar radiation reaching it.

With the temperature impact of volcanic eruptions deriving from the fact that the volcanic emissions absorb and reflect solar radiation, the direct impact lasts only as long as the emissions remain in the atmosphere. Any particulate matter that makes it only into the troposphere descends fairly quickly, but the emissions that reach the stratosphere can remain aloft for 1 to 3 years or sometimes somewhat longer. Hence, the large-scale weather impact from a volcanic eruption tends to last about 1 to 3 years and to be greatest for volcanoes that explode vertically with great force, injecting massive amounts of material into the stratosphere, and do so in the low latitudes, from where the spatial spread of the emissions tends to be greatest. Eruptions whose emissions instead explode out horizontally tend to do greater immediate local damage to the surrounding landscape but have less impact on global weather and climate.

THE OCEANS

The oceans cover approximately 71% of the surface area of the Earth and contain approximately 1.4 billion cubic kilometers of water. This constitutes greater than 96% of the Earth's total known water content and provides a living area for ocean-based life that exceeds by many times the living area available on Earth for land-based life. The oceans serve as the habitat for hundreds of thousands of species, including vibrant ecosystems in the depths of the ocean far from the reach of solar radiation as well as the more familiar species and ecosystems closer to the surface.

An important aspect of ocean water is that, in addition to pure water (H_2O), it contains dissolved materials collectively contributing to its "salinity" and typically constituting about 35 parts per thousand of its mass. Over 85% of this natural dissolved material consists of chloride and sodium ions, with the remainder largely being sulfate, magnesium, calcium, potassium, bicarbonate, and bromide ions. The salt content is more dense than the H_2O itself, and therefore higher salinity tends toward higher density. Salinity is one of two main factors affecting surface seawater density, the other being temperature. The temperature effect is such that a given sample of seawater becomes more dense as the temperature drops. Seawater density is of particular interest because of its role in ocean circulation. Dense water overlying less dense water forms an unstable situation, leading to sinking of the dense water and propagating impacts on further water flows.

Of the many important roles played by the oceans in the Earth system, two of the most important ones for the subject matter of this book are their role in the chemical and other exchanges between the ocean and atmosphere and their role in distributing heat around the planet.

Ocean/Atmosphere Exchanges

The majority of the water in the atmosphere arrives there through evaporation from the oceans. Some of this water returns from the atmosphere to the oceans directly through precipitation, but some instead returns by a much more circuitous route, traveling in the atmosphere to locations over land before precipitating and getting transferred from the atmosphere to the ground, vegetation, glaciers, or other land-based entities. Almost all the water precipitating over land eventually returns to the oceans, but the time scale on which this occurs varies greatly. Some of the water returns quickly to the oceans through rivers or streams, and some instead takes decades, centuries, or even millennia before returning after extended stays in a glacier, groundwater, or other land-based feature. All these water flows are part of the Earth's "water cycle."

Water might be the best known of the transfers between ocean and atmosphere, but it is only one among many. As winds blow on the water and generate waves, they are transferring momentum from the atmosphere to the ocean. When steam is seen rising from the ocean or other water body into a chilly atmosphere, heat, as well as water, is being transferred from the ocean to the atmosphere. Conversely, when the air is warmer than the water, heat gets transferred from the atmosphere to the ocean.

In addition to water, other chemical transfers between the ocean and atmosphere include oxygen (O_2), carbon dioxide (CO_2), dimethylsulfide (CH_3SCH_3, more commonly abbreviated as DMS), and chlorofluorocarbons (CFCs). Depending on time and location, major flows of O_2 and CO_2 can occur either upward from the ocean or downward into the ocean, but the flows of DMS are predominantly from the ocean to the atmosphere, and the flows of CFCs are predominantly from the atmosphere to the ocean. DMS is produced by marine algae, and CFCs are produced artificially by humans and generally released to the atmosphere rather than the ocean. The transfers of CO_2 from the atmosphere to the ocean and similarly those from the atmosphere to the land have helped reduce the buildup of CO_2 in the atmosphere and are important factors in the consideration of atmospheric CO_2 increases. Changes in the capacity of the ocean and/or land to serve as sinks for atmospheric CO_2 could have important impacts on the magnitude of future global warming.

Heat Redistribution: Ocean Circulation and the Ocean Conveyor

Because of the geometry of the Earth/Sun relationship, most of the solar radiation entering the Earth system enters in the low-latitude regions. Both the atmosphere and the oceans redistribute a portion of that energy. In the case of the oceans, probably the best-known heat transfer is the transfer northeastward in the Gulf Stream, bringing warm waters to the vicinity of Great Britain and transferring heat to the atmosphere in that region because of the relative warmth there of the ocean versus the atmosphere. This leads to northern European climates being much milder than they would otherwise be.

The Gulf Stream is one segment of a far larger ocean circulation. "Ocean conveyor" is an effective, memorable term brought into use in the late twentieth century by geochemist Wally Broecker of Columbia University's Lamont-Doherty Earth Observatory to refer to the very large-scale general circulation pattern of the global oceans. In this pattern, surface and near-surface waters flow northeastward across the North Atlantic in the Gulf Stream, sink to the depths of the ocean in the northern North Atlantic, flow southward at great depth from the northern North Atlantic through the North and South Atlantic to the Southern Ocean, rise and then sink again in the Southern Ocean, proceed at great depth to the Indian and Pacific oceans, then rise to the upper levels of the ocean and return as upper-level waters to the South Atlantic, from there heading northward to the North Atlantic, from whence the sequence repeats.[7] Influenced in large part by the temperature and salinity of the water, the ocean conveyor is also called the "thermohaline circulation." Alternatively, a third term is the "meridional overturning circulation."[8]

The reason that temperature and salinity are so important to the ocean circulation is that they are the two primary factors affecting seawater density. Horizontal density differences drive water from the high-density toward the low-density regions, compounded by land boundaries and the Earth's rotation. Vertically, ocean waters are stable where the denser waters underlie the less dense waters, which is the case for most of the ocean most of the time. However, there are circumstances in which surface waters become denser than the waters below them, and in those circumstances the dense surface waters will tend to sink, producing overturning in the water column and helping to drive ocean circulation. This happens largely in just one narrow and two broad regions of the global oceans. The narrow region is where very salty and hence dense water from the Mediterranean enters the Atlantic through the Strait of Gibraltar. The broader regions are in the high latitudes of the northern North Atlantic and the Southern Ocean. In both of these broad regions, the surface waters can become dense by loss of heat to the atmosphere, hence becoming colder, or by addition of salt as sea ice forms and ages, releasing salt content to the waters below. The resulting overturning of waters in the northern North Atlantic and Southern Ocean are primary processes in the ocean conveyor circulation. Wherever it occurs, sinking of surface waters to the depths of the ocean is termed "deep-water formation." No water is actually formed during deep-water formation, but water that had been at or near the ocean's upper surface gets transferred to great depths.

The ocean conveyor flow, which might take water from the northern North Atlantic 1,000 years to return to the North Atlantic, affects climates around the world, a particularly prominent case being Great Britain and the rest of northern Europe. The warmth brought to northern Europe because of the Gulf Stream is a central reason why northern Europe has such a mild climate relative to the climates of other regions at comparable latitudes. This becomes important later in the book because changes in the ocean conveyor (and especially the Gulf Stream portion of it) are prime candidates for explaining some dramatic climate changes in the past (chapter 3) and hence enter also into concerns about the future (chapter 5).

The ocean conveyor concept captures very important large-scale flow patterns in the global ocean. Ocean circulation also has many smaller-scale patterns, including regional-scale clockwise surface circulations covering much of the North Atlantic and North Pacific, matched by counterclockwise circulations in the Southern Hemisphere (South Pacific, South Atlantic, and Indian oceans) and many much smaller eddy patterns. As with the atmosphere, there are day-to-day, season-to-season, and year-to-year variabilities.

Sea Level

Over the natural course of Earth's evolution, sea levels have risen and fallen, ecosystems have adjusted, and life and the Earth have gone on without major added complications from the sea level changes. For humans, however, sea level changes—in particular sea level rises—can be among the most troublesome of the changes in the Earth system for the unequivocal reason that humans have built up very unnatural structures along the coasts that cannot easily adapt to rising seas. Today, with political boundaries, a lack of freely available land, and the substantial infrastructure that accompanies major human settlements, when sea levels rise it is not a simple matter for those living in low-lying areas to adjust, as whole communities cannot easily pick up their belongings and move inland. Hence, sea level rise becomes a much bigger issue to humans than to the Earth system.

Exactly where sea level is depends on many factors. Over the course of Earth's evolution (chapter 2), the location and topography of the ocean basins have changed significantly, with significant impacts on sea level. However, over the course of a human life span, those changes tend to be minimal as far as sea level is concerned.

Instead, the two factors controlling sea level on the scale of a century or so tend to be 1) the amount of water in the ocean and 2) the density of the ocean water, less dense water taking up more space and therefore raising sea level. Increasing global temperatures contribute to sea level rise by affecting both those factors: 1) rising temperatures cause more water to be added to the oceans, as some of the ice that had been on land flows into the ocean, either as ice or as meltwater, in either case raising the ocean's top surface by adding to the amount of substance in the ocean, and 2) as the water temperatures increase, the water expands, decreasing the water density and hence causing the water to take up more volume, raising the top surface without adding mass. This latter process of water expansion as temperatures increase is called "thermal expansion." In the opposite case, when global temperatures are decreasing, there is likely to be growth of land ice coverage, thermal contraction of the water, and consequent sea level fall.

THE EARTH'S ICE COVER

Much to most people's surprise, current ice volume on Earth equals approximately one-third of the ice volume that existed at the peak of the last ice age. Most people rarely if ever see or think about this ice, as it lies overwhelmingly in the Antarctic ice sheet, far from any human population centers. However, both the portion of the Earth's ice on the land and the portion in the sea play several important roles in Earth's climate, and the land-based ice has major implications for potential sea level rise.

Land Ice

Land-based ice exists because water has evaporated from the oceans, has been transported over the land, has precipitated out as snow, has piled up, and has consolidated into ice. After piling up sufficiently, the ice flows outward and downward under the force of gravity. When the ice melts or flows outward over the oceans (sometimes calving off as icebergs), it adds mass to the oceans, contributing to sea level rise. An "ice shelf" is a portion of an ice sheet or other glacier that remains attached to the land-based ice but extends out over liquid water (generally the ocean).*

The Antarctic ice sheet is awesome in its size, covering almost the entire Antarctic continent and having depths exceeding 2 miles of ice over wide expanses. It

* There also exist ice shelves formed from sea ice and held fast to the land (Jeffries 1992; Vincent et al. 2001), but the glacially fed ice shelves are the best known and are the only ice shelves discussed in this book. The two largest ice shelves on Earth today, the Ross Ice Shelf and the Ronne-Filchner Ice Shelf, are both in Antarctica.

contains approximately 25.7 million cubic kilometers of ice, which is enough ice to raise sea level around the globe by approximately 60 meters. Like other ice sheets and glaciers, the Antarctic ice sheet gains mass through snowfall accumulation and loses mass through melting and the calving of icebergs.

By far the second-largest ice mass on Earth today is the Greenland ice sheet, estimated as containing 3 million cubic kilometers of ice. This pales in comparison with the Antarctic ice sheet but is still a very substantial amount of ice, with approximately 7 meters of sea level rise potential. The rest of the Earth's land-based ice is also significant, almost completely covering some islands and providing spectacular scenery, recreational opportunities, and freshwater supplies in mountainous regions around the world. Still, the volume of all the rest of the ice is swamped by the volume of ice in the Antarctic and Greenland ice sheets.

Many of the glaciers around the world have lost volume in the past few decades, and this has contributed to sea level rise. However, because the volume of ice contained in Antarctica and Greenland overwhelms all the other ice on Earth, when it comes to the possibility of catastrophic rather than gradual sea level rise, attention focuses on those two remaining continental-scale ice sheets. Whether or not they are likely to cause rapid sea level rise is closely tied to their stability.

The bulk of the Antarctic ice sheet, located largely in the Eastern Hemisphere and named East Antarctica or the East Antarctic ice sheet, has generally been thought to be relatively stable and unlikely to experience a rapid retreat. More concern exists regarding Greenland, which averages much lower latitudes than Antarctica does, placing it at greater risk of catastrophic decay from warmer temperatures, and the West Antarctic ice sheet, which has elicited concerns for decades regarding its stability.[9] The West Antarctic ice sheet, located largely in the Western Hemisphere and divided from the East Antarctic ice sheet by the Transantarctic Mountains, has a volume of ice comparable to the volume of the Greenland ice sheet and, most important, lies on land that is predominantly below sea level, partly because of the land's having been pushed down by the weight of the ice on top of it. This grounding below sea level increases the susceptibility of the West Antarctic ice sheet to surges of the ice that might speed its flow into the ocean. In the extremely unlikely event of a total collapse of either the Greenland or the West Antarctic ice sheet, sea level would rise approximately 7 meters.

Because of their size, the Antarctic and Greenland ice sheets further affect weather and climate by affecting wind flow in their vicinities, and because of the high albedo of ice, they affect weather and climate by reflecting away most of the solar radiation incident on them.

Sea Ice

Of much lesser volume than the Earth's land ice but still of tremendous significance is the Earth's sea ice coverage. Sea ice forms in the sea (or ocean) and floats on the liquid water, thereby being in equilibrium with it and not affecting sea level as it expands or retreats. Compared with ice on land, it is quite thin, typically being less than 5 meters thick. However, it spreads over a vast area in both polar regions. In the Arctic, the areal extent of sea ice ranges from about 5 million square kilometers in late summer to about 15 million square kilometers in late winter, and in the Antarctic its extent ranges from about 3 million square kilometers in the late austral summer to about 18 million square kilometers in the late austral winter. For comparison, the area of the United States is 9.4 million square kilometers, and the area of Europe is 10.5 million square kilometers.

Despite being thin, sea ice forms a significant barrier between the ocean and the atmosphere, restricting many exchanges of heat and chemicals and momentum that would occur between the ocean and the atmosphere in its absence. Further, because of its high albedo, it reflects away most of the sunlight incident on it, removing that radiation from the polar environment. The ice is also a major factor in the habitat of polar animals, impacting not just the frequently featured polar bears and penguins but dozens of additional polar species as well.

POSITIVE AND NEGATIVE FEEDBACKS IN THE CLIMATE SYSTEM

The sharp contrast between the albedos of ice and liquid water is one of the primary reasons why major changes in the polar sea ice covers have important impacts on the rest of the climate system. For instance, if warming leads to a substantial reduction in sea ice (which is likely because of less ice formation and more ice melt), this will lessen the reflectivity of the surface, as part of it changes from highly reflective ice to highly absorptive liquid water. The lessened reflectivity means that more of the incoming solar radiation will stay within the Earth system, thereby leading to further warming. Similarly, as glaciers and snow cover retreat, their reflective white surfaces are replaced by much less reflective land surfaces so that again more solar radiation is kept within the Earth system rather than being reflected back to space, thereby encouraging the system to warm further. In scientific parlance, such a feedback, which enhances the change that initiated it, is termed a "positive feedback." Here the word "positive" has nothing to do with being favorable or unfavorable, so its use can sometimes be confusing to people unfamiliar with it. Much better terms would be "enhancing feedback" or "amplifying feedback," both of which remove the confusing element. However, "positive feedback" is the term used throughout the scientific literature. The positive feedbacks involving the ice and snow reflectivities are frequently referred to as the "ice-albedo" and "snow-albedo" feedbacks and reappear later in this book, sometimes with warming being enhanced because of a reduction of ice and snow coverage and other times with cooling being enhanced through the opposite sequence of cooling leading to more ice and snow, hence greater reflection of solar radiation and further cooling. Whether it is warming being enhanced or cooling being enhanced, the albedo effect brings about a positive feedback, as the initial change (warming or cooling) is being enhanced through the albedo change.

The ice-albedo and snow-albedo feedbacks are only two examples of the concept of a positive feedback. There are many other important positive feedbacks in the climate system, and several of these appear in later chapters. Much attention is given in climate studies to positive feedbacks because these feedbacks could significantly accelerate the continued warming of the planet or, in the event of a reversal in temperature trends, could significantly accelerate a cooling.

In addition to positive feedbacks, the Earth system also has counteracting negative feedbacks. A "negative feedback" is one that dampens rather than enhances the original forcing and hence hinders a small change from spinning out of control to produce a much larger change. As with "positive" in "positive feedback," the use of the word "negative" here has nothing to do with whether the feedback is desirable or undesirable, favorable or unfavorable, and, as a result, in the same way that "enhancing feedback" would have been a better term to use than "positive feedback," "dampening feedback" would have been better than "negative feedback." Nonetheless, "positive feedback" and "negative feedback" are the two widely used terms.

For an example of a negative feedback in the Earth system, consider the chemical weathering of rocks, a process that involves carbon dioxide. As carbon dioxide rises in the atmosphere, this can encourage more chemical weathering, but the chemical weathering consumes CO_2, thereby dampening rather than enhancing the original forcing of increased CO_2. This example is expanded on in chapter 5.

THE INTERGOVERNMENTAL
PANEL ON CLIMATE CHANGE

Because of the importance of the Intergovernmental Panel on Climate Change (IPCC) in discussions of climate change and climate prediction, IPCC documents are referred to often in this book. Hence I include here a brief description of this very large and influential group.

The IPCC was established by the World Meteorological Organization and the United Nations Environment Programme in 1988 and was charged with assessing wide-ranging aspects of climate change issues and formulating realistic response strategies.[10] The IPCC examines and consolidates projections from a large number of published scientific studies and quickly became an important force in climate change discussions. The first of the IPCC's major assessments was completed in 1990, with subsequent major assessments released in 1995 (the Second Assessment Report), 2001 (the Third Assessment Report), and 2007 (the Fourth Assessment Report [AR4]). These assessments have involved hundreds of scientists from around the world, a wide range of data sources, and many different computer model simulations. (I am among the hundreds of scientists involved, although with an extremely minor role in the context of the effort as a whole.) The IPCC assessments aim to establish a broad scientific consensus and as such are of tremendous value for reporters, politicians, and others in determining the mainstream scientific viewpoint.

However valuable, the IPCC reports are not easy reading. For one thing, they are long. For instance, the core of the 2007 AR4 report includes three large volumes, the first consisting of 996 pages on "The Physical Science Basis,"[11] with an 18-page "Summary for Policymakers"[12] and a 73-page "Technical Summary,"[13] the second consisting of 976 pages on "Impacts, Adaptation and Vulnerability,"[14] and the third consisting of 851 pages on "Mitigation of Climate Change."[15] Furthermore, it is unlikely that the authors of these reports would be accused, on the whole, of erring in them on the side of oversimplification. In the interest of high scientific integrity, the IPCC documents include many qualifiers and many results from a large range of studies. For instance, the chapters on "Global Climate Projections"[16] and "Regional Climate Projections"[17] present results from many different models simulating the future under various scenarios representing different possible levels of population growth, economic growth, and transition to more efficient technologies. It all adds up to a sophisticated scientific effort but not an easy read.

Like many (preferably all) of the participants in the IPCC process, I am aware that the consensus viewpoint is not necessarily correct. Nonetheless, the consensus probably provides as sound a starting point as any for examining the bulk of the current scientific evidence. Furthermore, it also provides a solid starting point for understanding why people are concerned, as the IPCC projections (incorporating the efforts and published results of many scientists and research groups) are widely accepted as scientifically based, and they include many far-from-desirable predicted conditions, as elaborated further on in chapter 5.

The IPCC effort was given a rousing vote of confidence on October 12, 2007, when the Chairman of the Norwegian Nobel Committee announced that the Nobel Peace Prize for 2007 would be awarded in equal parts to the IPCC and former U.S. Vice President Albert Gore Jr. "for their efforts to build up and disseminate greater knowledge about manmade climate change and to lay the foundations for the measures that are needed to counteract such change."[18]

PREVIEW

This book attempts to explain why so much confusion exists regarding climate change issues; to lessen that confusion by describing relevant aspects of the Earth system, its evolution, and the efforts to study and forecast it; and to make the case that we do not know enough to warrant undertaking potentially dangerous massive geoengineering schemes aimed at counteracting anticipated but uncertain future climate changes. Issues surrounding predicted upcoming climate change have become highly contentious in recent years, with the discussion at times degenerating into far-from-civil exchanges. Throughout this book, I try to be fair and respectful to all involved, as I do believe that each side has important points to make that deserve to be treated respectfully. Here is a preview of the chapter-by-chapter sequence.

Chapter 2 provides a summary in a few dozen pages of the enormous changes the Earth has undergone during its 4.6-billion-year existence. This places the changes witnessed in our lifetimes into a much longer-term context. Among much else, it shows that the Earth system has experienced much hotter as well as much colder conditions than today's, a fact that is relevant to understanding the climate system and the confusion regarding it yet does not affect the potential seriousness of future climate change. The chapter also makes explicit the interconnectedness of the Earth system, the great uncertainty in our understandings of the Earth's past, and the fact that our understandings continue to evolve.

Chapter 3 narrows down from the very broad overview of Earth history given in chapter 2 to consider the fact that on numerous occasions prior to the past several thousand years, the Earth experienced very abrupt climate changes. This is a sobering reality, as it indicates that we could be in for much greater and more rapid climate change than modern humans have experienced so far. In fact, we could unwittingly precipitate an abrupt climate change.

Chapter 4 summarizes many of the impacts that humans have had and are having on the Earth system. This includes the limited impacts in the distant past, the increased impacts with the invention and spread of agriculture, and the much more severe impacts since the start of the industrial revolution.

Chapter 5 explains why some people are so concerned about future climate, based on mainstream projections of continued global warming and its numerous unfavorable consequences, and why other people do not share those concerns. Consideration of climate-altering massive geoengineering proposals derives in large part from the troubling nature of the climate predictions.

Chapter 6 cautions that good intentions—including the intentions behind the proposals for massive geoengineering—do not necessarily lead to good results. This sad reality is illustrated with cases ranging from the misguided killing of cats and dogs during the 1665 London Plague to the purposeful addition of lead to gasoline and paint.

Chapter 7 summarizes several of the most-discussed geoengineering proposals, with comments about the dangers involved. This could be the book's scariest chapter for those concerned about the future of our planet, as some of the proposals include potentially quite damaging activities.

Chapter 8 discusses earlier, smaller attempts to modify the environment and how they have fared, focusing in particular on attempts at rainmaking, hail suppression, and taming hurricanes. The lack of success in these smaller efforts should be cautionary when it comes to considering the much larger-scale geoengineering proposals.

Chapter 9 warns through historical examples that even the support of experts and a strong scientific consensus by no means guarantee that a position is correct. Science is constantly evolving, and it is part of the regular course of scientific development that consensus views sometimes get overturned.

Chapter 10 describes some of the imperfections in the global climate models used to create the climate projections that have caused such consternation about the future and that have led to considerations of massive geoengineering efforts. These models are the best tools we currently have for climate projection, and certainly their results should be weighed heavily in considerations about future climate, but it is important to recognize their limitations.

Chapter 11 examines the social pressures that make the consensus view of the scientific community perhaps seem to be a stronger consensus than it really is.

Chapter 12 points out some of the alternatives now available or on the horizon that offer possibilities for correcting or reducing human-induced climate change with far less risk than massive geoengineering and with a far greater chance of addressing a broader range of human-induced impacts than the typically narrow global warming focus of many of the geoengineering schemes.

Chapter 13 concludes with a plea summarizing the argument that we are not yet knowledgeable enough to undertake massive geoengineering without substantial risk to the Earth system and the life within it. I do not argue that geoengineering should never be undertaken, only that we need to be extremely cautious about it. Most important is that we avoid doing even greater damage in the attempt to correct for anticipated damage.

CHAPTER 2

4.6 Billion Years of Global Change

The Earth has been undergoing climate change ever since its formation 4.6 billion years ago. It has experienced gigantic changes that have taken place with an intricate and incompletely understood intertwining of life, climate, internal tectonic forcings, and external forcings from the Sun and, occasionally, elsewhere. Through quite extended periods, it has been considerably warmer than it is now, but those periods were all long ago. Amazing as it is that we know as much as we do, through piecing together a wide range of theory and observations, our understanding of Earth history remains laden with uncertainties.

ALTHOUGH written records extend back only a few thousand years, scientists have been able to combine a wide range of evidence and theory to piece together a somewhat coherent picture of Earth history, revealing the ever-changing nature of our very dynamic planet. In this chapter, I attempt to summarize this postulated history while recognizing that there is quite considerable uncertainty and disagreement regarding almost all the dates and many of the other details. Despite the uncertainties, the basic thrust that the system is continually evolving and has many important and diverse influences is valid, and the long-term perspective is both interesting and relevant to examining the human impacts of today. This one-chapter overview of billions of years highlights that climate change has been occurring ever since the Earth formed, with gigantic changes taking place over the course of the 4.6-billion-year existence of our home planet both before and after the emergence of humans. (See the following two pages for an overview of Earth's geologic time line.) For a quick preview of the topic of most interest to many people, that is,

---

ABBREVIATED GEOLOGIC TIME LINE

4,600 to 2,500 m.y.a.	Hadean and Archean eons		
2,500 to 542 m.y.a.	Proterozoic Eon, ending with the Cryogenian (850 to 630 m.y.a.) and Ediacaran (630 to 542 m.y.a.) periods		
542 m.y.a. to the present	Phanerozoic Eon		
	Paleozoic Era 542 to 251 m.y.a.		
	542 to 488 m.y.a.	Cambrian Period	
	488 to 444 m.y.a.	Ordovician Period	
	444 to 416 m.y.a.	Silurian Period	
	416 to 359 m.y.a.	Devonian Period	
	359 to 299 m.y.a.	Carboniferous Period	
	299 to 251 m.y.a.	Permian Period	
	Mesozoic Era 251 to 65 m.y.a.		
	251 to 200 m.y.a.	Triassic Period	
	200 to 145 m.y.a.	Jurassic Period	
	145 to 65 m.y.a.	Cretaceous Period	
	Cenozoic Era 65 m.y.a. to the present		
		Paleogene Period 65 to 23 m.y.a.*	
		65 to 56 m.y.a.	Paleocene Epoch
		56 to 34 m.y.a.	Eocene Epoch
		34 to 23 m.y.a.	Oligocene Epoch
		Neogene Period 23 m.y.a. to the present*	
		23 to 5.3 m.y.a.	Miocene Epoch
		5.3 to 1.8 m.y.a.	Pliocene Epoch
		1.8 m.y.a. to 11,800 years ago	Pleistocene Epoch
		11,800 years ago to the present	Holocene Epoch

m.y.a. = million years ago

A NOTE ON DATES: Despite the many uncertainties in Earth history, considerable work done since the late eighteenth century has established a detailed geologic time line dividing this history into four eons subdivided into eras that are in turn subdivided into periods and those into epochs. Some of these, such as the Cambrian and Jurassic periods and the Miocene and Pleistocene epochs, have what are by now familiar names. Dates and other details on the geologic time line vary from source to source, and almost certainly many of the dates will change as further research uncovers new data or develops new dating techniques. Still, even imperfect dates can be valuable in terms of providing an approximate time scale, which accounts for the many dates throughout this chapter. On the advice of Professor Andrew Knoll from the Department of Earth and Planetary Sciences at Harvard University, I have chosen to use the time line presented by the International Commission on Stratigraphy (ICS), downloaded from their website at http://www.stratigraphy .org/chus.pdf on January 20, 2008, for the dates from 3.6 billion years ago to the present (the downloaded ICS time line does not include dates prior to 3.6 billion years ago). Because of the uncertainties involved and for simplification, I have rounded the numbers to the nearest million years for dates older than 10 million years ago and to two significant figures for younger dates until the last 15,000 years, when more precision is often warranted. The interested reader is referred to the ICS website (http://www.stratigraphy .org/chus.pdf) for the unrounded dates and also for error bars. The geologic time line includes a great many names and dates, but this abbreviated version highlights the portion of relevance for this chapter, reaching down to the period level starting 850 million years ago and one step further, to the epoch level, starting 65 million years ago.

---

* The Cenozoic Era is sometimes instead divided to the Tertiary and Quaternary periods, the former extending from the Paleocene through the Pliocene epochs and the latter consisting of the Pleistocene and Holocene epochs. The use of the Paleogene and Neogene here is for consistency with the ICS. As of this writing, in 2008, controversy exists regarding whether to give official status to the term "Quaternary" on the time line. A factor in its favor is that the term has been extensively used. However, when it was originally proposed, it was the fourth of four divisions labeled Primary, Secondary, Tertiary, and Quaternary, and the first three terms are almost never used anymore so that the term "Quaternary" has lost its original context. Readers interested in the controversy can find more about it in Kerr (2008).

how temperatures today compare with temperatures in the past: the near-surface atmosphere today is, on a global average, probably warm relative to most of the past million years but cold relative to most of the past 500 million years.

Throughout its history, two main drivers of the Earth's evolution have been the radiation reaching the Earth from the Sun and the Earth's internal energy, the latter consisting of both the internal energy of radioactivity and the Earth's primordial heat. Today and for at least the past 3 billion years, the radiation from the Sun helps to power the atmosphere and oceans and thereby also to drive the weathering and erosion processes on land and the deposition processes within the oceans, while the Earth's internal energy maintains plate-tectonic motions and affects global atmospheric composition and local soils through volcanic eruptions, when materials from within the Earth are blasted out into the atmosphere and onto the landscape. The Earth's crust is continually being replenished from the mantle at ocean-spreading centers such as the midocean ridges, and the additional crust being created there is offset by the recycling of old crust back to the Earth's interior at subduction zones where one gigantic tectonic plate sinks beneath another. The continual movement of crustal material contributes to the slowly changing configuration of land and ocean, which is evolving concurrently with the ever-changing atmosphere, oceans, and plant, animal, and other life forms. More dramatic changes in the Earth's topography occur when plates collide with sufficient force to cause the formation of major mountain ranges and associated valleys. But whether violent or gradual, the processes moving the Earth's crust and upper mantle produce the Earth's ever-changing topography, including the midocean ridges, mountain ranges, ocean basins, and continental and island boundaries. As detailed later in this chapter, both this changing geography and the changing composition of the Earth's atmosphere have major impacts on the planetary climate. But before the plate-tectonic processes could begin, the Earth had to settle down from its initial state, as the Earth did not burst into existence with a crust and mantle and plate tectonics already in place.

EARTH'S FIRST 2.1 BILLION YEARS, ENCOMPASSING THE HADEAN AND ARCHEAN EONS (4.6 BILLION YEARS AGO TO 2.5 BILLION YEARS AGO)

According to the current scientific consensus, the Earth formed approximately 4.6 billion years ago, roughly 9 billion to 10 billion years after the hypothesized big-bang origin of the universe. The main components of the Earth system, including a metallic core (which produces the Earth's magnetic field), an overlying convective mantle, a crust, an ocean, an atmosphere, and the beginnings of a biosphere (eventually including the plant and animal life on Earth as well as bacteria, fungi, and protoctists), all are thought to have appeared within the Earth's first billion years. None of these, however, has remained constant or even near constant since the Earth's creation, with all major components of the Earth system instead undergoing a continual and continuing evolution. The atmosphere, oceans, and land/sea boundaries and topography are of particular interest because of their immediate impacts on climate and climate change.

The atmosphere that evolved into today's atmosphere is believed to have derived not from the initial gases surrounding the Earth, which are presumed to have been blown off early by the solar wind, but instead from gases spewed out from the Earth's interior by volcanic activity. Such activity was likely considerably greater during the Earth's early history than it is now, because of the greater internal energy provided by the higher initial temperatures and the greater radioactivity provided by the younger atoms within the Earth's interior. By now, billions of years later, many of the radioactive atoms have decayed to stable forms, reducing the radioactivity of the system as a whole. Still, volcanic eruptions have continued sporadically throughout Earth's history to jolt the chemical and particulate composition of the atmosphere, sometimes with noticeable climatic consequences.

Because it formed from volcanic activity, the early atmosphere that evolved into today's atmosphere was probably dominated initially by carbon dioxide (CO_2) and water vapor (H_2O), the two major constituents of volcanic outgassing. Although volcanic gas composition today varies from one volcanic eruption to another and very possibly varied much more strongly from the distant past to the present, it is likely that the early Earth's atmosphere contained a percentage composition much closer to the average composition of today's volcanic gases than to the average composition of today's atmosphere. Despite the variation in emissions from one volcano to another, four gases almost always dominate, and those are water vapor (H_2O) at 50% to 90% of the volume of the gas emissions, carbon dioxide (CO_2) at 1% to 40% of the emissions, sulfur dioxide (SO_2) at 1% to 25% of the emissions, and hydrogen sulfide (H_2S) at 1% to 10%, with far lesser amounts of a variety of other gases, including carbonyl sulfide (COS), carbon disulfide (CS_2), hydrogen chloride (HCl), hydrogen bromide (HBr), hydrogen fluoride (HF), ammonia (NH_3), and methane (CH_4).[1] For a rough indication, average gas composition from modern volcanic eruptions includes approximately 65% to 70% water vapor and 20% carbon dioxide. Even more roughly, these numbers can be used as estimates for the approximate percentages of H_2O and CO_2 in the gas composition from the eruptions during Earth's early history as well. Clearly, other processes than volcanic outgassing must have additionally played crucial roles in the evolution of the atmosphere to account for today's predominance of nitrogen (at least 75%; 78% in the dry atmosphere, i.e., ignoring the water vapor content) and oxygen (at least 20%; 21% in the dry atmosphere), a relatively minor amount of water vapor (up to 4%), and a miniscule amount of carbon dioxide (approximately 0.04% in the early twenty-first century). As will become clear through the course of this chapter, the evolution from the early atmosphere, dominated by volcanic gases, to today's atmosphere is integrally tied to the evolution of the oceans and life, establishing in the very long term the huge importance of the highly interconnected nature of the Earth's atmosphere, oceans, and life.

The presumed high concentration of the greenhouse gases H_2O and CO_2 in the Earth's early atmosphere would have kept atmospheric temperatures considerably higher than they otherwise would have been. This greenhouse tendency for warming the system helped counteract the fact that during the Earth's first billion years the Sun was radiating only about 75% as much energy as it is now, as the Sun's energy output has gradually increased by about 30% over the course of its 4.6-billion-year history.[2] The dilemma of how the early Earth maintained above-freezing temperatures despite the far less radiant Sun has been called the "faint young Sun paradox."[3] In 1970, Carl Sagan and George Mullen from Cornell University proposed a super-greenhouse effect from massive amounts of methane and ammonia in the early atmosphere as a solution to the paradox,[4] and although the favored greenhouse gases later became water vapor and carbon dioxide, the super-greenhouse effect remains a prime candidate for resolving the earlier apparent paradox. The smaller amount of solar radiation reaching the early Earth was also counteracted by the initially higher internal Earth temperatures.

By about 3.8 billion years ago, assuming approximately correct dating of these very distant events, the hot surface rocks of the Earth had cooled below the boiling point of water, 100°C, and the water vapor in the Earth's atmosphere had begun to precipitate out, forming liquid water at the Earth's surface and a resulting shallow ocean covering the majority of the planet, interspersed with scattered volcanic islands. Liquid water almost certainly has continued to be present on the Earth ever since. Eventually, most of the atmospheric water vapor condensed, greatly reducing the water vapor contribution to the atmosphere from what might have been roughly 65% to 70% very early on to today's much lower values, averaging no more than 4%. However, by forming the oceans, the condensed water vapor also created a permanent source of moisture for the atmosphere, whose water vapor content has probably never decreased to zero during the past 3.8 billion years, and a new and vital, massive component in the water cycle of the planet. The oceans further affected the early atmosphere by absorbing much of the atmospheric carbon dioxide, some of which was also removed by interactions with sedimentary rocks, as chemical weathering of the rocks consumes carbon dioxide. As the percentages of water vapor and carbon dioxide in the atmosphere decreased, the percentage of nitrogen in particular rose.

By about 3.8 billion years ago also, with rocks cooled and liquid water at the Earth's surface, the chemistry and climate of the Earth had become amenable to the development of life. Surface temperatures are thought to have been in the range of 0°C to 50°C, and organic compounds, such as amino acids, nucleotides,* and the sugars glucose ($C_6H_{12}O_6$) and sucrose ($C_{12}H_{22}O_{11}$), likely existed in abundance, created as products of the effects of lightning and the Sun's ultraviolet radiation on atmospheric methane (CH_4), ammonia (NH_3), water vapor (H_2O), carbon dioxide (CO_2), and hydrogen (H_2). Many of the amino acids and nucleotides were eventually transferred from the atmosphere to the global ocean, providing the ocean with at least some of the building blocks of life.

Harvard University's Edward O. Wilson estimates the timing of the origin of life at 3.9 billion to 3.8 billion years ago, emerging "spontaneously from prebiotic organic molecules" and consisting of microscopic single-celled organisms, with stromatolite ecosystems appearing by 3.5 billion years ago.[5] Although Wilson does not explicitly say where these first organisms were, many people traditionally have thought that life on Earth originated in the upper ocean, while some have speculated that it might instead have first arisen in land-based muds. After the discovery in the 1970s of abundant life at hydrothermal vents at the bottom of the ocean, speculation arose that life might have developed in the deep ocean under dark, hot conditions very different from those at the surface[6] and that simple organic molecules might have first been synthesized in these hydrothermal vents rather than in

---

* Amino acids are organic acids containing nitrogen and constituting building blocks for proteins. Nucleotides are organic compounds consisting of a pentose sugar, base, and phosphate group and constitute building blocks for deoxyribonucleic acid (DNA) and ribonucleic acid (RNA).

the atmosphere.[7] The issue remains unresolved, and the possibility exists that life arose independently at different locations.

In any event, wherever life originally arose, the first life presumably was "prokaryotic," meaning that the cells lacked a well-defined nucleus. The recognized prokaryotes are divided into bacteria and archaea, sometimes alternatively termed "eubacteria" and "archaebacteria," respectively.[8] Many archaea are life forms that have thrived in what humans would consider extreme conditions, such as at hydrothermal vents or in other locations that are extremely hot, extremely salty, under extreme pressure, or characterized by some other extreme. It is not certain whether the first organisms were archaea or bacteria or "neither simply Archaean nor simply bacterial."[9] As mentioned earlier, the history of the Earth and of life on the Earth is full of uncertainties. Nonetheless, with current dating methods, the fossil record suggests that prokaryotic life had emerged by 3.5 billion years ago.[10]

At some point, primitive living cells began photosynthesizing, using solar energy to manufacture food for their continued survival. This was a momentous step, as they were now harnessing a consistently reliable and massive energy supply, that is, the energy from the Sun.[11] The first photosynthesizers likely dissociated hydrogen sulfide (H_2S) rather than water (H_2O), but eventually other life forms emerged that performed the photosynthesis reaction most common today, with sunlight acting on carbon dioxide (CO_2) and water (H_2O) to break the bonds uniting oxygen to carbon and hydrogen and create the simple sugar carbohydrate glucose and the waste product of oxygen: $6CO_2 + 6H_2O + \text{light} \rightarrow C_6H_{12}O_6 + 6O_2$. The cyanobacteria first performing this reaction are the forerunners of all green plants on the planet today.[12] Such reactions likely began by 3.5 billion years ago.

By 3 billion years ago, again if the current scientific understanding is correct, microscopic organisms swarmed in the oceans, while the land remained largely lifeless, except possibly within land-based muds. With almost no oxygen in the atmosphere except the oxygen bound into water vapor, carbon dioxide, and sulfur dioxide, there was no ozone layer in the upper atmosphere to protect potential life at the surface—either life in the upper layers of the oceans or life venturing out of the oceans—from the Sun's ultraviolet radiation.

Living organisms at the time were mostly single-celled and prokaryotic. Single cells, however, are highly vulnerable to destruction so that as cells began to group together, initially probably by accident, the multicelled life forms that developed had some survivability advantages over the more primitive single cells.

PROTEROZOIC EON
(2.5 BILLION YEARS AGO
TO 542 MILLION YEARS AGO)

As the global population of cyanobacteria increased, photosynthesis converting CO_2 and H_2O into glucose and oxygen provided an increasing supply of oxygen, a waste product as far as the photosynthesizers were concerned but vitally important to the future evolution of life on Earth. A small amount of oxygen (O and O_2, respectively, for the atomic and molecular forms) likely accumulated in the atmosphere by 2.5 billion years ago both from photosynthesis and from abiotic processes, but evidence suggests a rise in atmospheric oxygen between 2.45 billion years ago and 2.22 billion years ago that was significant enough to label the rise in that period as "the Great Oxidation Event."[13]

In the oceans, prior to about 2 billion years ago, much of the oxygen being generated by photosynthesis combined with ferrous iron, oxidizing the ferrous iron to form ferric oxides.* Ferrous iron had become abundant in the oceans, having been weathered from land-based iron-rich rocks and then transported to the oceans in rivers, streams, and groundwater flow. The ferrous iron, being soluble, mixed well with the ocean waters. After oxidation, however, the resultant insoluble ferric oxides settled to the ocean floor. Many of the iron formations formed by this process still exist and are observable in the geological strata dated to the period between 2.5 billion years ago and 1.8 billion years ago, hence providing evidence for this presumed sequence of what took place billions of years ago.[14] In fact, these iron formations form extensive sequences of alternating, broken layers of black or red iron-rich material and light gray iron-poor material. Many of these banded iron formations are hundreds of meters thick, although with the individual iron-rich and iron-poor layers only 1 millimeter to several centimeters in thickness. These iron formations constitute the Earth's most substantial reserves of iron ore, in some locations extending horizontally for 300 kilometers or more. Many of the banded iron formations contain small spherical bodies about 30 micrometers in diameter that are visible in petrologic thin sections and are perhaps the remains of primitive phytoplankton producing the oxygen that led to the iron formations.

---

* Ferrous iron has a valence number of +2, meaning that it has two electrons to lose as it combines with other elements; ferric iron has a valence number of +3.

Eventually, seemingly about 2 billion years ago, the ferrous iron in the oceans became sufficiently depleted by this process (and was not sufficiently rapidly replaced by new ferrous iron from the land) that a fundamentally important change occurred. Namely, much of the oxygen released during photosynthesis was no longer merged with the ferrous iron but instead remained as molecular oxygen. This oxygen could then accumulate in the oceans and atmosphere. With free oxygen (i.e., oxygen not chemically bound with other elements) present in the atmosphere, the iron in rocks and soils reacted with atmospheric oxygen to produce insoluble ferric oxides, which remained in the rocks and soils; that is, the land iron became oxidized in place, with the result that very little dissolved iron continued to flow into the oceans. This brought a near halt to the formation of the iron bands at the ocean bottom and additionally produced redbeds and soils more akin to modern soils than anything that had existed previously. By this time also, other aspects of the environment were probably becoming similar to today's as well, including a blue sky and a blue-gray sea.

Free oxygen molecules would likely have been poisonous to the earliest photosynthesizers, but by the time free oxygen began to accumulate in the atmosphere and oceans, some of the photosynthesizing organisms were manufacturing enzymes that counteracted the poisonous effects of oxygen. Such photosynthesizers began to dominate once significant amounts of oxygen had accumulated in the ocean and atmosphere, sending the unadapted organisms either to extinction or to withdrawal to anaerobic (i.e., without oxygen) environments.

The accumulating oceanic and atmospheric oxygen allowed the evolution of aerobic organisms able to use the oxygen to provide a much more efficient metabolism than any previous organisms had. Some primitive, single-celled prokaryotic aerobic organisms likely appeared as early as 2.8 billion years ago, when the atmospheric oxygen concentration was still extremely low, likely well under 1%. The new type of single-celled organism, able to take in oxygen and use it as its main energy source, contrasted markedly with the photosynthesizing cells. In fact, the new, oxygen-breathing cells were forerunners of the first animals. Oxygen allowed a rapid acquisition and utilization of energy, eventually helping animals become much more physically active than bacteria, plants, and other nonanimal life forms and increasing the chance that a capacity for advanced thought, which takes considerable energy, would develop.

The Earth's ecosystems continued through over 2 billion years, from about 3.8 billion years ago to about 1.8 billion years ago, to be dominated by single-celled prokaryotes, specifically bacteria and archaea. Multicellular prokaryotes probably also existed although in much smaller numbers. At some point in this time frame, perhaps as early as 2.7 billion years ago, perhaps not until 1.8 billion years ago, a new type of cell, the eukaryotes, arose.[15] The central characteristic distinguishing the eukaryotes from the more primitive prokaryote organisms was the existence of a cell nucleus and the enclosing of their deoxyribonucleic acid (DNA) in a nuclear membrane. A leading theory by the end of the twentieth century for the origin of eukaryotic cells involves prokaryotic cells devouring but not digesting other prokaryotic cells in a symbiotic merger. This "endosymbiotic" theory was developed in substantial part by Lynn Margulis of the University of Massachusetts, building on the idea of symbiogenesis, in which new organs and organisms are formed through symbiotic mergers, originally proposed by the Russian Konstantin Merezhkovsky (1855–1921).[16] The continuing role of symbiogenesis throughout life's evolution is a strong component of Margulis's theory although is less widely accepted than is the original role of symbiogenesis in the origin of eukaryotic cells.[17]

The first eukaryotes presumably contained only a single cell, but multicellular eukaryotes with increasing complexity eventually evolved. The vast majority of familiar organisms today are eukaryotic, the only exceptions being bacteria and a few other microscopic life forms. Nonetheless, prokaryotic descendants of the earliest bacteria still survive and thrive and do so in abundance although in anoxic environments such as beneath the sea floor, in marshes, and in the guts of humans and other animals. Sequencing of the genome of the human intestinal parasite *Giardia lamblia* in the early twenty-first century[18] is yielding new information and raising new questions about the origin of the eukaryotes,[19] providing yet another strong likelihood of eventual further revisions to current scientific understandings.

As photosynthesizers continued to evolve and spread, they added more exhaled oxygen to the atmosphere. It is uncertain exactly when the atmosphere was thereby transformed from reducing to oxidizing, although it likely occurred between 1.8 billion years ago, the estimated date of the laying down of the last of the known massive sedimentary banded iron formations, and 0.7 billion years ago, the date of the earliest known fossilized worms. Over that 1.1-billion-year period, the oxygen composition of the atmosphere likely increased from the very low levels of the early atmosphere—Wilson suggests a 1% oxygen concentration by 1.8 billion years ago—to approximately the same 21% concentration that exists today.[20]

By about 1 billion years ago, the oxygen levels in the atmosphere likely had risen sufficiently to allow the creation in the upper atmosphere, through photochemical reactions powered by the Sun, of a layer rich in the triatomic form of oxygen known as ozone (O_3). Ozone, well known now especially for protecting life at the Earth's surface from heavy exposure to the Sun's ultraviolet radiation (e.g., chapter 1), is formed after normal molecular oxygen (O_2), with two oxygen atoms, is broken apart into individual oxygen atoms, which then combine with remaining molecular oxygen: $O + O_2 \rightarrow O_3$. Once established, the protective ozone layer blocked much of the incoming solar ultraviolet radiation from reaching the Earth's surface, thereby increasing the chances that life could survive on the land as well as in the oceans. Other atmospheric constituents, such as smog products from the decomposition of methane and/or hydrogen sulfide, might also have reduced the ultraviolet intensity at the Earth's surface.

In addition to adding exhaled oxygen, another important impact of photosynthesizers on the evolving atmosphere was the removal of atmospheric carbon dioxide. Carbon dioxide, entering the atmosphere at that time predominantly through volcanic eruptions, also continued to be removed from the atmosphere by the nonbiological means of chemical weathering of exposed rock and the production of carbonate sediments. Biological photosynthesis and rock weathering together probably accounted for the majority of the decrease in atmospheric CO_2 concentration from its high early values to its current approximately 0.04% level. The lessened greenhouse effect as a result of the decreasing CO_2 and H_2O concentrations helped the Earth maintain a relatively stable global temperature despite the increasing energy output from the Sun, increasing gradually by approximately 30% over the course of Earth's history so far.[21] The relative temperature stability has allowed liquid water to be continually present in the Earth system for the past 3.8 billion years, although the temperature fluctuations have contributed to significant fluctuations in the amount of solid water, that is, ice.

Ice probably covered a considerable portion of the Earth's surface during some of Earth's early history, although the dating and extents of the early ice covers remain quite uncertain and controversial. Macdougall suggests three extended periods with major ice coverage in Earth's first 4 billion years, the first at about 2.9 billion years ago, the second at about 2.4 billion to 2.2 billion years ago, and the third at about 850 million to 550 million years ago.[22] Others eliminate the first and provide a different dating for the third. For instance, Kirschvink and colleagues highlight two periods, at about 2.4 billion years ago and at 800 million to 600 million years ago,[23] and Hoffman and colleagues date the approximately 700-million-years-ago ice cover at about 750 million to 550 million years ago.[24] The International Commission on Stratigraphy includes in its geologic time line the Cryogenian Period, with significant ice coverage, at 850 million years ago to 630 million years ago.[25] Regardless of the exact dates, an extended glaciation event at about 700 million years ago is thought to have been the most severe of all the Earth's glacial periods, with some speculation that the ice coverage may have extended globally on both land and ocean surfaces, creating a "Snowball Earth."

The term "Snowball Earth" was coined by American geobiologist Joseph Kirschvink in the early 1990s, although there had been precursor speculations about a near-global glaciation about 600 million years ago as early as 1964 based on examination of tillites from around the world.[26] Likely at least two and perhaps four or five major glacial episodes occurred between 850 million years ago and 550 million years ago,[27] with each successive pair separated by an interglacial episode with considerably less ice. During the glacial episodes, the global oceans might have had pack ice to thicknesses of 500 to 1,500 meters (versus today's spatially much more limited polar pack ice that is almost all thinner than 5 meters), and global average surface temperatures might have plummeted to −20°C to −50°C.[28] As for the cause of such massive glaciation, Russian climatologist Mikhail Budyko reasoned early on that it might have been in large part due to the land/sea configuration at the time, about 700 million years ago, when the continents are thought to have been clustered around the equator. With land rather than water covering much of the equatorial region, more sunlight would have been reflected back to space, promoting a cooling that would have amplified once glaciers and sea ice coverage, with their high reflectivities, started to grow.[29]

Among the considerations regarding the possibility of a Snowball Earth, concerns have been raised that a fully glaciated Earth would have been irreversible, that is, that the Earth could not have recouped from a runaway glaciation.[30] However, in rebuttal, those propounding the Snowball Earth hypothesis suggest the following mechanism for ending the iced-over state. With all the land and all the photosynthesizing life buried under the ice, atmospheric carbon dioxide would not have been depleted through chemical weathering or photosynthesis. This would have allowed carbon dioxide (still likely added to the atmosphere through volcanism even though the eruptions would have had to blast through some ice) to build up in the atmosphere and eventually warm the planet enough to melt the ice.[31] Hoffman and colleagues even suggest that volcanic outgassing might have raised atmospheric carbon dioxide amounts to a level 350 times the twentieth-century values, leading to warming from extreme greenhouse conditions.[32] Still, although many find the hypothesis intriguing and Hoffman and colleagues offer supporting geologic evidence,[33] there remains considerable uncertainty regarding whether a Snowball Earth ever existed. Modified scenarios have been hypothesized, with the Earth largely but not entirely covered by ice and snow, for instance with ice-free oceans at low latitudes and with atmospheric CO_2 amounts closer to four times rather than 350 times the twentieth-century values[34] or with scattered regions of open water that could have served as refugia for early life.[35]

Whether or not there was a Snowball Earth, glaciation appears to have been quite significant during the Cryogenian Period, with global impacts and with even tropical glaciers extending to sea level at least in some locations if not throughout the tropics.[36]

The Cryogenian Period was followed by the Ediacaran Period, lasting for the remainder of the Proterozoic Eon, from 630 million years ago to 542 million years ago. Although no longer at the level of a Snowball Earth, major glaciation episodes continued to occur well into the Ediacaran Period, with the last being the Gaskiers event at about 580 million years ago.[37] Geologically speaking, it was not long after the end of the Gaskiers glaciation that a host of new life forms emerged on the scene. Many of these also needed, in addition to the removal of the massive ice cover, an atmosphere amenable to substantial oxygen breathing, and this was indeed in place by that time.

By 600 million years ago, in the early Ediacaran Period, the chemical composition of the atmosphere had evolved to approximately its present state, with nitrogen constituting about 78% and oxygen about 21% of the dry atmosphere. These are generally thought to have remained the approximate nitrogen and oxygen percentages to the present time. The increase of nitrogen from its negligible early concentration came about substantially through biochemical processes during which ammonia (NH_3) was decomposed into hydrogen and nitrogen, with the nitrogen by and large accumulating in the atmosphere. The chemical changes in the atmosphere over the past 600 million years, some of which probably had very significant effects on the Earth's climate, have come largely in the remaining 1% chemical composition, involving the trace gas species, including CO_2.

If the photosynthesizers had proceeded unchecked, they might have eliminated the carbon dioxide component of the atmosphere altogether. A large part of the reason that the atmospheric composition could remain so stable over the past 600 million years is that the plant and other photosynthesizing species, inhaling carbon dioxide and exhaling oxygen to the atmosphere, were counterbalanced by animal species inhaling oxygen and exhaling carbon dioxide. The earliest animals, well prior to 600 million years ago, did not require much oxygen because the gas could diffuse across the walls of their cells. Hence, for them, the atmospheric oxygen concentration could have been (and might have been) as little as 0.1%. However, for most larger animals, considerably more oxygen is needed for an efficient aerobic respiration allowing fully active animal life.

Prior to about 575 million years ago, that is, for the first approximately 87% of Earth's existence, likely the only life forms were single-celled and small multicelled life. These were bacteria, archaea, algae, fungi, and some small marine animals with soft bodies. Then, from approximately 575 million years ago to 542 million years ago, in the late Ediacaran Period, there arose the Earth's oldest currently known complex macroscopic life forms, with indications of over 270 species found in over 30 locations on several continents.[38] It was a fascinating abundance of new life forms, including the first "large, architecturally complex organisms in Earth history."[39] In contrast to remains of earlier life forms, the Edicaran fossils are no longer exclusively microscopic, instead many having lengths of centimeters or decimeters, with some over a meter long, occasionally even close to 2 meters. Some structures are disc shaped, others leaflike, others segmented, and some unlike any in the modern world, labeled as "'failed experiments' in animal evolution."[40] The "failed experiments" included organisms with modular construction consisting of highly fractal elements and having some similarities to modern suspension-feeding animals. Many of the more familiar-looking organisms resembled the two modern animal phyla of Porifera (sponges) and Cnidaria (corals, jellyfish, and so on), although the Ediacara biota did not contain all the characteristics of the modern phyla. Among the firsts believed to have occurred during the Ediacaran Period are the first calcified metazoans, the first skeletal reefs, the first animals with mobility, and the first predators. Where they existed, the Ediacara populations were not always small, as densities have been found of 3,000 to 4,000 individuals per square meter, comparable to highly productive modern ecosystems.[41]

Major as the changes were during the Ediacaran Period, few of the new life forms survived into the following period, the Cambrian,[42] some believe because of the effectiveness of the first predators in eliminating other species.[43] In hindsight, the Ediacaran organisms were a prelude to the greater changes about to come in what is frequently referred to as the Cambrian "explosion" of life.[44] However advanced the Ediacaran organisms were versus those that preceded them, they still lacked macroscopic sensory organs such as compound eyes and antennae, and they still lacked legs, claws, or similar appendages, leaving a significant gap between their capabilities and those of many of their successors.[45]

PALEOZOIC ERA
(542 MILLION YEARS AGO
TO 251 MILLION YEARS AGO)

The end of the Ediacaran Period marked the end not only of a geologic period but of the entire Proterozoic Eon and the start of the Phanerozoic Eon, which continues to the present time. The Phanerozoic is divided into three eras, the Paleozoic, the Mesozoic, and the Cenozoic, roughly translated as ancient life, middle life, and new life, respectively, despite the far more ancient life that had existed earlier. During the entire Phanerozoic Eon, the Earth has had an atmosphere and oceans amenable to life as we know it, and indeed the changing suite of macroscopic life forms, as determined from the fossil record, is a major component of how the Phanerozoic is divided and subdivided.

The Phanerozoic began with the Cambrian Period (542 million years ago to 488 million years ago) and its previously noted "explosion" of life forms. To obtain a sense of the magnitude of this explosion, consider the number of genera* at the start and end of the Cambrian Period. At the start of the Cambrian, fewer than 100 genera are estimated to have existed, but by the end, the number of genera was probably at least 1,200, with much of this 12-fold increase in diversity occurring between about 530 million years ago and about 515 million years ago.[46]

With that said, however, identifying when any particular species or genus or other category of life (or equally any characteristic, such as compound eyes or fingernails) first appeared is extremely difficult. No matter how carefully one date for a "first appearance" might be established, there is always the possibility that another study examining and dating yet older rocks could uncover an earlier occurrence of that category or characteristic. Indeed, DNA analyses suggest that some and perhaps most of the animal phyla thought to have originated in the Cambrian, based on the fossil record, might have originated considerably earlier.[47] Nonetheless, the record so far indicates abundant firsts for the Cambrian time frame, including "essentially all of the readily fossilizable animal body plans," although some "lightly skeletonized" forms had appeared in the Ediacaran Period's final 5 million years.[48]

---

* In classifying life, species are grouped into genera and those in turn into families, then orders, then classes, then phyla, and then, finally, in the largest standard grouping in the classification of life, kingdoms. For a concrete example, humans are in the species *sapiens*, the genus *Homo*, the family Hominidae (for Hominids, including all the great apes), the order Primates, the class Mammalia (for mammals), the phylum Chordata (for Chordates, the human subphylum being Vertebrata, for vertebrates), and the kingdom Animalia (for animals).

With the Cambrian came the first widespread appearance of animals with hard parts, specifically shells and/or skeletons. The shells and skeletons made these animals much more likely to be fossilized than the earlier, softer animal forms and hence much more likely to be known about today, with remains available for examination by inquisitive modern scientists. As far as the record shows so far, the first vertebrates (animals with a spinal column, or backbone) appeared in the Cambrian in the form of jawless fishes, as did the first corals. Clams, starfish, jellyfish, seaweeds, mollusks, brachiopods (animals generally attached to the sea floor and filter feeding), and sponges all existed, as did echinoderms (invertebrates with five rays, such as starfishes and sea urchins), the latter in considerable abundance. Perhaps most abundant in the second half of the Cambrian were the trilobites, which seem to have first arisen about 525 million years ago.[49]

Following the Cambrian, the other periods in the Paleozoic Era were the Ordovician (488 million years ago to 444 million years ago), the Silurian (444 million years ago to 416 million years ago), the Devonian (416 million years ago to 359 million years ago), the Carboniferous (359 million years ago to 299 million years ago), and the Permian (299 million years ago to 251 million years ago).

By the end of the Ordovician, some fishes inhabited the oceans,[50] and some photosynthesizers had adapted to freshwater rivers, lakes, and ponds. Sometime in the Ordovician or Silurian, the first land plants appeared, and by the end of the Silurian, true vascular plants, with roots, stems, and efficient water-routing systems, were populating at least some land areas.[51] The initial colonization of land by plants was greatly facilitated by the symbiotic relationship between plants and fungi. The soils of the time would not likely have been readily hospitable to complex organisms. However, higher plants, then as now, allied themselves with fungi able to absorb nutrients from the soil and transfer some of the nutrients to the plants that provided them shelter and carbohydrates. Land animals in the form of invertebrates likely burrowed in the primitive soil.

By the early Devonian, the evolving land ecosystems included such arthropods (today constituting the largest of the animal phyla, including insects and other species) as scorpions, mites, and spiderlike trigonotarbids.[52] All land plants at the time were seedless, and among them were ferns, which expanded and proliferated throughout the Devonian. Although now far from dominant, seedless plants have survived (with evolutionary changes) to the present, the ferns continuing to be among the best-known examples. By about 400 million years ago, plants had spread widely over the continents, generally as thick mats and low shrubbery, housing swarms of spiders, mites, centipedes, and insects.[53]

In the Devonian oceans, there were immense coral reefs, some extending for hundreds of miles, and a wide diversity of fishes, dominated through much of the Devonian by jawless fishes (specifically, agnathans) and placoderms (named for their "plated skin" and containing a neck joint and jaws). The dominance of the jawless fishes and placoderms gave way in the late Devonian to cartilaginous and bony fishes, and few placoderms survived into the Carboniferous Period. In fact, placoderm remains are limited almost exclusively to the Devonian.[54] During the Devonian, fish were so dominant that the period is sometimes labeled "the age of fishes."[55] However, it was also during the Devonian that amphibians first appeared, likely evolved from bony fishes adapting to a local habitat that was becoming increasingly dry. Like the most familiar modern amphibian, the frog, the early amphibians were four-legged animals breathing air and needing water to keep their skin moist. Whether today's three major groups of amphibians—frogs, salamanders, and limbless caecilians—all evolved from a common ancestor back in the Devonian remains uncertain.[56]

The amphibians were so well suited to the environment at the time and became so abundant that they are sometimes said to have "ruled the Earth" during the Carboniferous Period, following the Devonian.[57] Also alive during the Carboniferous were giant insects and the earliest known reptiles, the latter perhaps evolving from the amphibians. Winged insects might first have appeared in the early Carboniferous and were certainly abundant by the late Carboniferous. Millipedes were up to 2 meters long and some of the dragonflies had wingspans comparable to those of today's seagulls. Cockroaches and beetles also swarmed over the land. Dominating in the oceans were sharks and bony fishes, although those shared space with giant coral reefs, brachiopods, starfishes, gastropods, sea urchins, algae, and a wealth of other life.[58]

> As coal is burned today, it returns to the atmosphere carbon dioxide that had been taken from the atmosphere by living plants between 359 million years ago and 299 million years ago.

Plant life in the Carboniferous has a relevance today that goes far beyond intellectual curiosity. Tall lycophyte trees, mosses up to 40 meters tall, and a large variety of ferns contributed to a profuse vegetative cover on Earth's land areas.[59] The lush forests from this, the Carboniferous Period, consumed CO_2 from the atmosphere, with much of the carbon then being incorporated into soil or peat. Much of the peat was buried and, very slowly, converted into extensive coal deposits, formed from the compaction of fossilized plant remains and thus of necessity formed well after the banded iron formations, which were not dependent on fossilized plants. As the resulting coal is burned today, it returns to the atmosphere carbon dioxide that had been taken from the atmosphere by the living plants between 359 million years ago and 299 million years ago.

The Carboniferous coal deposits are particularly prominent in North America and western Europe, where in many locations they appear in layers. Specifically, layers of coal alternate with layers of marine sedimentary rocks, suggesting that the swamps that formed the peat that led to the coal were repeatedly inundated by seawater.[60] A logical explanation for the layering would be a sequence of alternating glacials and interglacials during the Carboniferous, with major flooding occurring at the glacial/interglacial transition because of the melting of the ice, and indeed it does appear, from additional evidence, that the Carboniferous experienced such glacial cycling.

After the massive ice coverage in the Cryogenian and early Ediacaran periods from 850 million years ago to 580 million years ago, the next extended span of time with compelling indications of major glaciation* began at about 340 million years ago, in the Carboniferous, and lasted about 80 million years, until well into the Permian Period.[61] This is also a time frame during which the movement of tectonic plates was bringing most land areas together into one gigantic continent called Pangaea, and indeed the two extended events could well be connected.

The evidence for the Carboniferous/Permian glaciation comes from Antarctica and southern South America, South Africa, India, and Australia, all of which were located in the southern portion of the supercontinent Pangaea existent by the end of the Permian. In fact, the glacial markings in these areas helped scientists piece together the presumed configuration of Pangaea. The land areas that eventually became North America and western Europe, where the coal deposits were being formed, were in the equatorial regions during the Carboniferous, while the glaciated land areas were in the high southern latitudes and included the areas that subsequently separated into Antarctica, Australia, India, South Africa, and South America, the ice radiating from a central dome.[62] The land/sea configuration was crucial to providing a geography conducive to glaciations. In fact, after the Carboniferous/Permian glacial/interglacial cycling, Pangaea migrated out of the central south polar region, and there followed 200 million years believed to have been largely devoid of ice.[63]

Quite likely a contributing factor for the Carboniferous/Permian glaciation was the massive uptake of CO_2 from the atmosphere by the plants that would, with time, create today's coal deposits.[64] Further, it was about 310 million years ago that

---

* Lesser glaciations appear to have occurred at several times during the interval between 580 million years ago and 340 million years ago.

the tectonic plates carrying what would become Europe and Asia were slowly but forcefully pressed against each other, uplifting the Ural Mountains.[65] The resulting chemical weathering of the newly exposed rocks likely removed significant additional atmospheric CO_2. Assuming no counteracting greenhouse gas increases, the CO_2 consumption would have lessened the greenhouse effect and its warming influence, hence encouraging cooling and eventual glaciation. With cooling, water vapor also would be expected to decrease, from reduced evaporation, further reducing the greenhouse effect. In this instance as in others, the highly interconnected nature of the various major components of the Earth system is apparent.

By the end of the Carboniferous, the ferns and other seedless plants were facing competition from a relatively new plant type, the gymnosperms. Gymnosperms, such as conifers, develop seeds and pollen, providing survival advantages over seedless plants, particularly in dry environments. The gymnosperms likely first arose either in the late Devonian or in the early Carboniferous Period. In any event, they seem to have become the dominant land flora by the end of the Permian. Although they have long since lost their dominance, being overshadowed today by flowering plants, some gymnosperms have survived quite well, present examples including pine, spruce, and fir trees, each still abundant in some localities.

In the oceans, both the Carboniferous and the Permian periods had extensive coral reefs and their associated ecosystems, with arthropods, worms, starfish, nautiloids, fishes, and others. Other marine ecosystems were centered on filter feeders attached to muds at the bottom of shallow waters. Most of the filter feeders—and the ecosystems they anchored—did not last beyond the Permian.[66]

In the air, a major new entry in the Permian was the earliest flying vertebrate— the gliding reptile *Coelurosauravus*. Flapping flight was still far in the future, but *Coelurosauravus* had a membrane of skin stretched over ribs, allowing for gliding and, through gliding, preying on airborne insects.[67]

While hundreds or thousands of species were coming into existence and evolving from the Cambrian through the Permian, others were losing in the battle for survival and going extinct. Just as it is difficult (or even impossible) to specify definitively when one or another life form arose, it is equally difficult to specify when it went extinct or, with modern life, whether it even is extinct. Still, at times in the past 500 million years, enough species appear to have gone extinct at approximately the same geologic time that these are considered "mass extinctions." Of the mass extinctions, five stand out as particularly severe and are labeled the five "great mass extinctions."

In a review of the mass extinctions during the Phanerozoic Eon, American paleontologist Richard Bambach mentions previous publications that include up to 61 mass extinction events.[68] Bambach narrows the list to 18 marine extinction events that he would consider to be truly mass extinctions, with three in particular standing out. Two of these three mark the boundaries between the Paleozoic and the Mesozoic eras and between the Mesozoic and the Cenozoic eras; that is, they mark the boundaries separating the three eras of the Phanerozoic Eon. The third of Bambach's big three occurred much earlier, at the end of the Ordovician Period, and the other two of the frequently labeled five great mass extinctions occurred in the Devonian Period and at the end of the Triassic Period. Each of the five great mass extinctions is discussed further, in turn chronologically, as are a few of the particularly noteworthy smaller extinctions.

A great deal could be said about the record of mass extinctions. Bambach estimates the average number of published papers concerning the fossil record of extinctions to have been 7.25 per year for the years 1954–1957 and to have risen decade by decade since then, attaining a rate of over 330 papers per year by the early years of the twenty-first century.[69] Out of necessity, I mention only some highlights, this simply to give a flavor of the reality of extinction events throughout the history of life on Earth. The interested reader can find

> One caution to keep in mind: The literature is filled with contradictions.

massive amounts of additional information about this and all other aspects of Earth history by probing the scientific literature, a reasonable place to start being the 2006 article by Bambach[70] and the other articles referenced in this chapter as well as the further references listed in those articles. One caution, however: the literature is filled with contradictions, as there are both disagreements in interpretation of the data and revisions in understanding as new data and models become available.

Although the current understanding is likely to be revised eventually, the first of the recognized five great mass extinction events occurred about 444 million years ago, marking the end of the Ordovician Period. John Sepkoski Jr., a paleobiologist at the University of Chicago, estimates that at least 70% of ocean species went extinct at this time. Included among them were many species of brachiopods, echinoderms, ostracods (microscopic crustaceans), and agnostids, although new species of some of the same groups arose late in the Paleozoic.[71] The second of the great mass extinctions occurred about 375 million years ago, in the second half of the Devonian Period. Trilobites, placoderms, and specific groupings of sponges, corals, and brachiopods were particularly hard hit, with extinctions of more than 65% of the genera of those groupings. This was followed by a lesser extinction event at the end of the Devonian that wiped out over 70% of the remaining genera of trilobites and placoderms, among others.[72]

The third and greatest of the great mass extinctions occurred about 251 million years ago, marking the end of the Permian Period. Two prime candidates for what might have caused this most major of catastrophes are an impact from an extraterrestrial body (asteroid, comet, or meteorite)[73] and volcanic eruptions.[74] In a book focused on the Permian extinctions, paleontologist Michael Benton of the University of Bristol reviews the many arguments and considerable evidence and weighs in on the side of volcanic eruptions.[75] More specifically, the extinctions may have been connected with a vast volcanic outpouring of basalt known to have occurred in Siberia and the disruption that this outpouring might have caused in the Earth's atmosphere and water cycle. Ian Campbell of the Australian National University and colleagues from Australia, the United States, and Russia suggest that the critical process was the injection of sulfur dioxide and dust into the atmosphere, which would have led to sulfate aerosols and dust blocking out the Sun's radiation and consequently cooling the climate.[76] Others, however, argue that the geologic record indicates global warming rather than cooling and speculate that the warming was from the greenhouse effects of the sulfur dioxide and carbon dioxide emissions connected with the volcanic outpourings, perhaps further enhanced by the release of the greenhouse gas methane from frozen gas hydrates as they melted after the warming began.[77] In other words, the volcanic outpourings might have led to cooling causing the extinctions or might have led to warming causing the extinctions. The basalt outpourings were massive, with the remains still spreading over approximately 340,000 square kilometers[78] and the original basalt estimated to have covered 3.9 million square kilometers to a depth of 400 to 3,000 meters.[79] In the early 1990s, attributing the extinctions to the Siberian basalt outpourings was greatly complicated by the fact that several attempted datings of the basalt outpourings placed them somewhat after rather than before or synchronous with the extinctions.[80] However, the dating issues were resolved, with Renne and colleagues dating both events at 250 million years ago,[81] a date later slightly revised to 251 million years ago.[82] Other hypothesized possible causes for the late Permian mass catastrophe include nutrient collapse, hypercapnia (high CO_2), hydrogen sulfide toxicity, anoxia (lack of oxygen), or oceanographic gas eruption.[83]

> Approximately 90% of all species on Earth went extinct at the end of the Permian.

Whatever the cause of the extinctions at the end of the Permian, the result is believed to have included the extinction of at least 90% of all marine species and approximately 70% of land-based vertebrate families,[84] including nearly 75% of amphibian and reptile families.[85] Benton indicates that 60% to 70% of all families (land or water based) and approximately 90% of all species disappeared, with the losses perhaps being even more severe on land than in the sea despite earlier estimates to the contrary.[86] Among the casualties: all remaining trilobites, most of the radiolarians (microscopic plankton with skeletons generally made of silica), most species of fusulines (the dominant foraminiferans at the time), most corals, most echinoderms (including most species of sea lilies, starfish, and sea urchins), ammonoids (free-swimming cephalopods), and perhaps 99% of all species of brachiopods.[87] With so much of the plant life reduced to decaying organic matter, fungi thrived, and a resulting fungal spike appears in the geologic record at the Permian/Triassic boundary at many locations around the world.[88] In terms of estimated percentage losses, this was the most severe of all mass extinctions, concluding the Paleozoic Era that had begun with the great Cambrian explosion of life.

MESOZOIC ERA (251 MILLION YEARS AGO TO 65 MILLION YEARS AGO)

The Paleozoic gave way to the Mesozoic Era and the first of its three periods, the Triassic. By this time, much of the coal-forming vegetation dominant in the Carboniferous Period had died out, except the ferns, which, along with conifers, cycads, and cycadeoids, dominated land vegetation. Animals entering into the empty niches left by the Permian extinctions included two-legged carnivores, the first dinosaurs, the first modern insects, and, by about 210 million years ago, the first mammals.

During the Triassic (251 million years ago to 200 million years ago), reptiles became the dominant land animals, beginning in the early Triassic with *Lystrosaurus*, a moderate-sized dicynodont, about 1 meter long, that survived the Permian extinctions and was widespread, abundant, and so dominant in the early Triassic that they might have constituted as much as 95% of all land animals at the time.[89] The dominant reptile types varied through the period, but those that gained ascendancy by the late Triassic—when ecosystems were back to full robustness—became known as the "ruling reptiles." The latter included dinosaurs, pterodactyls (winged reptiles), crocodiles, and alligators. The dinosaurs lasted for nearly 200 million years before finally becoming extinct about 65 million years ago, a longevity swamping the longevity so far of humans, which, under even the most generous of definitions of what might constitute a "human," would be no more than a few million years. Furthermore, for much of their history, the dinosaurs were the dominant land animals on Earth, and the entire period from approximately 251 million years ago to approximately 65 million years ago is often popularly referred to as the Age of Reptiles, more formally as the Mesozoic Era, with its three periods, the Triassic, the Jurassic (200 million years ago to 145 million years ago), and the Cretaceous (145 million years ago to 65 million years ago).

Of course, many other animals existed in the Mesozoic in addition to the reptiles, and with hindsight it is clear that the progress of mammals during this era set the stage for the succession of the Age of Reptiles by the Age of Mammals. Even during much of the Age of Reptiles, the early mammals tended to have larger brain size relative to body size than the contemporary reptiles. Still, throughout the Mesozoic, the mammals were fairly inconspicuous, likely with none growing to a size larger than a modern small dog.[90] Warm blood helped the mammals, then as now, to maintain a near constant body temperature, aiding in the accurate functioning of the brain, although many dinosaur species were likely warm-blooded as well and considerably more intelligent than the picture portrayed of them even as late as the mid-twentieth century.

Dinosaurs were not all large, but the largest were gigantic, with some having thighbones longer than the height of the tallest of modern humans, some having skulls over 2.5 meters long, lengths over 40 meters, and weights possibly exceeding 100 tons. Of course, they did not start at such sizes, all being far smaller when they first emerged. By the late Triassic, the largest was *Plateosaurus*, which abounded in large numbers in central Europe and grew to lengths of up to 8 meters,[91] but much larger dinosaurs, with lengths sometimes exceeding 40 meters, arose in the early Jurassic.[92] The largest, the giant sauropods, had emerged by the late Jurassic and have species names such as *Diplodocus*, *Brachiosaurus*, *Supersaurus*, *Ultrasaurus*, and *Seismosaurus*. Not as large but well known from their many remains in North America were the Cretaceous dinosaurs *Triceratops* and *Tyrannosaurus rex*.[93]

Along with the dinosaurs and mammals, the Triassic Period witnessed the first (or the earliest with remains that have been found and classified) turtles, first crocodilians,[94] and first flying vertebrates, reptiles called pterosaurs.[95] Also abundant at the time was the much smaller (on the order of 0.5 meters long) tuatara (*Sphenodon punctatum*), notable because the species survived all the extinctions and other gyrations of the past 200 million years, with living samples today thought to be nearly identical in appearance to the individuals alive 225 million years ago, although today they are found only on a few small islands off the two main islands of New Zealand.[96]

The Triassic ended approximately 200 million years ago with the fourth of the five great mass extinction events. This event marked the end of the conodonts, a category of primitive fishes,[97] as well as the extinction of at least 67% of the genera of calcareous sponges, scleractinian corals, nautiloid cephalods, ammonites, and some categories of brachiopods.[98]

In the Jurassic, seed-bearing gymnosperms were the dominant trees, and marine animals included the ammonites and belemnites (both being groups of mollusks), the former existing in abundance and having coiled shells used for buoyancy control and sizes ranging in diameter from 1 centimeter to 2 meters. New were the teleosts, a type of bony fish with a new jaw apparatus that allowed them to manipulate their food. Teleosts today include salmon, tuna, eels, and seahorses. Among the animals feeding on the Jurassic teleost fishes were the plesiosaurs, also new in the Jurassic. The plesiosaurs were up to 4 meters in length and were probably in turn eaten by the much larger pliosaurs, which ranged to 12 meters in length.[99] The Earth's first bird was likely *Archaeopteryx*, evolved from dinosaurs in the late Jurassic.[100]

In the Cretaceous, key new additions were the flowering plants, or angiosperms, the oldest fossils of which are dated at about 130 million years ago. Within about 50 million years, the angiosperms had outstripped both the seedless land plants (existent since the Ordovician or Silurian) and the gymnosperms (existent since the late Devonian or early Carboniferous), and by the end of the Cretaceous, the angiosperms included magnolia, oak, palm, birch, walnut, fig, holly, sycamore, and willow trees. The flowering plants remain dominant today in the plant world. New also in the Cretaceous—and some quite helpful to the flowering plants—were bees, moths, ants, butterflies, and termites.[101] The flowering plants and pollinating insects proceeded to coevolve in mutually beneficial ways, becoming heavily dependent on each other.[102] Deposits laid down in the Cretaceous from all this abundance of life have become major chalk beds, coal fields, and oil reserves.[103]

During the Age of Reptiles, global temperatures were considerably higher than in the past few million years, and during much of the era the Earth was likely without any major glacial conditions: no Antarctic or Greenland ice sheets and no glacial conditions in North America or Eurasia. This was probably due in part to high greenhouse gas concentrations—the Intergovernmental Panel on Climate Change (IPCC) suggests that CO_2 concentrations were perhaps above 3,000 parts per million in the Jurassic[104] (versus about 386 parts per million in 2008)—but was also greatly facilitated by the land/sea geography at the time. In particular, there were no major landmasses in the central polar region of either hemisphere and water could move freely from low to high latitudes and vice versa. Much of the land was still in the supercontinent of Pangaea (formed during the late Paleozoic Era) during the Triassic and Jurassic, but this mass had rotated, drifted, and otherwise adjusted considerably from where it had been during the Carboniferous/Permian glaciations. In the mid-Jurassic, the portions eventually to become South America and Africa were largely in the Southern Hemisphere (although straddling the equator), the portion to become North America was in the low and midlatitudes of the Northern Hemisphere, much of Eurasia was in the midlatitudes of the Northern Hemisphere (except India, which was situated well south in the Southern Hemisphere), and Antarctica and Australia were predominantly in the midlatitudes of the Southern Hemisphere, joined to each other and, on the Antarctic side, to South America, Africa, and India, and extending south to about 75°S. Later in the Jurassic, the continents began to separate and drift apart, a process that has continued and is ongoing today.[105] The rupturing included the splitting of Pangaea first into two large landmasses of Laurasia in the Northern Hemisphere and Gondwana (or Gondwanaland) in the Southern Hemisphere, this split occurring in the Jurassic, then later further splitting into the more familiar continents of today. Gondwana began to separate into smaller pieces about 120 million years ago, with South America, Africa, the combined Antarctica/Australia, and the subcontinent of India all drifting apart by the end of the Cretaceous.

By the late Cretaceous, North America and Eurasia were joined in the north, roughly at the landmass that became Greenland. Eurasia and Africa were separated by a narrow waterway, and North and South America were separated by a much greater expanse of water. India was about midway on its slow trek from Antarctica to its eventual protracted collision, centered about 35 million years ago, with southern Asia. Antarctica was roughly in its current south polar location.[106]

The last of the five completed great mass extinctions occurred about 65 million years ago, ending the Mesozoic Era (the Age of Reptiles) and its final period, the Cretaceous. At this time, about half of the then-existent species of plants and animals became extinct, the most famous being the dinosaurs. Most dinosaur species had probably died out earlier, but the remaining ones became extinct as a part of the great mass extinction event 65 million years ago. In contrast to some extinction events, this one has a solid prime candidate for its cause. Specifically, a large asteroid (on the order of 10 kilometers in diameter and 1,000 billion tons) striking the Earth just off the Yucatán Peninsula near Chicxulub, Mexico, is thought to have triggered the extinctions.[107] The hypothesis that an asteroid was the root cause was presented by Luis and Walter Alvarez, Frank Asaro, and Helen Michel of the University of California, Berkeley, in 1980, based on anomalously high iridium concentrations found in deep-sea sediments in the vicinity of Italy, Denmark, and New Zealand.[108] Although forcefully denounced by some geologists in the 1980s,[109] the asteroid hypothesis gained substantial support after an appropriately sized and dated crater was found near Chicxulub.[110] As for the mechanism by which the asteroid precipitated the disaster, the favored hypothesis appears to be that it sent dust and other particles into the atmosphere, both causing a disastrous sudden cooling and also sufficiently restricting sunlight transmission to the surface to seriously curtail photosynthesis.[111] Other speculations include that the impact might have caused a sudden severe greenhouse warming by sending CO_2[112] or steam into the atmosphere or might have caused raging fires. A prime nonextraterrestrial hypothesis is that the extinctions might have been caused—or at least influenced—by massive volcanic eruptions occurring in India and now evidenced by the resulting Deccan Traps covering much of northern India.[113] Many other theories of the dinosaur extinctions have been offered over the years, in fact so many that Benton lists 100 of them. For a flavor, speculations put forth as early as the decade of the 1920s include topographic changes, gradual global cooling, disease, volcanic eruptions, and consumption of dinosaur eggs by mammals.[114]

Whatever the reason for the extinctions, both the reign of the dinosaurs and the Age of Reptiles came to an end. Among the extinctions: the dinosaurs, pterosaurs, mosasaurs, plesiosaurs, many species of plankton, and some families of birds, marsupial mammals, teleost fishes, ammonites, and belemnites.[115] Also gone: over 60% of the genera of bivalve mollusks, ammonites, and marine reptiles.[116] Animals that survived the extinctions include beetles, flies, bees, wasps, ants, moths, butterflies, oysters, clams, salamanders, turtles, lizards, crocodiles, some primitive snakes, and numerous species of sharks and small mammals. Ferns once again proved resistant to extinction, and while many species of flowering plants also survived the extinction episode, others did not. As with the other extinction episodes, some of the largest and most dominant animals (e.g., in this case the dinosaurs) proved the most susceptible, having gained their dominance through attaining a high level of specialization that made them less able to adapt to altered conditions.

CENOZOIC ERA (65 MILLION YEARS AGO TO THE PRESENT)

After the extinction of the dinosaurs, mammals gradually rose from their fairly insignificant status during the Age of Reptiles to become the dominant medium-sized and large-sized land animals on Earth, resulting in the entire period since 65 million years ago being popularly labeled the Age of Mammals. More formally, it is the Cenozoic Era, divided into the Paleogene and Neogene periods and subdivided, in order, into the following epochs: the Paleocene (65 million years ago to 56 million years ago), Eocene (56 million years ago to 34 million years ago), and Oligocene (34 million years ago to 23 million years ago) in the Paleogene, and the Miocene (23 million years ago to 5.3 million years ago), Pliocene (5.3 million years ago to 1.8 million years ago), Pleistocene (1.8 million years ago to 11,800 years ago), and Holocene (11,800 years ago to the present) in the Neogene. Among the mammals, primates first appeared apparently about 80 million years ago, survived the extinctions 65 million years ago, and gradually increased in numbers, sizes, and types, with apes estimated to have first emerged between 55 million years ago and 38 million years ago.

Global climate is thought to have warmed during the first 10 or so million years after the extinctions 65 million years ago, during the Paleocene and early Eocene. Within this span, a particularly prominent warming, by several degrees Celsius, occurred in about 1,000 to 10,000 years circa 55 million years ago,[117] at which time the central Arctic Ocean is believed to have been free of ice and to have had sea surface temperatures rising to approximately 24°C (75°F).[118] A leading candidate for the cause of the warming 55 million years ago is the possibility that there was a large release of the greenhouse gas methane from the ocean depths, described as a "megafart" and the Earth's "biggest fart ever."[119] (In fairness to the gas methane, flatulence contains many gases, often including nitrogen, oxygen, carbon dioxide, and hydrogen sulfide as well as methane.) An estimated trillion tons of methane, from methane clathrates that broke down because of ocean warming, are believed to have risen from the ocean into the atmosphere, causing the atmosphere to warm globally by between 6°C and 10°C as a result of the greenhouse effect. Furthermore, at least one other greenhouse gas is believed to have been prominent at this time as well, with atmospheric CO_2 concentrations estimated to have been over five times as high near the transition from the Paleocene to the Eocene as at the end of the twentieth century. Along with the greenhouse-induced warming near the Paleocene/Eocene transition and perhaps connected to it, there was a mass extinction of deep-sea organisms.[120] An estimated 50% of all the species of foraminifera are believed to have perished. In their wake, the fossil record of 55 million years ago shows the first appearance of many modern mammals, from rodents to individual primate species, and the first hooves, first claws, and first stabbing canines.[121]

> In the last 55 million years, the global temperature trend has basically been downward, in other words, global cooling.

In the 55 million years since the relatively brief warming spurt about 55 million years ago, the global temperature trend has basically been downward, that is, global cooling, although compounded by many fluctuations. Among the fluctuations, the basic trend over the past 18,000 years has been a warming; but however significant those 18,000 years have been to humans, they are quite minor in the time frame of the past 55 million years, with the result that today's temperatures, overall, are lower than the temperatures for most of the period from 55 million years ago to 2 million years ago.[122]

The large-scale, long-term cooling of the Cenozoic is associated with the break-up of the supercontinent Pangaea and the migration of the resultant smaller continents into the polar regions. Antarctica had already migrated into the south-central polar region, and North America and Eurasia moved into high northern latitudes. In both cases, the movement of land into the polar regions made conditions ripe for gradual increases in snow and ice cover on land. The resultant increase in planetary albedo (reflectivity) from the increased ice and snow cover further accentuated the cooling trend, as more of the Sun's radiation was reflected back toward outer space. The ice cover increased planetary albedo both directly, because of the very high reflectivity of ice and snow, and indirectly, because when water was transferred to the land-based ice, sea level was lowered, replacing some ocean surfaces along the continental boundaries by more reflective land surfaces.

The long-term cooling that began about 55 million years ago is also associated with decreasing atmospheric CO_2 levels. At 55 million years ago, CO_2 is estimated to have constituted about 2,000 parts per million of the atmosphere; by 40 million years ago, it was down to about 700 parts per million; and by about 24 million years ago, it had decreased to below 500 parts per million, perhaps for the first time in hundreds of millions of years. All this puts the much lower early twenty-first-century value of about 380 parts per million into a different perspective than when comparing it only with the past few millennia.[123] The current CO_2 levels, however much of a concern, are considerably lower than the CO_2 levels for most of the period from 55 million to 2 million years ago.[124]

> The current atmospheric CO_2 levels, however much of a concern, are considerably lower than the CO_2 levels for most of the period from 55 million years ago to 2 million years ago.

Greater than half the drop from the early Eocene's atmospheric CO_2 concentrations of 2,000 parts per million is thought to have occurred approximately 50 million years ago. Evidence from deep-sea cores drilled in the Arctic during the 2004 Arctic Coring Expedition offers a possible explanation, as these cores include a substantial layer dominated by the plant *Azolla* and dated at about 50 million years ago. *Azolla* is a fast-growing freshwater fern that has been used in Asia for hundreds of years as a natural fertilizer in rice fields because of its absorption of atmospheric nitrogen. But *Azolla* also absorbs atmospheric CO_2. In fact, *Azolla* absorbs atmospheric CO_2 at a rate high enough that widespread floating *Azolla* mats in the Arctic, as suggested from the deep-sea cores, are considered likely to have been a significant contributing factor in the reduction of atmospheric CO_2 about 50 million years ago.[125]

Another possible strong, perhaps dominating contributing factor to the decreasing atmospheric CO_2 levels in the Cenozoic is the formation of the Himalayan Mountains. Following its slow drift northward after breaking off from Gondwanaland, about 50 million years ago India began to collide with Asia, forcing the slow uplift of the Tibetan Plateau and the formation of the Himalayan Mountains. Chemical weathering through tens of millions of years in this vast mountain region removed CO_2 from the atmosphere and thereby surely contributed to a lessened greenhouse effect and resultant cooling since the early Eocene.[126] Some regard it as likely not only a cause but rather the primary cause of the cooling.[127]

The continental migrations, lessened CO_2, and associated cooling eventually led to increased ice coverage in both polar regions. In the south, although Antarctica might have already been positioned in the south polar region for tens of millions of years, major glaciation on the continent seems to have begun about 34 million years ago.[128] Modeling studies by Robert DeConto and David Pollard of the University of Massachusetts and Pennsylvania State University, respectively, suggest that cooling brought about by CO_2 decreases played the dominant role in the emergence of the Antarctic ice sheet, although the specific timing suggests that another strong triggering factor might have been the opening of the water passageways between Antarctica and Australia and between Antarctica and South America.[129] These passageways isolated Antarctica from the more equatorward landmasses, perhaps by the end of the Eocene, 34 million years ago.[130] Furthermore, the resulting circumpolar current surrounding the continent likely reduced the flow of warm waters from lower latitudes through to the Antarctic continent, hence also contributing to the emergence of major Antarctic glaciation approximately 34 million years ago. Not until much later, with further CO_2 decreases, further cooling, and further drifting of the northern continents, did comparably major Cenozoic glaciations begin in the Northern Hemisphere.

During the extended period of general cooling after 55 million years ago, mammals continued to evolve, with apes prominent in some regions by 24 million years ago. Remains of Miocene apes (23 million years ago to 5.3 million years ago) have been found in Africa, where they are thought to have first emerged,[131] and in Europe and Asia. These apes seem to have flourished in forested areas on all three continents in the early and mid-Miocene.[132] However, as the regions continued to cool and dry, the habitat became less favorable, with the result that most of the Miocene ape species went extinct by the end of the Miocene Epoch. Apes had not yet arrived in the Americas, but during the Miocene North America reached close to its modern complement of amphibian and reptile families, including the addition of three new genera of salamanders, seven genera of lizards, and nine genera of snakes.[133]

It is estimated that somewhere between 8 million years ago and 5 million years ago, the last common ancestors of humans and chimpanzees roamed the African homeland.* Naturally, at the time of the split the two creatures looked extremely similar to each other and distinctly different from modern humans. Although perhaps not apparent at the time the split occurred, an eventual distinguishing feature of the group whose descendants included humans was upright walking. This feature took time to develop, but some signs of upright walking appear in remains of individuals from as early as 6 million years ago or perhaps even 7 million years ago, and by 4 million years ago all species of the genus *Homo* and all species of the genus *Australopithecus* were likely walking upright. The period from about 6.5 million years ago to about 5.4 million years ago is thought to have been unusually cold and dry, perhaps forcing our ancestors to adapt to less wooded habitats than their ancestors had been used to.[134] Readers seeking a readable account of the highly contentious search for the earliest humans and the discoveries over the last decades of the twentieth century and the first few years of the twenty-first century are referred to *The First Human: The Race to Discover Our Earliest Ancestors* by Ann Gibbons.[135] Other early hominid species include *Ardipithecus ramidus*, which likely still moved easily in trees as well as being able to walk on two legs.[136]

For a while, Australopithecines were thought to have been ancestral to humans, but many scientists in the early twenty-first century instead believe that all Australopithecine species went extinct and were not among our direct ancestors. Nonetheless, several of the Australopithecine remains have become well known, in part because of the early thought that they were in our direct ancestral line.[137] Among the

---

* Until the 1990s, it was thought by many that the Hominidae family of primates, eventually including all species of the human genus *Homo* (of which only modern humans, *Homo sapiens*, survives) and *Australopithecus*, split from the great apes, consisting of chimpanzees, gorillas, and orangutans. DNA studies in the 1990s, however, suggest that chimpanzees and humans are so closely related that chimpanzees belong in the Hominidae family. In any case, the split between the ancestors of chimpanzees and humans is thought to have occurred between 8 million years ago and 5 million years ago and to have taken place in Africa.

most famous: 1) the Taung Baby, discovered by Raymond Dart in 1924 in Taung, South Africa, and dated to between 1 million years ago and 2 million years ago; 2) Zing (nicknamed the Nutcracker Man), discovered by Mary and Louis Leakey in 1959 at Olduvai Gorge in Tanzania (within the Great Rift Valley of eastern Africa) and dated to 1.8 million years ago; 3) Lucy, discovered by Donald Johanson in 1974 in Hadar, Ethiopia (also within the Great Rift Valley) and dated to between 3.1 million years ago and 3.2 million years ago; and 4) Abel, discovered by Michet Brunet's Mission Paléoanthropologique Franco-Tchadienne team in Koro Toro, Chad, in 1995 and dated to 3.5 million years ago.

By the time of the most famous of the Australopithecines, the Miocene had been succeeded by the Pliocene (5.3 million years ago to 1.8 million years ago). Although continental movements continued and still continue today, by the start of the Pliocene all the Earth's continents had reached approximately their current configurations. However, land ice was not as abundant as currently, and as a result sea level was higher, by an estimated 15 to 25 meters.[138] The mid-Pliocene (about 3.3 million years ago to 3.0 million years ago) is thought to be the most recent period when global temperatures exceeded those of today by a substantial amount for a sustained period, with estimates by the IPCC that global temperatures at that time were 2°C to 3°C higher than in about 1800.[139] Somewhat ironically, given the current concerns about global warming, this period, thought of as a possible analogue to the predicted future warmer Earth, is often called the mid-Pliocene "climatic optimum," as though warmer were better.[140]

By 3 million years ago, the long-term cooling under way since 55 million years ago and the continuing evolution of land/sea boundaries through continental drift had reached a temperature and land/sea configuration amenable to ice accumulation in the Northern Hemisphere, providing the precursor conditions for the Northern Hemisphere glacial/interglacial cycling to come. The ice sheet formation starting in the Arctic about 3 million years ago[141] was likely the first large-scale appearance of ice in the Northern Hemisphere during the Cenozoic Era, meaning that abundant Northern Hemisphere ice came tens of millions of years after ice had become well established in Antarctica.* This marked the beginning of an ice age sequence that started in the Pliocene and continued in the Pleistocene, with ice sheets repeatedly forming and then decaying in North America, Greenland, and

---

* Preliminary results from the August 2004 Integrated Ocean Drilling Program Arctic Coring Expedition include suggestions that the Arctic Ocean may have been ice covered tens of millions of years earlier than had been thought previously, with the timing of the onset of glaciation perhaps being comparable in both polar regions (Moran et al. 2006). Revisions to our understandings of this and other aspects of Earth history are bound to continue to occur as more records are obtained and analyzed.

Eurasia.[142] The growth/decay cycles at first lasted approximately 41,000 years, although as of 900,000 years ago they shifted to a 100,000-year cycling,[143] a topic that is discussed in greater depth in chapter 9.

Also about 3 million years ago, the combination of continental drift, falling sea level, and perhaps a mountain-building episode resulted in the joining together of North and Central America with South America at the isthmus of Panama. With that connection, quite suddenly land animals could readily move between the two continents, and indeed armadillos, porcupines, giant sloths, tortoise-like glyptodonts, and others moved north from their region of origin in South America, while dogs, cats, horses, elephants, deer, and the ancestors of the modern llamas traveled south into South America.[144] Just as important, the waters and marine animals of the tropical Pacific and Atlantic oceans could no longer readily move between those oceans. Blocking the exchange of waters allowed diverging characteristics in the two oceans, notably in salinity, with the average salinity of the tropical Atlantic increasing over that of the tropical Pacific as the fairly persistent east-to-west-flowing trade winds carry water evaporated from the Atlantic westward to the Pacific, freshening the Pacific waters through rainfall while making the Atlantic waters more saline.

Although apes had been in Europe, Asia, and Africa for many millions of years, hominids appear to have evolved in Africa and to have remained only in Africa until more recently than 2 million years ago. During this period, Africa was undergoing climate changes that likely played a role in hominid development. In particular, the region went through cycles of cool and dry conditions alternating with warm and wet intervals, with the larger trend being toward cooler and drier. During the cool/dry periods, the forests shrank and hence the overall trend toward cooler and drier meant overall less tree coverage and more coverage of grasslands.[145] As a result, the local forest-dwelling hominids were forced to adjust either to a decreasing habitat area or to life at least in part away from the trees. Those that adjusted to less dependence on the trees emerged with important evolutionary advantages. Although they likely had become bipedal before the Pleistocene cool/dry periods severely reduced the size of the forests, that reduction encouraged the transition toward becoming ground dwellers rather than tree dwellers. The transition to the ground, with bipedal locomotion, had a very significant advantage in freeing the hands for uses other than moving from place to place, for instance, for communicating, carrying infants and other items, throwing, and making tools.[146]

Upright walking developed by about 4 million years ago, well in advance of the expanded brain size that also characterized later *Homo* species. Tool use, with primitive stone tools, was likely common by about 2.5 million years ago or soon thereafter.[147] Several *Homo* species prior to *Homo sapiens* have been identified, including *Homo rudolfensis*, *Homo habilis*, *Homo erectus*, *Homo ergaster*, *Homo heidelbergensis*, and *Homo neanderthalensis*, although Bernard Wood of George Washington University and Mark Collard of the University College London argue that *Homo rudolfensis* and *Homo habilis* should be removed from *Homo* status and probably reclassified in the genus *Australopithecus*.[148] *Homo rudolfensis*, alive in Africa from about 2.4 million years ago to about 1.8 million years ago,[149] was perhaps the first species to use tools extensively.[150]

Fossil remains suggest that *Homo habilis* arose between 2.3 million years ago and 1.9 million years ago and lasted until about 1.4 million years ago, *Homo ergaster* arose about 1.9 million years ago and lasted until about 1.5 million years ago, and *Homo erectus* arose about 1.9 million years ago and lasted until about 200,000 years ago; all three species likely first arose in Africa.[151] Although *Homo erectus* was thought by many to have evolved from *Homo habilis*, that evolutionary line was called into question in 2007 with the announcement of the discovery of new fossils in Kenya's Koobi Fora Formation, east of Lake Turkana. These fossils suggest that *Homo habilis* and *Homo erectus* coexisted in this region for almost half a million years and that neither was ancestral to the other.[152] Regardless of ancestry, groups within *Homo erectus* eventually appeared in Asia, likely although not definitively as a result of migrating from their original homeland in Africa. Remains of *Homo erectus* were discovered in 1891 by Eugene Dubois in Trinil, Java, dated to between 0.8 million years ago and 1.2 million years ago, and in 1929 by Davidson Black in Peking. These early discoveries of what have become known as Java Man and Peking Man, respectively, led many in the early twentieth century to feel that the origins of humanity were in Asia, prior to attention shifting in the mid-twentieth century quite decisively to Africa as the continent of human origins.[153] Some scientists in the early twenty-first century feel that the evidence unveiled so far is too spotty to be certain and are reconsidering where the genus *Homo* originated, the prime candidate other than Africa being Asia. Africa in any case retains a critical role, as the proposed sequence for the Asian origin seems to be that prehuman hominids from Africa migrated to Asia and that their descendants evolved into humans in Asia.[154] However, even though the evidence is spotty, the predominance of the evidence in the early twenty-first century supports an African origin.

In line with the rest of Earth history prior to the past few thousand years, huge uncertainties exist regarding almost all aspects of early human development. Take, for instance, the illustrative issue of when humans first starting cooking and the importance of that to the evolutionary enlargement of the human brain, a topic of much interest at the Primatology Meets Paleoanthropology Conference held on April 17–19, 2007, at the University of Cambridge in the United Kingdom. As summarized by Gibbons,[155] Richard Wrangham of Harvard University suggests that cooking began between 1.9 million years ago and 1.6 million years ago and was fundamental to future human development because the chemical changes that occur during heating reduce the internal human energy required for digesting food. This frees a significant amount of the internal human energy to be used elsewhere, notably in the brain, hence leading to tying the development of cooking to the doubling of the brain size that is thought to have occurred between about 1.9 million years ago and 1.6 million years ago. Others, however, noting the lack of any evidence for controlled fire by 1.6 million years ago, argue that cooking may not have arisen until 500,000 years ago or even later. Even if so, John Allman of the California Institute of Technology suggests that the development of cooking could still be connected with brain enlargement, although with the further expansion of the brain that is thought to have occurred in the past 500,000 years rather than with the doubling of brain size between approximately 1.9 million years ago and 1.6 million years ago. Whenever it arose, cooking certainly helped us out—chimpanzees are said to spend an average of five hours a day chewing their food! But it is by no means the only candidate trigger for the apparent brain enlargement between 1.9 million years ago and 1.6 million years ago. In 1995, Leslie Aiello of University College London and Peter Wheeler of Liverpool John Moores University proposed instead that increased meat eating, regardless of cooking, provided greater energy per volume of food. Hence, guts could shrink and less energy was needed for digestion, again leaving more energy for the brain. The Aiello/Wheeler theory has received some support from studies of other primates, birds, and fish. Others speculate that upright walking and running could have been brain-enlargement triggers, and certainly it is quite likely that multiple triggers all played important roles.

The importance of controlled fire extended far beyond cooking, as it could be used to scare animals away, provide light in caves, and provide warmth, the latter being especially important as people moved to harsher climates in Europe and Asia and as the regional temperatures cooled, especially leading into ice ages.[156] Although it is not known exactly when hominids first became capable of generating and using fire for their own purposes, it is believed to have certainly happened by 500,000 years ago.

Also contentious is the timing of the emergence of our modern human species, *Homo sapiens* (anatomically modern humans).[157] Some feel that *Homo sapiens* might have emerged by about 500,000 years ago,[158] while others place the date much later, for instance, at about 150,000 years ago[159] or about 120,000 years ago.[160] Dennell and Roebroeks indicate that *Homo sapiens* existed in eastern Africa by about 200,000 years ago and in Asia (specifically Israel) at least by about 115,000 years ago and perhaps much earlier.[161] Regardless of the timing, although groups of earlier *Homo* species had previously ventured out of Africa, *Homo sapiens* is now widely believed, like the first *Homo* species, to have originated in Africa, perhaps a descendant of *Homo ergaster* rather than, as many earlier believed, a descendant of *Homo erectus*.[162] With the evidence still quite slim, new findings could eventually overturn some or all of these current understandings.[163]

In the context of the 4.6-billion-year history of the Earth, the emergence of humans is a quite recent event, not occurring until well after 99.9% of Earth's history had elapsed (at least according to current scientific dating methods). Despite our late arrival, we have certainly grown to become a potent global force. From Africa (presumably), migrating *Homo sapiens* reached the Middle East, then India and elsewhere in Asia, and then Europe. Some had ventured as far as Australia by 65,000 years ago, and some had reached New Guinea by 45,000 years ago.[164] In Europe, they encountered *Homo neanderthalensis* (Neanderthals), which had arisen by about 250,000 years ago (perhaps much earlier), and lived in proximity to them for thousands of years before the Neanderthals became extinct about 35,000 to 30,000 years ago.[165] Neanderthals are notable as the first humans known to have left behind convincing evidence that they cared for those who were sick and buried their dead. They did not, however, have as large and sophisticated a set of tools as the Cro-Magnons, and that placed them at a distinct disadvantage.

By the time of the earliest *Homo sapiens*, the Earth was well into its most recent cycling of glacial and interglacial episodes. Deep-sea cores suggest that these cycles have been occurring for the past several million years, in which case they began before the start, at 1.8 million years ago, of the Pleistocene Epoch. Ice cores confirm the cycling for the past 740,000 years, during which time frame a full glacial/interglacial cycle has tended to last approximately 100,000 years, with about 70,000 to 90,000 years of each cycle experiencing glacial conditions, characterized by major ice sheets spreading over much of northern North America and Eurasia. The interglacials typically last about 10,000 to 30,000 years, and the warming at the transition from glacial (or ice age) to interglacial conditions typically occurs mostly in about 5,000 years.[166]

Records collected from locations scattered around the world, many from ice cores and deep-sea cores, confirm that the globe as a whole experienced coherent climate changes during these glacial/interglacial cycles. In the Antarctic region, the glacial/interglacial temperature contrasts were about 10°C, as determined from analysis of the Vostok ice core,* drilled vertically through 3,623 meters of the Antarctic ice sheet at the Russian Vostok station in East Antarctica and reaching back 420,000 years.[167] In the tropical oceans, the glacial/interglacial temperature contrasts were about 3°C to 4°C, and on a global average they were about 5°C.[168] So when people speak of a potential future global warming of 5°C, they are speaking of a warming equal in magnitude to the contrast between full-scale ice age conditions and today, only in the direction of further warming rather than in the direction of a return to an ice age.

> When people speak of a potential future global warming of 5°C, they are speaking of a warming equal in magnitude to the contrast between full-scale ice age conditions and today.

As the Northern Hemisphere ice sheets cyclically advanced southward, ecosystems were forced to adjust. Near the ice, tundra and taiga likely dominated, whereas farther equatorward, remains suggest an interleaving of species that lived in the region during the interglacials with species invading from the north as the ice covered the northern habitat. When the ice retreated, as a glacial episode transitioned to an interglacial, the landscape initially emerging in the newly deglaciated regions contained a mixture of sterile mud, silt, sand, and gravel. Among the key pioneer species stepping in to establish a foothold in these less-than-ideal conditions were spiders, voles, willows, and a white and yellow rose called the "dryas," preparing the way for the reestablishment of more complete ecosystems.[169]

As the glaciations of the early Pleistocene, 1.8 million years ago to 400,000 years ago, lowered the sea level sufficiently for a Bering land bridge to exist between Asia and North America, some of the important animals crossing that bridge from Asia and entering the Americas were the mammoth, modern horse, hare, saber-toothed cat, and rodents. About 400,000 years ago, the first bison to enter North America are believed to have taken the Bering land bridge route.[170]

Over the glacial/interglacial cycles, atmospheric CO_2 concentrations—well below what they had been earlier in the Cenozoic—have typically varied from a low of approximately 180 parts per million during the depths of the glaciations to a high

---

* The Vostok ice core is additionally of interest because of a large lake, Lake Vostok, underlying the ice in this region of Antarctica. Lake Vostok has an area of approximately 14,000 square kilometers, with a water volume estimated at 5,600 cubic kilometers. The many microbes found in the accretion ice overlying the lake (at the bottom of the Vostok ice core record) suggest that Lake Vostok may house a fascinating and complex (although microbial) ecosystem underneath the vast frozen Antarctic ice sheet (D'Elia et al. 2008).

of approximately 300 parts per million during the midst of the interglacials,[171] and atmospheric methane concentrations have typically varied from about 400 parts per billion during the glaciations to about 700 parts per billion during the interglacials.[172] The record from the Vostok ice core suggests a strong correlation extending back at least 420,000 years between atmospheric temperatures and each of these greenhouse gases.[173] Notably, however, the greenhouse gas changes lag behind the temperature changes (by an estimated several hundred years) and hence are not the instigator of them,[174] although they could well provide feedbacks that serve to amplify the temperature changes once begun.

There are several compelling reasons why atmospheric methane concentrations rise and fall in concert with the glacial/interglacial cycles and hence with somewhat of a lag behind the atmospheric temperature increases and decreases. One reason, explaining at least a portion of the methane decreases during glaciations, is that any wetlands overridden by the ice sheets would no longer be emitting methane to the atmosphere. Another, related reason is that the dry conditions thought to have predominated elsewhere around the globe during glaciations would also have reduced the extent of wetlands and hence their methane emissions to the atmosphere. With the retreat of the ice during deglaciation, wetlands could be reestablished, in the process reestablishing that source for atmospheric methane. Another methane source during deglaciation would have been thermokarst (thaw) lakes.[175] Thermokarst lakes form when water accumulates in the depressions produced in the landscape as permafrost in the underlying ground decays. Methane from below bubbles up through these lakes, and some of this methane makes its way into the atmosphere. Katey Walter from the University of Alaska and colleagues estimate that 33% to 87% of the high-latitude increase in atmospheric methane during the deglaciation following the last great ice age (the ice age starting about 116,000 years ago and peaking about 21,000 years ago) came from Northern Hemisphere thermokarst lakes.[176] These various factors—ice overriding former wetlands, the reduction of wetlands due to drying during glaciations, and the absence of thermokarst lakes during glaciations—together provide a quite reasonable explanation of why atmospheric methane would be lower during major glaciations than during interglacials.

Turning from methane to CO_2, two postulated explanations for why atmospheric CO_2 decreases as the temperatures drop and the Earth goes into a glaciation are as follows: 1) The dry, arid conditions in low latitudes during times of major glaciation lead to more and larger deserts, hence generating enhanced dust storms. The dust storms transport iron-rich mineral dust across the oceans, fertilizing the oceans for plankton, which then consume atmospheric CO_2.[177] 2) The cold conditions during major glaciations also induce increased ocean absorption of CO_2 through a pure temperature effect, as cold water has a greater capacity to absorb CO_2 than warmer water does. Both of these processes would encourage reduced atmospheric CO_2. However, at the same time, the huge areas covered by ice during a major glaciation would cover plant life and rocks that would, if uncovered, potentially be taking CO_2 out of the atmosphere. Hence, before the explanation for why there is reduced atmospheric CO_2 during glaciations can be solidified, there is a need to quantify that the extra ocean consumption of CO_2 because of the enhanced dust storms and the colder oceans would more than compensate for the reduced land-based consumption of CO_2 caused by the burial of plant life and rocks under the ice.

However sound or unsound the explanations might be, both the decreasing CO_2 and the decreasing methane are well recorded in the ice core records. Although lagging behind and hence not instigating the coolings, both the CO_2 and the methane decreases would have contributed to enhancing the cooling through the greenhouse effects of those two gases.

With greenhouse gas changes lagging the temperature changes, other mechanisms must have been in play to cause the glacial/interglacial cycling. Among the mechanisms put forward as possibilities, the one that has been favored since the mid-1970s is forcing by changes in the Earth's orbit around the Sun, specifically the version first detailed by Milutin Milankovitch in the first half of the twentieth century. This version incorporates changes in the tilt of the Earth's axis, the eccentricity of the Earth's elliptical orbit, and the precession of the equinoxes, all with long cycles of tens of thousands of years (see further discussion and details in chapter 9). This Milankovitch orbital forcing is presumed to come into play with respect to ice age cycles only once the land/sea geography is amenable to glaciations, hence explaining why these glacial/interglacial cycles do not occur continually throughout Earth's history.*

The Milankovitch theory places an emphasis on summer insolation, with the ice ages coming in times of low summer insolation and hence less likelihood of ice and snow melting away during the summer season. With this theory being the current favored explanation for the ice age cycles (see chapter 9), it becomes relevant that the next large reduction in Northern Hemisphere summer insolation is calculated to come 30,000 years from now.[178] This would suggest that, in the absence of

---

* In addition to Milankovitch orbital forcing during the Pleistocene, deep-sea cores reveal Milankovitch forcing, especially the eccentricity cycles, influencing Earth's climate during the Oligocene (34 million years ago to 23 million years ago), when the initial growth of the modern Antarctic ice sheet began (Pälike et al. 2006). Earlier periods of major ice coverage on Earth also seem to have undergone glacial/interglacial cycles, as identified earlier in this chapter.

disruption of the ice age cycle by human activities, the estimated timing for the next ice age would be approximately A.D. 32,000. Hence, in terms of natural forcings, the next ice age does not appear to be imminent.

The last major interglacial, termed the "Eemian" (or alternatively the "Sangamon"), lasted from approximately 130,000 years ago to approximately 116,000 years ago and is thought to have had global conditions somewhat warmer than today (perhaps about 2°C warmer[179]), with less land ice than today and with sea level approximately 4 to 6 meters higher than today.[180] Two leading contenders for where the land ice might have been less than today's ice coverage, accounting for a portion of the higher sea levels, are the West Antarctic ice sheet[181] and the Greenland ice sheet.[182] Supporting the suggestion of a reduced Greenland ice sheet, pollen records from marine sediment cores off southwestern Greenland suggest dense fern vegetation over southern Greenland during the Eemian (and forest vegetation over southern Greenland during the earlier and longer interglacial at approximately 400,000 years ago).[183]

The Eemian was followed by the last major ice age, often called "the last great ice age." This most recent ice age started about 116,000 years ago, peaked in mid- and high northern latitudes about 21,000 years ago[184] and in the tropics and subtropics between 18,900 and 17,500 years ago,[185] and was largely concluded by 11,800 years ago, the date that marks the end of the Pleistocene and the start of the Holocene Epoch. At its peak, the last great ice age had massive ice sheets in North America and Eurasia as well as Antarctica and Greenland, with ice covering approximately 30% of the global land area, approximately three times today's land ice coverage.[186] In addition to the ice sheets, the last great ice age was characterized by cold and dry conditions over much of the Earth. Tropical climates were probably cooler and more variable than now,[187] and tropical rain forests covered a considerably smaller area in both Africa and South America than they do today or did during the preceding interglacial, remaining only in relatively small refuges.[188] However, within these generalities, the climate of the ice age varied considerably over the tens of thousands of years of the glacial episode. Some of the variation is mentioned further in chapter 3, on abrupt climate change.

Because of all the water contained in the ice of the massive land-based ice sheets, sea level was significantly lower during the midst of the last ice age than it is today, with continental boundaries located considerably farther outward around the world. This was particularly relevant at locations where the extended continental boundaries resulted in a land connection joining landmasses that are separated today and had been separated during the last interglacial. Particularly noteworthy in this regard is the region of the Bering Strait, where the waters are sufficiently shallow that the lowering of sea level during the ice age produced a land bridge, called the

Bering land bridge, between northeastern Eurasia (now Siberia) and northwestern North America (now Alaska) from what had been (and again is) the shallow ocean bottom. This land bridge also existed in some of the earlier glacial periods, facilitating earlier land animal migrations. Other land bridges during the last ice age were important in connecting England to continental Europe and New Guinea to Australia, but the Bering land bridge has particular significance because it is by walking across this strip from Siberia to Alaska that humans likely first arrived in North America and, from North America, eventually journeyed south to populate Central and South America as well. Humans had continued to survive in Eurasia and Africa and to expand their capabilities and tools throughout the glacial/interglacial cycles. By about 50,000 years ago, spears and harpoons had been invented, allowing the hunting of large animals from a distance, as had sewing,[189] helping improve the quality of clothing and the ease of surviving harsh weather conditions.

Estimates of when humans first arrived in the Americas range from about 50,000 years ago to about 13,500 years ago, and speculated routes for the first arrivals also vary. Genetic evidence from nuclear gene markers, mitochondrial DNA, and Y chromosomes are helping to resolve some of the uncertainties and controversies. Using the genetic evidence in conjunction with archaeological evidence, human skeletal remains, and previous studies, Ted Goebel and colleagues conclude that the earliest Americans descended from a single Siberian population that migrated toward the Bering land bridge between about 30,000 and 15,000 years ago and first colonized the Americas about 15,000 years ago, preceding by at least 1,800 years the Clovis populations that flourished from about 13,200 to about 12,800 years ago, based on revised carbon-14-based dates.[190] From Alaska, the migrating humans spread southward through North, Central, and South America, reaching the southern tip of South America by about 12,000 years ago.[191] Much later, and far away, humans spread outward across the Pacific from western Polynesia, settling Hawaii by A.D. 800, New Zealand by A.D. 1000, and Easter Island by A.D. 1200.[192] Like most of the topics in this chapter, the details of human expansion around the globe are not known with certainty, and likely our current understandings will eventually be revised in light of new data or reconsideration of old data.

While the human species continued to advance, other species suffered an extinction episode near the end of the last ice age. These extinctions were weighted strongly to large land mammals, with birds, fish, reptiles, amphibians, and small mammals all faring much better than the large mammals. Large animals typically fare less well during extinctions (recall the dinosaurs), being less able to adjust readily to changing conditions than many smaller animals. However, in this case an additional factor was involved, namely, human hunters in sufficient numbers and with sophisticated enough tools to make an impact (see chapter 4).

In addition to affecting sea level and continental boundaries, the ice of the massive ice sheets overlying much of northern North America and Eurasia had lasting impacts on the underlying landscape, many of which remain quite evident today. The ice carried soil and rocks from Canada south to the United States; it gouged U-shaped valleys and in mountainous regions amphitheater-like settings called cirques; as it carried rock debris along, it polished underlying rocks, often leaving them grooved and striated in the direction that it had passed over; and as it melted and retreated, it left boulders and unsorted debris called tills, with sizable tills along the sides or end of the retreating ice called lateral moraines and end moraines, respectively.[193] The meltwater streams from the ice were effective sorters of materials, laying down sand deposits separate from gravel deposits, with both types of deposits thousands of years later being collected in mass by modern societies for construction materials.[194] Silt from the meltwater streams became the valuable, easily cultivated soil of the farm belt of the United States.[195] Furthermore, with the removal of the massive ice cover, which had depressed the land underneath it, the land rebounded and in fact is still rebounding today, for yet another continuing impact of the past ice coverage.

As is clear from the lack of ice covering northern North America and Eurasia today, considerable warming and deglaciation has occurred from the peak of the last ice age to the present. This has by no means been a steady process, but, overall, warming since the last glacial maximum is estimated to have been about 4°C to 7°C.[196] Globally averaged, warming is thought to have predominated from 18,000 years ago to about 15,000 years ago, followed by cooling and advance of the recently retreating ice sheets, warming at about 14,500 years ago, cooling about 14,000 years ago, and warming about 13,000 years ago, followed by a prominent temporary reversal to further cooling about 12,800 years ago in what can be viewed as the last gasp of the last great ice age. Through these ups and downs, the overall trend was toward warming and the associated deglaciation. More on some of these major fluctuations is included in the next chapter, on abrupt climate change.

As the temperatures rise and fall and the ice sheets contract and grow, consequences are felt worldwide, with sea level changes being among the most notable. Since the peak of the last major ice age about 21,000 years ago, sea level is estimated to have risen 120 meters,[197] giving an average rate of sea level rise of nearly 0.6 meters per century. The rate has not been even close to uniform, however, as most of the deglaciation and hence most of the sea level rise from the peak of the last ice age until today occurred prior to 10,000 years ago.[198] More specifically, for several thousand years near the start of the major deglaciation, from about 20,000 years ago to about 14,500 years ago, sea level rise averaged about 1 meter per century, for a total sea level rise of approximately 55 meters over that period. Then for a 400-year period from 14,500 to 14,100 years ago, the rate averaged about 5 meters per century, for a total sea level rise, from land ice melt and ocean thermal expansion, of about 20 meters in 400 years. Some speculation exists that the very rapid 400-year sea level rise might have been in part due to a collapse of the West Antarctic ice sheet.[199] These rates of sea level rise far exceed the estimated rate of 0.17 meters (17 centimeters) per century for the twentieth century (see chapter 4), a fact that illustrates how much larger natural variability in sea level can be than what has been experienced recently.

> For a 400-year period from 14,500 to 14,100 years ago, the rate of sea level rise averaged about 5 meters per century, making the twentieth century's estimated 0.17-meter sea level rise appear trivial in comparison.

The period of cooling about 14,000 years ago, following the very rapid deglaciation 14,500 to 14,100 years ago, is called the Older Dryas; and the period of cooling starting about 12,800 years ago is called the Younger Dryas. The Older and Younger Dryas periods are named after the white and yellow Arctic rose called the dryas, from which remains exist in European bog sediments from those two periods and from an earlier cold period called the Oldest Dryas.[200] The Younger Dryas is one of the most studied periods of abrupt climate change and as such is discussed further in chapter 3. After the Younger Dryas, warming and deglaciation resumed. Enough deglaciation had occurred by 11,800 years ago that the Pleistocene is dated to have ended at that time, initiating our current epoch, the Holocene.

Well into the Holocene, about 8,200 years ago (6200 B.C.) the northern North Atlantic vicinity abruptly returned to much colder conditions and remained relatively cool for about 400 years.[201] This cool period was followed by about 2,500 years of greater warmth, with global temperatures perhaps comparable to temperatures at the start of the twenty-first century.[202]

In the centuries since the abrupt cooling about 8,200 years ago, climates have tended to be relatively stable, meaning that in the past 8,200 years the Earth's continually changing climate system has not experienced any large-scale abrupt changes on the order of the Older or Younger Dryas, 14,000 and 12,800 years ago, or the lesser event at about 8,200 years ago. It is during this period of relative stability in the past 8,200 years that humans have become a dominant force on our planet (chapter 4). The fact that we have benefited from the relative stability is perhaps best demonstrated by how seriously impacted human societies can be even by climate changes that might pale in comparison to some of those in the paleoclimate record. Some of the relatively small later changes have been implicated as significant contributing factors in the demise of quite a few civilizations, including the Akkadian Empire, the Roman Empire, the T'ang dynasty in China, the Angkorian Empire in Southeast Asia, and the Classic Mayan civilization in Central America.[203]

Among the twists and turns of climate over the most recent millennia, a medieval warm period lasted from about A.D. 800 to about A.D. 1300,[204] known originally largely from European records and of uncertain global extent. Cooling then took over, resulting in a relatively cold period from the fourteenth or early fifteenth century until the mid-nineteenth century, in a period termed the Little Ice Age. Here too, the records were originally largely from Europe, although in this case much additional evidence has been found elsewhere as well, and the cooling is actually more prominent in ice cores from tropical Peru than in polar ice cores.[205] Enough records from around the globe exist to lead British climate historian and glaciologist Jean Grove (and others) to declare the Little Ice Age a global phenomenon. Grove's book titled *The Little Ice Age* details the global nature of the overall cooler conditions while also making it clear that considerable spatial variation existed, along with many fluctuations, with, for instance, glaciers at times retreating temporarily before again advancing.[206] Volcanic eruptions could well have been among the causes, if not in initiating then at least in perpetuating the cooler conditions, as there were at least five major eruptions in the seventeenth century, beginning with a massive eruption in the Peruvian Andes in 1600.[207] Each of these poured substances into the atmosphere that reflected sunlight and otherwise blocked sunlight from getting through to the Earth's surface.

Since the Little Ice Age, the Earth has experienced, overall, warming until the 1940s, cooling until the 1970s, and then continued warming.[208] The temperatures in the early twenty-first century are possibly, on a global average, the warmest of the past 1,000 years, although, like so much else, this possibility is not without controversy. Because the recent warming is tied to the enhanced greenhouse effect from human activities, I delay discussion of this controversy to chapter 4, on human impacts.

SYNOPSIS

The brief sketch given in this chapter of Earth history over the past 4.6 billion years provides a framework for understanding the long-term evolution of the Earth system, but almost every aspect of it could require revision at some point as new information becomes available. Indeed, revisions to our understandings continue at a rapid pace, with at least a few stories coming out each year regarding significant new developments and uncertainties regarding them. However, despite all the uncertainties, it is still relevant to understand the basics of Earth history as scientists currently believe it to be. With that in mind, the following provides a quick summary of some of the most relevant climate aspects described in this chapter: For the total temporal scale, the Earth has been in existence for approximately 4.6 billion years. There have

> Compared to Earth history as a whole, today is probably unusually icy and cold.

been a few periods when it was colder and considerably more ice existed than exists now, but likely for the majority of Earth's history the global temperatures were higher than now, the concentration of greenhouse gases in the atmosphere was higher than now, and there was considerably less ice than now. During many of the periods with little or no ice, it is probable that no large landmasses existed close to the geographic north and south poles and that the oceans extended in some regions from the equator to the pole. Under those circumstances, no land was most appropriately positioned for major ice sheets to grow, and the polar oceans could maintain temperatures above the freezing point through the ocean transport into the polar regions of warm water from lower latitudes. Consequently, snow and ice tended to exist in those periods only on high mountain peaks. During the past 500 million years, it is likely that the only substantial periods in which major ice sheets existed were during the past 35 million years, with ice in both polar regions, and during an extended period (lasting about 80 million years) centered at about 300 million years ago, with ice predominantly in the south polar region. Hence, compared to Earth history as a whole and more specifically to the past 500 million years, today—with major ice sheets in both Greenland and Antarctica—is probably unusually icy and cold. However, if considering just the past million years, then today is quite warm, with global temperatures well above the temperatures during the glacial periods that have dominated the past million years. These glacial periods have alternated with much shorter interglacials, which, like today, retain some ice but far less than the amount of ice during the glacial episodes.

A former comforting concept regarding the climate changes that Earth has experienced during the past 4.6 billion years is that these changes were thought, by and large, to have occurred extremely slowly. Slow changes are something to which humans and other species can generally adjust. Of course, the demise of the dinosaurs and all the other extinctions are sobering, establishing that even quite dominant animals have encountered changes that they were unable to withstand, but those were thought to be highly unusual, being due to dramatic events like asteroids striking the Earth or major and prolonged volcanic eruptions. Normal climate change was thought to occur slowly. However, in the late twentieth and early twenty-first centuries, evidence has been unveiled showing that climate change not forced by asteroid collisions or other extraterrestrial or deep-Earth events can also play out much more abruptly than earlier believed. This has particular relevance because, whether natural or human caused, an abrupt climate change could devastate human societies far more now than when humans were sparsely distributed and lacked the political boundaries and massive infrastructures that now hinder ready migrations to more favorable environments. The next chapter examines some of the fairly recently recognized abrupt climate changes in the paleoclimate record.

CHAPTER 3

Abrupt Climate Change

The Earth system has repeatedly experienced climate change at a rate that far exceeds anything that has occurred over the past few thousand years. In other words, the system is demonstrably capable of abrupt changes at a level that humans have not had to deal with since the development of cities and agriculture but that they likely might have to deal with at some point in the future.

THE OVERVIEW of the Earth/atmosphere system's 4.6-billion-year history provided in the previous chapter mentions some huge changes that have occurred during that extended period. Among these are that the percentage contribution of nitrogen in the atmosphere has risen to about 78%, from a starting point likely no higher than 1.5%; land/ocean configurations have changed continually and dramatically, for instance transitioning from a single supercontinent surrounded by ocean hundreds of millions of years ago to today's configuration of seven continents and many islands; dominant life forms have changed from microorganisms to dinosaurs to mammals; and the geographic coverage of ice has perhaps varied all the way from near 100% during a very ancient hypothesized Snowball Earth to near 0% during extended portions of the reign of the dinosaurs. However, until recently, a fundamental aspect of the modern scientific thinking about these and many other changes was that, by and large, they took place extremely slowly, over extremely long periods of time. The belief that large, long-lasting changes could occur only slowly changed in the late twentieth century, when evidence from Greenland ice cores and elsewhere strongly suggested that some large changes have happened very quickly.

The fact that major local and regional changes can happen abruptly, for instance when caused by earthquakes, volcanic eruptions, hurricanes, or meteorite or asteroid impacts, has probably never been in dispute. Further, at least since the early nineteenth century, there has been some recognition that short-term hemispheric or global-scale changes can also happen quickly. In particular, temporary widespread atmospheric cooling has been attributed to massive volcanic eruptions, as discussed in chapter 1 and epitomized by the large-scale coolings after the 1815 eruption of Mount Tambora and after the smaller but more recent 1991 eruption of Mount Pinatubo.

> Various paleoclimate data have now revealed that some changes previously thought to have occurred over a span of centuries or millennia might have taken place within a few years, perhaps even a single year.

Still, despite the recognition of the several-year impact that a volcanic eruption might have on hemispheric or global temperatures, it was widely thought that the larger, longer-lasting climate changes in the Earth's past took place over very long expanses of time. This view had been solidified in the mid-nineteenth century when the geological "uniformitarians" essentially won out over the "catastrophists" following a many-decades-long controversy between the two groups, the first emphasizing slow, gradual changes and the principle that the forces operating on the Earth in the past are the same as those operating today and the second emphasizing abrupt catastrophic changes.* Hence, in the late twentieth century, when the Greenland ice cores and other paleoclimate records suggested that some prominent climate changes took place quite rapidly, this came as a shock to many scientists as well as others. Various paleoclimate data have now revealed that some changes previously thought to have occurred over a span of centuries or millennia might have taken place within a few years, perhaps even a single year. In addition to opening up the exciting new research area of abrupt climate change, this realization

---

* For details on the controversy between the uniformitarians and the catastrophists, the reader is referred to Hallam (1983). Once the controversy ended in victory for the uniformitarians, the catastrophist concepts were relegated largely to the history books. As is often the case, a better resolution would have been an acceptance of valid aspects of the arguments on both sides, recognizing that some changes happen slowly and others rapidly, similarly to the acceptance of the particle/wave duality theory of light in the early twentieth century. The dual particle/wave theory of light emerged only after centuries of scientists believing light to be either a wave phenomenon or a particle phenomenon, with the wave theory and the particle theory each holding sway for extended periods and heated controversy between representatives of the two viewpoints in other periods.

also raises major concerns regarding how well humans could adapt if such rapid changes were to occur today. Further, it raises the additional concern that human-induced changes might at some point trigger an abrupt climate change.

GROWING AWARENESS OF ABRUPT CHANGES

In 1961, Henry Stommel from Harvard University wrote a theoretical paper on thermohaline circulation, showing the possibility of two distinct stable regimes in an idealized two-vessel experiment. Near the end of the paper, he suggested that different states might be possible in the ocean as well as in a two-vessel laboratory experiment and that, if so, the system might "jump" from one state to another, given a "sufficient perturbation."[1] Two decades later, Claes Rooth from Princeton University and the National Oceanic and Atmospheric Administration wrote that "catastrophic transitions in the structure of the thermohaline circulation are not only possible, but have probably occurred on many occasions in the earth's climatic past. . . . Drastic changes in thermohaline circulation regimes, accordingly, provide a likely mechanism for catastrophic shifts in climatic regimes."[2] Rooth explicitly mentioned in this context the possible importance of changes in the freshwater input to the North Atlantic both when the North Atlantic experiences a reduction in freshwater input in the initial stages of major glaciation and when the North Atlantic obtains a burst of freshwater input with the drainage of a meltwater lake during deglaciation.[3]

Among the most compelling data sources suggesting that abrupt climate change actually did take place has been evidence from long ice cores. These cores hold frozen within them a tremendous amount of climate information both in the ice itself and in everything embedded within the ice. The latter includes particulate matter that fell on the ice sheet and then got buried as more snow piled on over time and consolidated into additional ice; it also includes atmospheric gases that got captured in air pockets, first in the snow and then in the ice as the snow consolidated into ice. The gases embedded in the ice reveal a great deal about the composition of the atmosphere through time, including the carbon dioxide and methane content, and the particulate matter reveals deposits of extraterrestrial material, occurrence of volcanic eruptions, and occurrence of significant dust transport to the ice sheet. One aspect of the ice itself that has generated particular interest is the relative amount of different oxygen isotopes in the water (H_2O) that makes up the ice. Most of the ice—just like most of the water in the oceans—has the most abundant type of oxygen, with an atomic weight of 16, but a small percentage of the ice has oxygen with an atomic weight of 18, and an even smaller percentage has oxygen with an atomic weight of 17. Although the correspondence is not precise, scientists have been able to derive information about past temperatures from examining what proportion of the ice at different levels in the core contains the heavier oxygen (O18) versus the lighter oxygen (O16).* Overall, during colder periods, less of the water with O18 gets evaporated and deposited onto the ice sheets.

The theoretical concept (e.g., from the previously mentioned Stommel [1961] and Rooth [1982] publications) that changes in the climate system can happen abruptly was bolstered as ice cores from the Greenland ice sheet, in particular from a core drilled at Camp Century in northwestern Greenland in 1966[4] and a core drilled at Dye 3 in southern Greenland in 1981,[5] revealed that some changes at least in Greenland might have occurred rapidly. Furthermore, in line with the work of Stommel and Rooth, ice core researchers from Switzerland, Denmark, and the United States suggested modifications in ocean convection as the cause of the rapid changes.[6] Wally Broecker from Columbia University agreed and extended and publicized the ideas in a series of publications and talks, with particular emphasis on the importance of the North Atlantic and the ocean conveyor (see chapter 1).[7] Deep-sea and land-based bog records provided important additional support for the concept of rapid fluctuations,[8] although because stirring of the sediments by worms and other animals tends to blur brief events depicted in the deep-sea and bog records,[9] the ice core records remained of particular importance, especially when trying to narrow down just how fast the abrupt changes were.

While spared the worm difficulties of the deep-sea cores, the ice cores have difficulties of their own. One troublesome problem with the Camp Century and Dye 3 cores and with a third core drilled at Renland in far eastern Greenland in 1987 is that all three cores were drilled at sites where ice is flowing in (throughout the depth of the ice) from upstream, in the natural very slow outward and downward

---

* Thompson et al. (2003) provide a detailed overview of the temperature information that can be derived from oxygen isotope ratios (or proportions) in tropical ice cores, showing that while other factors such as precipitation also influence these ratios, the dominant factor is temperature. For more on the complications in the relationship between temperature and oxygen isotopes in tropical cores, see Thompson et al. (2000). For polar ice cores, both Alley (2000) and Steffensen et al. (2008) indicate that temperature is the dominant influence on the oxygen isotope ratios. However, at all latitudes, local circumstances also have an impact; for instance, while the oxygen isotopes in an ice core from the Guliya ice cap on Asia's Tibetan Plateau are, as elsewhere, used as a partial proxy for past temperatures, they are also influenced by the positioning of the high-pressure atmospheric system that is often quite persistent over the plateau (Thompson et al. 1997).

flow of ice under the force of gravity. This ice flowing into the location of the core greatly complicates the interpretation of the ice core data, as only the ice near the top of the core derived from snow deposited at the core's geographic location, the rest having been deposited farther upstream before flowing out and down. These complications contributed to considerable skepticism regarding the suggestions of rapid climate change revealed from these cores.

To eliminate the uncertainties regarding how much impact ice flow from up-stream of the core sites might have on the results, plans were finalized in the late 1980s to drill two long ice cores from the summit of the ice sheet, where no up-stream ice confuses the record.[10] One of these cores was from a primarily European effort named the Greenland Ice Core Project (GRIP), and the other was from a primarily U.S. effort named the Greenland Ice Sheet Project 2 (GISP2). The two groups coordinated their efforts, and both cores were drilled from the Greenland summit at 72.6°N, the GISP2 site 30 kilometers (18.6 miles) to the west of the GRIP site.

Coring at the GRIP site began in 1989 and was completed in 1992, when the core reached a depth of 3,028.8 meters (9,937 feet) of ice, which came close to but did not quite reach the bedrock underlying the ice. Annual layers of ice in the GRIP core were deemed sufficiently distinct to allow counting year by year through the ice for the past 14,500 years, reaching down to a depth of about 1,760 meters in the ice core. Dating becomes much more difficult below the level of being able to iden-tify individual annual layers of ice, but the ice flow model used in the initial analysis of the GRIP core yielded estimates that the ice at a depth of 2,788 meters is 110,000 years old and that the full ice core covers at least the past 250,000 years.[11] It soon became apparent, however, that the bottom 300 meters of the core were disturbed because of folding of the ice close to the bedrock, making the results for the bottom 300 meters unreliable and setting the length of the usable climate record from the core at about 110,000 years.[12]

Coring at the GISP2 site began in 1989 and finished on July 1, 1993, when the drill reached an ice depth of 3,053.5 meters (10,018 feet) and struck the bedrock beneath. As with the GRIP core, annual layers of ice can be distinguished in the top portion of the GISP2 core, with dating increasingly difficult as one progresses down the core. Despite the difficulties, the 3,053.5 meters of ice in the GISP2 core is thought to contain a climate record extending back at least as far as 110,000 years ago,[13] encompassing the last ice age. Two of the key scientists involved in the GISP2 effort—Paul Mayewski of the University of Maine and Richard Alley of Pennsylva-nia State University—have each written highly readable books about the GISP2 ice core and the surrounding science and results.[14]

Because of the difficulties at the bottom of the ice in the GRIP core and perhaps also in the GISP2 core, a new coring site was selected at 75.1°N and 42.3°W that was underlain by a flat bedrock. The North Greenland Ice Core Project (NGRIP) core was drilled there starting in 1996 and reached bedrock in July 2003. The NGRIP core is 3,085 meters (10,121 feet) long, with a record felt to extend back to 123,000 years ago, hence covering not just the Holocene and the entire last ice age but much of the last interglacial, the Eemian.[15]

Dating of the ice in an ice core is a meticulous, difficult activity. Hence revisions to an initial dating sequence are not unexpected, especially as new cores become available, providing additional information to cross-compare with the earlier cores. In this context, after the NGRIP ice core was obtained, a common ice core chronology was developed for a sizable portion of the Dye 3, GRIP, and NGRIP cores, necessitating some revisions to the earlier datings of the Dye 3 and GRIP cores and accounting for differences in dates that might be quoted for those cores.[16] Before the revisions, datings of the GISP2, GRIP, and NGRIP cores had agreed to within 750 years back to 40,000 years ago but had disagreed by several thousand years for some of the more distant periods. The revised ice core chronology for Dye 3, GRIP, and NGRIP had discrepancies of up to 2,400 years with the GISP2 time scale within the period between 40,000 and 60,000 years ago.[17]

Early results from the GRIP core confirmed the suggestions from the Camp Century, Dye 3, and Renland cores that the last ice age experienced numerous episodes of relatively warm conditions and suggested that these episodes lasted anywhere from about 500 to about 2,000 years and, most important, began "abruptly, perhaps within a few decades."[18] Furthermore, initial analysis farther down the core suggested that abrupt changes occurred not just during the last ice age but also during the preceding interglacial, the Eemian, with major climate changes occurring "perhaps even within a few decades,"[19] only 10 to 30 years being required to shift to an altered climate state that then lasted for anywhere from 70 to 5,000 years, all within the Eemian interglacial.[20] Even though the Eemian portion of the GRIP core was later called into question, the suggestion of abrupt change from that portion of the core added to the growing acceptance that climate can change rapidly and that rapid changes can occur during a warm period as well as during an ice age. Later analysis of the Eemian interglacial from the NGRIP core suggested an abrupt cooling at about 119,000 years ago and an abrupt warming at about 115,000 years ago, although with a gradual rather than abrupt cooling afterward, initiating the last ice age.[21]

As the GRIP, GISP2, and NGRIP cores continued to confirm abrupt climate changes, the skepticism that had attended the earlier results from the Camp Century, Dye 3, and Renland cores relaxed, and the possibility of abrupt changes in the climate system gained adherents and active researchers throughout the 1990s and early 2000s. By the early twenty-first century, the study of abrupt climate change had emerged as an important and vibrant field, involving a wide range of paleo-climate data sets and a variety of computer models. Evidence for abrupt changes has now been found from around the globe and in all types of environments, from deserts to coral reefs to glaciers.[22]

DANSGAARD-OESCHGER EVENTS/ OSCILLATIONS AND HEINRICH EVENTS

Formulating a coherent picture of the abrupt changes revealed in the paleoclimate records is far from straightforward, and the picture continues to evolve. However, sufficient consistencies and patterns have emerged for some categories of abrupt shifts to be named and grouped. In particular, throughout the approximately 100,000 years of the last ice age, there were numerous abrupt warming events, now termed "Dansgaard-Oeschger events," followed by a return to the colder conditions of the core ice age, the full warming and return sequence being termed a "Dansgaard-Oeschger oscillation." (The warm period is called an "interstadial," interrupting the colder "stadials" of the full ice age conditions.) The Dansgaard-Oeschger events/oscillations are grouped into "Bond cycles," in which a particularly strong Dansgaard-Oeschger event is followed by progressively weaker events until, at the end of the Bond cycle, there is a long cold stadial during which deep-sea cores show (informatively; see below) a sudden peak in ice-rafted debris accompanied by a further cooling, termed a "Heinrich event." After the Heinrich event, the next Bond cycle begins with another abrupt and major Dansgaard-Oeschger warming. Looking at the records from the ice cores and deep-sea cores, it is clear that each Bond cycle—and even each Dansgaard-Oeschger event and each Heinrich event—has its own idiosyncrasies, but the basic pattern tends to hold, at least through much of the last great ice age.

The Dansgaard-Oeschger events were named after Willi Dansgaard of the University of Copenhagen and Hans Oeschger of the University of Bern in honor of their extensive work on the early Greenland ice cores and their identification of abrupt changes revealed in the Camp Century and Dye 3 cores.[23] The Dansgaard-Oeschger events/oscillations are characterized by warmings within a few decades of about 5°C to 10°C[24] or perhaps even 8°C to 16°C,[25] followed by a slow cooling and then a more rapid cooling.[26] The interval between Dansgaard-Oeschger events tends to be about 1,470 years or a multiple of 1,470 years, and a possible cause of the events is thought to be systematic variations in the amount of radiation received from the Sun.[27] Although they were discovered through the Greenland ice core records, even within those records evidence exists that the Dansgaard-Oeschger events represent changes affecting a much broader region than just the North Atlantic vicinity, as, for instance, the ice cores show atmospheric methane (CH_4) rising substantially, sometimes by 50%, strongly suggesting expanded Northern Hemisphere wetlands during the Dansgaard-Oeschger events.[28]

The Bond cycles grouping the Dansgaard-Oeschger events are named after Gerard Bond from Columbia University's Lamont-Doherty Earth Observatory for his leadership in the effort first identifying the cyclical pattern. Through analysis of North Atlantic deep-sea cores, Bond and his colleagues confirmed the imprint of the Dansgaard-Oeschger events/oscillations outside of Greenland and proceeded to identify for the period from about 80,000 years ago to about 20,000 years ago the cyclical pattern later named the Bond cycle.[29]

Heinrich events are named after Hartmut Heinrich, who, in examining deep-sea cores from beneath the eastern North Atlantic Ocean in the 1980s, first identified six layers of sediment with anomalously high amounts of coarse rock fragments, all dated from the last ice age, and postulated that the source of the debris was melting icebergs.[30] Debris from the underlying ground is naturally entrained in the bottom of an ice sheet as the ice advances, and when the ice calves off into icebergs, it carries much of this debris with it, eventually releasing the debris in the ocean as the icebergs melt. Once released, the debris sinks to the ocean bottom, forming the debris layers found in the deep-sea cores.[31] Since Heinrich's 1988 publication, additional evidence for the Heinrich events has been found from deep-sea cores spreading eastward from North America to just off the coast of Portugal.[32] Most tellingly, the anomalous layers of ice-rafted debris are several meters thick in the Labrador Sea but are progressively thinner proceeding eastward across the Atlantic, reducing to less than 2 centimeters thickness by 10°W.[33] In view of the geographic pattern and the fact that geologists have determined that the rock fragments in the Heinrich layers came from the Hudson Bay region,[34] the high debris amounts most likely derived from anomalously large discharges of icebergs into the North Atlantic from the Laurentide ice sheet that covered much of northern North America during the last ice age.

Several different mechanisms have been proposed for the assumed repeated major episodes of iceberg releases resulting in the Heinrich layers. One is a "binge–purge" sequence in which the Laurentide ice sheet would grow while the sediment base on the floor of Hudson Bay and Hudson Strait was frozen; then, when the sediment thawed, a lubricated discharge pathway would emerge, and massive amounts of icebergs would be dumped into the Labrador Sea.[35] Other hypothesized mechanisms include massive flooding events (jökulhlaups) outward through Hudson Bay from a presumed Hudson Bay lake dammed by ice,[36] a buildup and collapse of ice shelves fringing the Laurentide ice sheet,[37] and a sea level rise trigger whereby after all the melting during multiple Dansgaard-Oeschger warming events, sea level would have risen sufficiently to destabilize some of the ice shelves surrounding the ice sheet, causing the ice shelves to crumble and break into massive volumes of icebergs.[38] In addition to the uncertainty regarding the mechanism, there are also differing estimates of how long the Heinrich events lasted and the amount of ice involved.*

Consisting as it did of very cold freshwater, the large amount of ice arriving in the North Atlantic during the Heinrich events necessarily cooled and freshened the North Atlantic surface waters. Although cold, the meltwater from the ice, without the relatively dense salinity content of salt water, was likely less dense than the preexistent Atlantic surface waters and the underlying waters. Hence, the inflow of the ice and its melt could well have temporarily severely reduced (and perhaps even halted) the regional convection that sends water from near the ocean's upper surface to great depths, in the North Atlantic's deep-water formation (see chapter 1). Severely reduced convection in the North Atlantic could in turn have shut down or nearly shut down much of the global ocean conveyor, in which case warm tropical waters from the South Atlantic would no longer have been transported northward to the northern North Atlantic. This could at least partially explain the various pieces of evidence suggesting that Europe and the North Atlantic cooled while the South Atlantic and Antarctica warmed during the Heinrich events. Additional evidence also suggests that during the Heinrich events, Europe was drier than normal, the South Asian summer monsoons were weaker than normal, the northwest tropical Pacific was saltier than normal, northern South America was drier than normal, and North America was cold and wet.[39] So, although the Heinrich events were first identified in North Atlantic deep-sea cores, by now the evidence for anomalous conditions during those periods is quite significantly more widespread.

---

* For a detailed review of the research on Heinrich events, the interested reader is referred to Hemming (2004).

THE YOUNGER DRYAS AND THE LATER
COLD EVENT ABOUT 8,200 YEARS AGO

The most recent and by far the most extensively studied Heinrich-type event, some-times labeled H0,[40] is the Younger Dryas cold period that occurred as the Earth was coming out of the last ice age and that temporarily reversed the warming trend associ-ated with the shift from glacial to interglacial conditions (see chapter 2). The Younger Dryas started approximately 12,800 years ago, with a temperature drop of 2°C to 8°C (depending on location) that occurred within a few decades and perhaps within a single decade. This plunged the Earth back into ice age conditions for about 1,200 years, until the Younger Dryas ended, likely even more abruptly than it began.[41] Alley indicates that the warming at the end of the Younger Dryas came in three steps spread over 40 years, with most of the change occurring in the middle step, when the sur-face temperature of Greenland, annually averaged, increased by approximately 8.3°C (15°F) in no more than 10 years. Regarding that middle step, Alley wrote, "I cannot insist that the climate changed in one year, but it certainly looks that way."[42]

By the early twenty-first century, a leading contender for what might have caused the Younger Dryas was the presumed emptying into the North Atlantic of a large lake formed from the meltwater of the retreating North American Lauren-tide ice sheet.[43] Named Lake Agassiz, this glacial lake formed well before the start of the Younger Dryas, but its drainage had originally been largely southward into the Gulf of Mexico. The speculation was that about 12,800 years ago, the bulk of the drainage shifted, suddenly draining not into the Gulf of Mexico but into the Great Lakes basin and from there into the North Atlantic. If so, this would be a case where the much-overused term "tipping point" might be fully appropriate, with the northward retreat of the ice reaching a position where the drainage suddenly shifted from going southward to going eastward or northeastward, where it could affect the deep convection in the North Atlantic.

If indeed the drainage from Lake Agassiz shifted at the start of the Younger Dryas, the high volume, cold freshwater influx into the North Atlantic might have been too low in density to sink, halting ocean convection in the northern North Atlantic and thereby triggering a shutdown of the ocean conveyor. The cold fresh-water influx from Lake Agassiz could also have cooled the northern North Atlantic surface waters sufficiently to freeze them, triggering further freezing through the ice-albedo feedback (chapter 1) and perhaps thereby further thwarting the con-tinued operation of the ocean conveyor. A shutdown of the ocean conveyor would have halted the flow of warm Gulf Stream waters northward and sent the northern North Atlantic and northern Europe plummeting into much colder conditions.

However satisfying the Lake Agassiz explanation of the Younger Dryas might be, by 2005 a major blow had been dealt this potential explanation as new evidence suggested that the Laurentide ice sheet might not have retreated sufficiently by the start of the Younger Dryas to shift the drainage of Lake Agassiz eastward that early.[44] Some scientists continued to support the Lake Agassiz explanation,[45] and even more continued to regard a weakening and perhaps stoppage of the ocean conveyor as a critical factor in the Younger Dryas.[46] Other possible causes for a weakened ocean conveyor, all involving freshwater inflow to the North Atlantic, include melting icebergs and increased river outflow[47] and, more specifically, a massive flood down the St. Lawrence River[48] or a major freshwater flow from the Arctic Ocean, through Fram Strait, in connection with meltwater/iceberg discharge from Canada to the Arctic Ocean basin.[49]

With a slowdown or stoppage of the ocean conveyor, less of the warm surface water of the South Atlantic would have moved northward to the North Atlantic. This would explain not only the records of Younger Dryas cooling in Northern Europe, the North Atlantic, and elsewhere in the Northern Hemisphere, including as far afield as the Tibetan Plateau,[50] but also the suggestive observational evidence, especially from the vicinity of Antarctica[51] and perhaps New Zealand,[52] that portions of the Southern Hemisphere warmed during the same period. It is less clear, however, how well it corresponds with the cooling that is thought to have occurred in tropical Peru, from an ice core drilled at Huascarán (9.1°S, 77.6°W).[53]

An alternative possibility is that the Younger Dryas was initiated by an extraterrestrial impact, evidence coming from carbon-rich debris layers found at various sites in North America and dated to approximately 12,900 years ago,[54] buttressed by what might be nanometer-sized impact diamonds.[55] This possibility remains quite controversial, but if indeed there was an extraterrestrial impact at that time, the impact might have contributed not just to the Younger Dryas cooling but also to the megafauna extinctions that are often attributed to human hunters.[56]

After the Younger Dryas setback, the Northern Hemisphere's prolonged emergence from the last ice age resumed. The next major abrupt event was the cooling at about 8,200 years ago (6200 B.C.) mentioned in chapter 2. Like the Younger Dryas, this event is thought likely to have been triggered by a massive flow of cold freshwater into the North Atlantic and a resulting slowdown of the ocean conveyor. In fact, the cold event 8,200 years ago might have been caused by drainage into the North Atlantic from Lake Agassiz, previously thought to have caused the Younger Dryas (certainly the possibility exists that Lake Agassiz had more than one major drainage episode[57]). More specifically, a massive flood might have emptied about

100,000 cubic kilometers of water from Lake Agassiz and Lake Ojibway to its east, draining first into Hudson Bay and through Hudson Bay into the North Atlantic, with much of this occurring in perhaps as few as 6 to 12 months.[58]

The cold event 8,200 years ago registers the strongest climate signal revealed in the Greenland ice cores for the past 10,000 years, although with an amplitude only slightly more than half that of the earlier Younger Dryas[59] and with a 350-year duration that is considerably shorter than the Younger Dryas.[60] The event might also have been more localized than the Younger Dryas,[61] although the Intergovernmental Panel on Climate Change (IPCC) mentions a widespread signature of the event from the tropics to the Arctic,[62] Thompson and colleagues describe abrupt changes in tropical Africa in the same time frame (perhaps 8,300 years ago),[63] and others connect the cold period with increased aridity in the Near East and the abandonment of farming villages in northern Mesopotamia.[64] Overall, the cooling over much of the Northern Hemisphere's mid- and high latitudes is estimated to have been about 1°C to 3°C, accompanied by a warming in the Southern Hemisphere.[65] Simulations with a global climate model and comparisons of the simulated results with paleoclimate records from multiple locations lend support to the concept that the cold event 8,200 years ago involved a slowdown by about 50% of the North Atlantic downwelling portion of the ocean conveyor, possibly caused by catastrophic drainage of Lakes Agassiz and Ojibway through Hudson Bay and eastward into the North Atlantic.[66]

THE GLOBAL ASPECT OF CLIMATE CHANGE

As is made clear in the IPCC reports[67] and elsewhere, we remain far from fully understanding the Dansgaard-Oeschger events, the Heinrich events, the Younger Dryas, the cold event 8,200 years ago, and the other extremely rapid climate changes revealed in the paleoclimate records. Many seem to have involved changes in the ocean conveyor circulation and particularly changes in deep convection in the northern North Atlantic. Hence, researchers have focused considerable attention on the Laurentide ice sheet and the many records available from the northern North Atlantic vicinity. However, abrupt climate changes are by no means confined to high latitudes. They have been found to occur, for instance, in mid- and low-latitude temperatures and glaciers[68] and in the South Asian monsoon.[69] The paleoclimate record reveals repeated instances of prolonged droughts with abrupt onsets in both mid- and low latitudes.[70]

For decades, from the late 1970s to the early twenty-first century, ice core research in the mid- and low latitudes was dominated by one individual: Lonnie

Thompson of the Byrd Polar Research Center at Ohio State University. While many researchers were concentrating on ice cores from Greenland and Antarctica, Thompson persisted in an extended effort to obtain and analyze the records preserved in the ice of lower latitudes.* Among the wealth of information that he and his team have obtained about past climates in the nonpolar regions, there are several indications of abrupt climate change. These include abrupt warming and drying at Sajama, Bolivia, 15,500 years ago, abrupt cooling at the same location approximately 14,000 years ago,[71] and abrupt cooling at Mount Kilimanjaro, Tanzania, about 5,000 years ago, the latter abrupt event also being recorded in a variety of additional tropical and midlatitude paleoclimate records.[72] (One of the secondary tragedies of the continued warming in tropical and midlatitudes is the disappearance of the climate records preserved in the local glaciers as these glaciers melt away, a point made well by Thompson and others.[73] Meltwater seeping through the glacier can damage the record long before the glacier itself is gone, but many of the glaciers are now in danger of disappearing altogether.[74] This increases the urgency that Thompson or others obtain the cores before it is no longer possible to do so.)

Although less frequently highlighted, the climate changes elsewhere than in the northern North Atlantic might have comparable or even greater importance than those in that vicinity, and indeed some of the abrupt events in the vicinity of the northern North Atlantic might have been triggered by events farther south. Even evidence from Greenland itself suggests that might be the case, as detailed analyses of the different sources of information in the high-resolution NGRIP ice core have led researchers to conclude that an abrupt Dansgaard-Oeschger warming of more than 10°C in only 3 years at about 14,700 years ago was likely triggered by a northward shift of the atmospheric Intertropical Convergence Zone (ITCZ) in the tropics.[75] Other studies have further suggested that Dansgaard/Oeschger events in general, although often attributed to changes in the ocean conveyor, might have been

---

* Obtaining ice cores from the very high altitudes of mid- and low-latitude glaciers is by no means an easy task, involving as it does a tremendous amount of logistics, international cooperation, physical endurance, and acclimatization to the lack of oxygen. Readers interested in the adventure as well as the science of Lonnie Thompson's spectacular career are referred to the book *Thin Ice: Unlocking the Secrets of Climate in the World's Highest Mountains* by Mark Bowen (2005) or, for those preferring a shorter account of an individual expedition, to the article "Thompson's ice corps" by the same author (Bowen 1998). I have known and been friends with Lonnie and his wife and primary colleague Ellen Mosley-Thompson since we were graduate students together in the mid-1970s, and I can say with great confidence that despite his tremendous accomplishments, Thompson remains consistently unassuming and thoughtful. In casual conversation, he does not let on in any way that he is "in the ranks of our great explorers," as once described by Wally Broecker (1995b, p. 212).

triggered by changes in tropical atmospheric patterns.[76] Of course, the ITCZ shift itself would have had a trigger, and this might have involved activities occurring even farther to the south. Blunier and colleagues suggest that, for the period from 47,000 to 23,000 years ago in the midst of the last ice age, climate change in Antarctica tended to lead rather than follow changes in Greenland, on average by about 1,000 to 2,500 years.[77] Stott and colleagues suggest that the warming initiating the end of the last ice age started in the mid- and high latitudes of the Southern Hemisphere before spreading to the tropics and then farther north,[78] and Thompson and colleagues suggest the warming in central Greenland began several thousand years after the warming in lower latitudes.[79] Further, in a separate study, Thompson and colleagues suggest that the abrupt event 8,200 years ago revealed in the Greenland ice cores was driven by abrupt changes in the tropics, particularly Africa.[80]

In view of the interconnectedness of the global system, major happenings in either hemisphere can be expected eventually to influence happenings in the other hemisphere. Once the paleoclimate record is more complete globally, a global synthesis likely will reveal long-term oscillatory patterns wherein all portions of the globe are important players and all major changes have causes that derive from elsewhere and consequences that spread elsewhere, with neither hemisphere either "leading" or "following" any more than the other.

CURRENT RELEVANCE

In addition to generating great excitement within the discipline of paleoclimatology, the recognition of past abrupt climate changes has also generated far broader interest, as it raises concern about the possibility of similarly abrupt climate changes in the future. Humans have been lucky so far. Assuming that the Greenland ice cores are being interpreted correctly, more than 20 abrupt temperature increases and additional abrupt temperature decreases occurred between 110,000 and 10,000 years ago in the Greenland vicinity, but none has occurred in that region in the past several thousand years, suggesting a much more stable climate system during the Holocene than during the previous 100,000 years.[81] Further, the much longer Vostok ice core from Antarctica indicates that the stability of the Holocene is unique even when compared with the entire preceding 410,000 years,[82] suggesting that humans have advanced from the beginnings of agriculture to today in a steadier climate than might be normal, at least over the past 410,000 years. If our luck runs out and climate suddenly changes as rapidly as it has in some of the most dramatic instances in the past, modern society will likely have to endure an extremely difficult transition period as it adjusts either to a new climate state or to a new level of instability in the climate system. Troublesome possible changes include an abrupt sea level rise, submerging coastal communities, and abrupt precipitation changes, causing flooding in some areas and drought in others, inconveniencing many and perhaps forcing the abandonment of what had been productive agricultural regions.

The climate system might, at least in the North Atlantic region, have different stable or semistable states that can exist for extended periods before a disruption occurs that causes the system to shift from one state to another, perhaps doing so with an abrupt climate change. Evidence from the Atlantic suggests three distinct ocean circulation modes in the past 120,000 years. In one mode, as today, a strong ocean conveyor operates with overturning and deep-water formation in the Nordic Seas in the northern North Atlantic. In such a case, as described in chapter 1, heat is transported northeastward across the North Atlantic by surface waters in the Gulf Stream, and consequently Europe and Greenland are warmer than they otherwise would be. A second mode, typified by presumed conditions during the Heinrich events, has both North Atlantic convection and the ocean conveyor halted or at least greatly slowed, with a resultant cold northern North Atlantic vicinity. The third mode, typified by normal conditions during the ice age stadials, has an operating ocean conveyor but with deep convection occurring geographically farther south than today, in the subpolar North Atlantic. Any of these three modes can operate for extended periods, but if a disruption causes a change from one mode to another, this change can be abrupt.[83] Andrey Ganopolski and Stefan Rahmstorf from the Potsdam Institute for Climate Impact Research in Potsdam, Germany, have succeeded in using a computer model to simulate different ocean conveyor modes and abrupt shifts from one to another. After examining how stable the different modes are in the model simulations, they conclude that the most stable mode is the circulation typical of cold stadial periods, with deep convection in the North Atlantic Ocean south of where it now occurs. The simulated mode similar to the modern situation is deemed "marginally unstable."[84]

Some comfort might be taken from the fact that the favored hypothesized mechanisms for many of the abrupt climate changes in at least the past 110,000 years are tied to the Laurentide ice sheet that formerly covered much of northern North America but now no longer exists. With that ice sheet gone, at least the planet is spared upcoming abrupt climate changes deriving from happenings in that particular ice sheet. However, this does not spare us other potential triggers of abrupt change, such as the shutting off or significant slowing down of the ocean conveyor for reasons unconnected with the Laurentide ice sheet; accelerated melting of the ice sheets of Greenland or Antarctica; massive release of methane to the atmosphere as permafrost thaws in the Arctic, enhancing the greenhouse effect; massive release of methane from ocean sediments; or accelerated emissions of greenhouse gases to the atmosphere from human activities. And, of course, abrupt change could also come quite independently of anything humans do or anything internal to the climate system, for instance, through a large asteroid impact or a sequence of particularly explosive volcanic eruptions.

Although no global changes on the order of a shutdown of the ocean conveyor have occurred in thousands of years, lesser but still visually spectacular changes have occurred, some recently enough to be captured in satellite imagery. Among these changes is the dramatic shattering of a portion of the Larson B Ice Shelf along the east coast of the Antarctic Peninsula from late January to early March of 2002.[85] This 2002 decay of the Larsen B Ice Shelf followed a 1995 decay of a more expansive Larsen B and of the Larsen A Ice Shelf directly to its north.[86] These 1995 and 2002 abrupt changes are part of a somewhat systematic southward sequence of ice shelf decays along the Antarctic Peninsula as the peninsula has warmed in recent decades. Each of these decays is believed to have occurred at least in part because of meltwater from the surface seeping through the ice and breaking the ice apart.[87] With expected continued warming, more meltwater and more ice shelf decay can be expected as well.

The potential importance of collapsing ice shelves goes far beyond the changes in the ice shelves themselves. Most crucially, the ice shelves could be buttressing the land ice masses upstream, in which case ice shelf decay could precipitate further outflow of land ice, adding to the mass of the oceans and thereby contributing to sea level rise. Both Terry Hughes and John Mercer of Ohio State University expressed concern about this possibility back in the 1970s, specifically regarding the West Antarctic ice sheet,[88] and although the importance of the buttressing mechanism has engendered some controversy,[89] evidence following the breakup of the Larsen A and Larsen B ice shelves supports the reality of the buttressing effect. Specifically, the glaciers that had flowed into the Larsen A Ice Shelf increased in velocity by up to three times in the 4 years after the January 1995 Larsen A collapse, and the glaciers flowing into Larsen B increased in velocity by two to six times soon after the January–March 2002 Larsen B collapse.*[90] Fortunately, the ice shelves

---

* Interestingly, the hypothesis that ice shelf collapses might have caused the Heinrich events developed partly in response to the marked ice shelf disintegration along the east coast of the Antarctic Peninsula in the 1990s and early 2000s, which provided a stark modern demonstration that such disintegration can occur.

along the Antarctic Peninsula buttress only the ice along the peninsula, not the far more massive West Antarctic ice sheet, which, like Greenland, contains enough ice to raise sea level by about 7 to 8 meters. However, the increased flow toward the sea of the ice in the peninsula area following the collapse of ice shelves there bodes poorly for what might happen should the larger ice shelves buttressing the West Antarctic ice sheet collapse.

In the 1980s, attention regarding West Antarctic ice sheet stability focused on the ice streams flowing into the largest ice shelves buttressing the West Antarctic, namely, the Ross Ice Shelf and the Ronne Ice Shelf.[91] Concern rose when evidence suggested a net loss of ice, estimated at 23 cubic kilometers of ice per year, from the region of the West Antarctic flowing into the Ross Ice Shelf.[92] However, further research on the flow into the Ross Ice Shelf lessened concerns about that region, and the flow rates into the Ronne Ice Shelf also appeared not to be a problem.[93] Concern then shifted to the Thwaites and Pine Island Glacier region (100°W–110°W) to the west of the Antarctic Peninsula.

As early as the mid-1970s, the Thwaites and Pine Island Glacier region was recognized by George Denton and Terry Hughes of the University of Maine (Hughes had moved in the mid-1970s from Ohio State) as potentially the "weak underbelly" of the West Antarctic ice sheet both because the Thwaites and Pine Island glaciers drain a sizable fraction of the West Antarctic and because these two glaciers do not end in buttressing ice shelves.[94] Pine Island Glacier alone discharges on average about 75 gigatons of ice each year, giving it the largest discharge rate of any West Antarctic ice stream.[95] However, measurements in the 1980s and early 1990s did not raise alarms, as the ice in this region did not seem significantly out of balance,[96] the accumulation through snowfall approximately balancing the ice loss through melt and calving. Interest increased when new measurements determined that Pine Island Glacier both receded and thinned during the 1990s.[97] Analyses of data from satellite radar altimetry suggest that the full Thwaites and Pine Island Glacier region experienced a net loss of several dozen gigatons of ice per year over the period from April 1992 to April 2001.[98] No one knows much about how these changes might compare to changes in previous decades, but the more significant unknown is whether the decreases will continue. Of special relevance is whether at some point they might accelerate and precipitate a major decay of the West Antarctic ice sheet, perhaps providing us with the initiation of an abrupt climate change. It could go either way, with concern about the Thwaites and Pine Island Glacier region either easing, as it did for the Ross Ice Shelf region, or getting greater, with an accelerated ice sheet decay.

Adding significantly to the concern that humans at some point may have to adjust to wide-scale abrupt climate changes is the fact that humans may inadvertently help to trigger such a change. Never before the twenty-first century has there been a human population of 6.8 billion people or a human population rising by over 85 million people per year, and never before have the Earth resources on which we depend been depleting as rapidly as they are currently. Because of our numbers and the multitude of things we do to the environment, we are hardly a benign presence on the planet. Although we cannot yet intentionally influence an asteroid impacting the Earth or an earthquake or a volcanic eruption, we do have a level of control over what we as humans do. It is therefore understandable that of all the possibilities of climate change, abrupt or otherwise, the ones receiving the most attention are those that could be caused in part or in whole by humans. Hence, I now turn to the topic of human impacts.

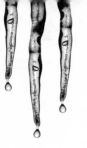

PART II

The Human Factor

CHAPTER 4

A Short History of Human Impacts

Humans have impacted local environments from our beginnings, but the impacts rose dramatically with the invention of agriculture 10,000 years ago and, even more so, with industrialization in the past 250 years. With agriculture, industrialization, and the continually increasing human population, our impacts are now global. Although by no means all negative, some of the impacts are cause for considerable concern.

THERE are many disagreements regarding exactly when and how humans first emerged, but whenever and however it was, we have had impacts from our beginnings. We trample whatever we walk on (or lie on or sit on); we change the composition of the atmosphere immediately around our mouths by the mere fact of breathing, as the oxygen we inhale takes part in chemical reactions within our bodies, with the result that carbon dioxide gets exhaled, thereby increasing the carbon dioxide amounts in the immediate atmosphere while decreasing the oxygen amounts; and we need to eat and drink and to deposit the waste from the eating and drinking, thereby necessarily having an impact on the local water supplies and plant and/or animal life.

The impacts of humans on the environment accelerated hundreds of thousands of years ago with, among other things, the intentional use of fire, the hunting of game, increasing tool usage and variety, and, perhaps somewhat later, the construction of dwelling places. Still, as long as the human population remained small, the impacts of humans remained local and minor in the context of the Earth as a whole. As we became more populous and skilled, however, the impacts rose considerably.

EARLY EXTINCTIONS PERHAPS CAUSED
IN PART BY HUMAN HUNTER/GATHERERS

A prime candidate for the earliest really substantial impact of humans would be the role that humans likely played in large animal extinctions starting in the latter part of the last ice age. In no case is it certain that humans were the prime cause of an individual extinction, but in many cases a strong likelihood exists. Humans killed animals for food, for clothing, and, in some instances, for the animal bones, which could be molded into tools. Humans were greatly aided by the spears and harpoons they had invented by about 50,000 years ago, allowing killing to occur from a distance and thereby greatly reducing the danger to the human hunter and providing a major advantage against the prey.

Although other factors may have been involved, in both Australia/New Guinea and the Americas, within a few thousand years of the arrival of humans, notable extinctions of large animals occurred on those continents. These extinctions, including 15-foot-tall mastodons, 750-pound saber-toothed cats, and the woolly mammoth in North America[1] and giant kangaroos and rhinoceros-sized marsupials called Diprotodons in Australia/New Guinea,[2] were likely caused at least in part by the human arrivals. More broadly, the extinction percentages around the world near the end of the last ice age provide a strong case that human hunters played a role. Specifically, in Australia, 94% of the genera of mammals weighing at least 44 kilograms were lost, and the corresponding extinction rates were 80% in South America, 73% in North America, but only 29% in Europe and only 5% in Africa south of the Sahara.[3] Notably, the high extinction rates were on continents where humans were relatively recent arrivals, bringing hunting technologies that could decimate the unsuspecting native mammals. Large mammals in Europe and Africa, in contrast, had evolved alongside humans for hundreds of thousands of years and hence had a chance to adjust more gradually to the technologies. Still, the 29% extinction percentage in Europe is quite large, appearing small only in comparison to the much higher rates in Australia and the Americas.

Much later, within a few hundred years of when humans arrived in Madagascar, about A.D. 500, the island's megafauna, including a bird almost 3 meters tall and a lemuroid as large as a gorilla, were all extinguished; and within less than 800 years of when humans arrived in New Zealand from Polynesia about A.D. 1000, numerous animal species were hunted to extinction, the most notable among them being all 13 or more species of large flightless birds called moas.[4] Some of the moas, killed probably predominantly for food, weighed over 200 kilograms.[5]

FROM AGRICULTURE TO INDUSTRIALIZATION

Serious as the impacts were from hunter/gatherers (and certainly the extinctions of hunted animal species were serious), they pale in comparison with what was to come. A new development with enormous consequences began about 10,000 to 12,000 years ago (8000–10,000 B.C.), when groups of humans started settling into permanent communities and turned at least in part from hunting and gathering to food production following the invention of agriculture. This began with the cultivation (deliberate growing) of wild plants and progressed eventually to plant domestication, whereby a plant undergoes genetic changes through human interference.[6]

Agriculture is believed to have arisen first in the Fertile Crescent region of southwestern Asia, perhaps just west of Diyarbakir in southeastern Turkey[7] about 8500 B.C., and to have spread from there and also from several other regions scattered around the globe where it arose independently although probably later.[8] Agriculture is not necessarily less exhausting physically than hunting and gathering, but one of the pressures that could have encouraged food production would have been a depletion of the population of wild game to hunt,[9] perhaps brought about by humans through unsustainable killing rates. New findings might change the specifics, but agriculture clearly arose and eventually spread (or arose independently) through much of the ice-free land regions of the globe. Through agriculture, individuals and communities could exert increasing control over their food supply, thereby greatly reducing a major day-to-day uncertainty regarding the source of the day's meals.

Quite logically, the early food crops were readily edible and fairly easily grown: emmer wheat, einkorn wheat, lentils, peas, and barley in the Fertile Crescent;[10] pepo squash and maize in Mexico and moschata squash, yams, lima beans, sweet potatoes, and peanuts in South America;[11] bananas, taro, and yams in New Guinea;[12] and millet and rice in China.[13] Much later, perhaps about 4000 B.C. in the Fertile Crescent, the domestication of fruit and nut trees began. The early tree domestications included olive, fig, date, and grape trees, all of which take sufficiently long to grow that their domestication necessarily awaited communities settled in one location for many years. In the meantime, in addition to food crops, the inhabitants of the Fertile Crescent domesticated flax, probably by about 7000 B.C., obtaining fibers from it for linen,[14] and cotton was domesticated in South America.[15]

The development of agriculture had a huge impact on the local landscapes, as humans vastly alter the landscape each time they clear areas for agricultural production. This effect would multiply as the centuries passed and more and more land came under agricultural production. Exactly when these effects were large enough to make a difference in terms of the global rather than strictly local or regional conditions is uncertain, but University of Virginia environmental scientist William Ruddiman argues that human impacts on global greenhouse gas concentrations could have been as early as 8,000 years ago. First to be affected were carbon dioxide concentrations, which began a slow rise 8,000 years ago that Ruddiman attributes to the deforestation in sizable regions of China, India, and Europe, as farmers cleared land for crops and pastures. Approximately 3,000 years later, global atmospheric methane concentrations started to rise, perhaps tied to the expansion of irrigated rice farming, especially in Southeast Asia, and, to a lesser extent, human raising of livestock and biomass burning.[16]

From an environmental standpoint and especially from the standpoint of the trees and other life displaced, agriculture can be seen as a highly unfavorable development. But from the standpoint of the development of human societies and civilization, it was fundamental. From early on, agriculture enabled enough food production to feed far more people than just those actively involved in producing the food, leaving time for others to pursue and become specialists in other arenas. All of us today who are not farmers benefit tremendously from the fact that agriculture enables a few producers to feed many, well beyond their immediate families. Agriculture also went hand in hand with a settled lifestyle, as the agricultural fields both allowed and encouraged people to stay put near those fields. Hence, with agriculture came firmly settled community life, spurring on all other developments that a settled lifestyle facilitates, including, notably, the accumulation of possessions beyond those that can readily be carried from place to place. It similarly allowed larger families, as infants no longer needed to be carried during frequent moves in a nomadic existence, and thus families could more readily accommodate more than one or two infants. This naturally spurred population increases.

Along with agriculture and the domestication of plants came the domestication of animals, including goats, sheep, cows, pigs, and cats in the Fertile Crescent; turkeys and dogs in the Americas; and pigs, dogs, and chickens in China. Domestic animals helped (and continue to help) humans in multiple ways. They pulled plows and performed other strength-related tasks; they provided a means of transportation by allowing humans to mount and ride them, with horses, donkeys, yaks, reindeer, and camels being especially helpful in this regard; they provided milk and fertilizer while alive and meat after death; and, along with domestic plants, they provided fibers for making clothing, blankets, rope, and other items. Another very substantial assistance was that some, particularly the cat, helped by feeding on rodents, which otherwise likely would have been an even greater nuisance than they were.[17] Water buffalo and horses, domesticated approximately 6,000 years ago, provided strength far exceeding what a comparable number of humans could provide by themselves and thus greatly increased the area of land that could be plowed by one or a small group of people.[18]

With a settled lifestyle and specialization, humans joined into larger and more formal groupings, from tribes to chiefdoms, kingdoms, and nations. As the groupings got larger and more formal, the human-caused changes to the landscape increased further, with the accelerated construction of buildings, roads, and other infrastructure. From a human standpoint, much of this was good, increasing comfort, security, movement, and progress despite the often unfavorable environmental consequences.

Major empires and other political entities, while consolidating and expanding their power and in many cases also refining and advancing human culture, quite often denuded the natural landscape over vast areas. The widespread clearing of forests has been in part to make way for other things, such as buildings, roads, and agriculture, and in part to make use of the wood, for all its many applications. The tropical deforestation that has captured considerable attention and concern in the latter half of the twentieth century and early twenty-first century is a recent continuation of what humans have done for thousands of years. Much of Europe and significant parts of Asia had been deforested through intentional human activities by the 1700s, followed by major deforestation of North America, predominantly east to west across the continent as European colonists and their descendants spread from the Atlantic to the Pacific coasts in the 1700s, 1800s, and early 1900s. Similarly, many islands and other regions around the world have been deforested by humans settling those regions.

Much of the deforestation and similar destruction of virgin grasslands arose because of the increased need for croplands and pastures in order to feed a growing population. Although the numbers are not known definitively, Kees Klein Goldewijk estimates a global increase of the area of croplands from 265 million hectares in 1700 to 401 million hectares in 1800, 813 million hectares in 1900, and 1,471 million hectares in 1990 and a global increase in the area of pastures from 524 million hectares in 1700 to 942 million hectares in 1800, 1,955 million hectares in 1900, and 3,451 million hectares in 1990. (A hectare is 10,000 square meters, equal to the area in a square with sides of 100 meters each; in acres, 1 hectare is 2.471 acres.) Regionally, the greatest increases in croplands from 1700 to 1950 were in the regions of the former Soviet Union, the United States, and southern Asia, but since

1950, the United States, Europe, eastern Asia, and Japan have all shown decreasing cropland areas, while the greatest cropland increases have been in South America, Southeast Asia, Oceania, and western Africa.[19]

Starting well before agriculture began and increasing markedly over time, humans have used the atmosphere, oceans, and land as dumping grounds for waste products from a vast array of human activities. When population remained sparse, people could, at least theoretically, destroy the environment in one area and then move to another area and repeat the destruction there without fear of running out of new places to migrate to and, in general, without destroying the environment beyond what nature could repair fairly readily once humans left the area. It is not so simple now, with people populating a larger portion of the Earth and with both political boundaries and extensive immovable infrastructure restricting large communal migrations. Global human population was probably under 10 million 12,000 years ago. It increased to about 300 million 1,000 years ago, to about 1 billion by 1800, and to 6.8 billion by the middle of 2009.[20] This is a strongly accelerating increase, with population now doubling in considerably fewer than 100 years. How many people the Earth can comfortably accommodate is an issue with opinions ranging from numbers well under the current global population to numbers several times the current population. In part, it depends on how we behave, but regardless of the "correct" number for comfortable accommodation, the current population has proven itself fully sufficient to have impacts that have moved far beyond being just local and regional in scale. Some of our impacts, such as the amount of carbon dioxide in the atmosphere, extend globally.

Pollution of the atmosphere, lakes, rivers, and other waterways became a problem locally in almost every locality where people settled permanently in large numbers. From early on, major efforts were made to transport the pollution out of major cities, as illustrated, for instance, by the extensive wastewater drainage system of the Romans, in place in at least a primitive state by several hundred years B.C. Later, chimneys became common to remove smoke and pollutants from dwellings and other structures, and as time went on, the height of industrial smokestacks was increased specifically to disperse the pollutants farther from their place of origin.

In England, increasing pollution of the air and waterways led to the British Parliament's passing, in 1388, a sanitation act forbidding the dumping of garbage and entrails in ditches, rivers, and other waterways.[21] Almost three centuries later, in 1661, John Evelyn addressed the issue of air pollution in a pamphlet addressed to King Charles II and titled *Fumifugium, or The Inconvenience of the Air and Smoke of London Dissipated*. Evelyn explained the damage to the lungs caused by the polluted air and identified specific means for improving the situation, including moving factories out of the cities and planting trees and shrubs.[22] He was largely ignored, however, and the air quality continued to worsen.

Already in Evelyn's day, pollution from coal usage was noticeable. In England, use of bituminous (soft) coal began in the 1200s as a decidedly cheaper means of heating homes and shops than burning wood,[23] and when James VI of Scotland became James I of England in 1603, he had cleaner-burning anthracite coal from Scotland imported to London.[24] (Coal had been used in China for hundreds of years prior to the 1200s.[25]) The economic and warmth advantages of coal outweighed the unpleasantness of the resulting airborne soot and sulfur dioxide, although by the time of Evelyn's pamphlet, episodes in London of what in 1905 would be coined "smog" (combining coal smoke and fog) were already notorious. Then as now, sulfur dioxide in smog was converted to sulfuric acid, which damages eye, nose, and respiratory tract membranes, becoming a cause of emphysema, bronchitis, bronchial asthma, and lung cancer as well as eating away at buildings and statues.[26]

THE INDUSTRIAL REVOLUTION

Coal usage hugely increased with the so-called industrial revolution, a term acknowledging the vastly increased industrialization of societies beginning in western Europe in the late 1700s and spreading around the globe in the next two centuries. As explained in chapter 2, the coal that we burn today took millions of years to form, much of it from decayed organic matter originating in the Carboniferous Period approximately 359 to 299 million years ago. Our burning the coal now returns to the atmosphere carbon that had been sequestered from the atmosphere hundreds of millions of years ago as plants breathed in carbon dioxide for photosynthesis.

The industrial revolution involved many factors, inventions, and societal changes, but one of the primary inventions sparking it was James Watt's steam engine. Patented in 1769, Watt's engine incorporated major improvements over the first practical steam engine, constructed by Thomas Newcomen in 1712.[27] Among the improvements, Watt's rotary engine, developed in 1784, replaced the vertical motion of the Newcomen engine by a circular motion.[28] The Watt steam engine had numerous direct applications that spurred further developments, such as the invention of the steamship, with its many consequences through easing and speeding the transport of humans and materials from place to place. For instance, when the steamship *Great Western* crossed the Atlantic from Bristol, England, to New York in 1838, it did so in 15 days 5 hours, smashing the previous record for an

England–to–New York transatlantic crossing and thereby allowing closer contact between the people on the two continents.[29] This increased the exchange of people, materials, and ideas, as did the rapid spread of railroad tracks around the world in the mid-nineteenth century, for instance in the United States increasing from 23 miles of track in 1830 to 30,626 miles of track in 1861.[30]

The exchange of ideas and information was further increased through vast advances in communication technology. Optical telegraph networks—based on one person signaling to another within the immediate line of sight and that person signaling onward—were constructed in France, Sweden, and Britain in the 1790s, and their success in speeding communications led to an expanding network throughout Europe over the next decades.[31] Across the Atlantic, it was optical signaling that relayed stock prices in the 1830s from New York's Wall Street to Philadelphia each business day, taking approximately 30 minutes to do so.[32] The optical telegraph networks were decidedly faster than runners or horseback riders hand carrying messages from place to place, but they were a far cry from what was to come.

The first practical electric telegraph systems were invented and demonstrated in the 1840s, followed in the next several decades by the construction of telegraph lines across continents and, one by one, the construction of submarine telegraph cables to link locations across masses of water. The telegraph was aptly described by Tom Standage as the "Victorian Internet" for the leaps and bounds it made in communications and for the social networks it engendered. Notably, there were more than 12,000 miles of telegraph lines in the United States by 1850,[33] England was connected by submarine cable to continental Europe in 1851, and, after several false starts, Europe was connected by submarine cable across the Atlantic to North America in 1866. In 1902, a submarine cable was completed across the Pacific Ocean from Vancouver, Canada, to Australia and New Zealand.[34] These were major steps toward allowing near instant communications around the globe, made vastly more widespread with the further inventions of the telephone and radio in the nineteenth century and television and the Internet in the twentieth century so that by the early twenty-first century, not just major events but even the most minor of occurrences and ideas can be routinely communicated almost instantaneously to people around the globe.

Society has benefited immensely from the increased speed of communications and from industrialization in general, which has created conveniences and luxuries that few people would now want to do without. Among these, central heating, widespread interior lighting, running water, and inexpensive newspapers (spurred by the 1820s invention of the rotary steam press), magazines, and books (expensive ones were available much earlier) all became common in middle-class houses during the first half of the nineteenth century.[35] Later, we obtained electric washing machines, dramatically reducing household labor, especially by housewives, and electric refrigerators, with all the benefits that refrigeration allows by lengthening the time that food can remain fresh and edible. The use of fossil fuels has also reduced the pressure on the world's forests, as it has meant less wood gathering for heat, at least in industrialized nations.

However, along with the tremendous benefits provided by industrialization came significant drawbacks. Treatment of factory workers could be horrendous, with rampant employment of women and children at low wages, long hours, and sometimes extremely unhealthy conditions. More directly relevant to this book, the industrialization that produced the horrifying working conditions also vastly increased the volumes of waste products spewed into the atmosphere by human activities, iconically illustrated in the nineteenth and twentieth centuries by massive black clouds emerging from smokestacks. Initially, these emissions were often viewed by local governments and peoples with pride as visible evidence of great industrial power and advance. The pride lessened as industrialization became more common and the dangers of the resulting pollution more obvious. Then an event occurred in December 1952 after which it became almost impossible not to recognize that something needed to be done to lessen some of the more serious negative consequences of industrialization.

Over the five days from December 4 to 8, 1952, a high-pressure atmospheric system that had stalled over London prevented the normal dissipation of the smoke pouring out of the city's millions of residential and industrial chimneys. This smoke combined with a persistent fog to produce a smog that is estimated to have killed 4,000 people and left 50,000 to 100,000 seriously ill. The smog was so thick that visibility was no more than a few feet, and in at least one reported instance, a doctor, unable to find his way through the smog, wisely and successfully enlisted the help of a blind neighbor to lead him to a sick patient's house.[36] Precursor events had occurred as early as December 1873, when 700 Londoners died from respiratory distress during a three-day toxic fog event,[37] but none of these matched the magnitude of the 1952 episode. Spurred on by the events of December 1952 and other instances, finally in 1956 the Clean Air Bill was passed by the British Parliament, and the burning of bituminous coal in London was prohibited.[38]

While London in the 1940s and early 1950s was experiencing fearful smog episodes, the people of Los Angeles were experiencing periods with many of the same symptoms. In Los Angeles, however, the prime initial suspect, sulfur dioxide, proved not to be the primary culprit. Specifically, instead of a smoke/fog "smog," the problem was a photochemical haze produced by the thousands of tons of hydrocarbons emitted into the air each day from cars and trucks. The situation in Los Angeles is compounded by the local geography and meteorology. In particular, the relatively cold marine air flowing onshore from the Pacific Ocean produces a low-lying cool air layer that is frequently capped by an overlying warm air layer from the desert to the east, hindering the upward flow of the photochemical haze, while at the same time the surrounding mountains prevent the flow of air outward around much of the city; in other words, the air gets stuck inside the city. Photochemical haze became prominent first in Los Angeles, largely because of the large numbers of cars and trucks and the complicating geography and meteorology, but soon took its toll in other cities as well.[39]

As industrialization spread in the nineteenth and twentieth centuries, the pollution intensified in the industrialized cities and expanded to affect larger geographic regions. By the second half of the twentieth century, it had become widely recognized that particulate matter that poured into the atmosphere through human activities was at least in some instances endangering human health and sometimes human lives. As a result, not only the British Parliament but many governments and communities took actions to restrict particulate emissions. Less obvious were the global impacts of the massive quantities of gases being poured into the atmosphere along with the particulate matter. These emissions had reached far beyond local and regional impacts, as they were now affecting the composition of the global atmosphere as a whole.

CHANGING ATMOSPHERIC COMPOSITION
AND ITS RELATIONSHIP TO TEMPERATURE

Some of the gases humans are adding to the atmosphere are pollutants and hence inadvisable from that perspective when retained at ground level. However, from the perspective of the Earth's climate, it is the human contribution to the trace gases affecting the greenhouse effect (chapter 1) that has garnered the most attention.

It is not clear whether humans are having a significant impact on the Earth's major greenhouse gas, that is, water vapor, and hence the discussions regarding the human contribution to the greenhouse effect generally revolve around other greenhouse gases, the ones that humans are known to be affecting. In particular, these include carbon dioxide (CO_2), methane (CH_4), nitrous oxide (N_2O), ozone (O_3), and other gases, such as sulfur hexafluoride (SF_6), that do not occur naturally in the Earth's atmosphere but are now there as a result of various manufacturing activities. All these gases are being increased in the lower atmosphere through human activities, although ozone is a special case in that ozone in the upper atmosphere is instead decreasing, also because of human activities and also with serious consequences. The ozone decreases in the upper atmosphere (specifically the stratosphere) constitute an entirely separate issue and are discussed later. Here the greenhouse gases being increased by humans, including ozone in the lower atmosphere, are discussed in turn, beginning with carbon dioxide, the Earth's second most important greenhouse gas.

Carbon dioxide is the trace gas that so far has produced the largest human-induced enhancement to the greenhouse effect. Estimates are that in 2000, 77% of greenhouse gas emissions from human activities were emissions of carbon dioxide.[40] By the start of the twenty-first century, through the burning of fossil fuels (i.e., fuels formed from fossilized remains of plants and animals; most prominently, these include coal, oil, and natural gas) and, to a lesser extent, the production of cement, humans were injecting about 25 petagrams of carbon dioxide into the atmosphere every year.[41] This 25 petagrams is shorthand for 25,000,000,000,000,000 grams, or 25,000,000,000,000 kilograms. In metric tons, it equals 25 billion metric tons, or 25 gigatons, annually. For those more interested in the amount of carbon than the amount of carbon dioxide, the approximately 25 petagrams of carbon dioxide divides to approximately 7 petagrams of carbon and 18 petagrams of oxygen. Unfortunately, these emission amounts keep rising. By 2005, the global emissions of carbon from fossil fuel burning and cement production had risen to approximately 7.8 petagrams, up from approximately 6.1 petagrams in 1990 and approximately 7 petagrams in 2000. Additional human activities, such as gas flaring, deforestation and other land use changes, and biomass burning, are also adding carbon dioxide to the atmosphere although at lesser amounts.[42]

Some of the carbon dioxide injected into the atmosphere through human activities accumulates in the atmosphere, contributing to atmospheric warming, while some gets transferred to the oceans or to the land biosphere. Estimates vary considerably and depend on the time period covered,[43] but estimates for the period of the past few decades indicate that about half the CO_2 inserted into the atmosphere by human activities has stayed in the atmosphere, about 30% has gone into the oceans, and about 20% has gone into the land biosphere.[44] Much of the CO_2 that goes to the land biosphere is used in photosynthesis and hence can have the favorable effect of stimulating vegetative growth. Some of the CO_2 that enters the oceans is also used in photosynthesis, in ocean plant life such as phytoplankton, whereas some instead takes part in a variety of chemical reactions within the water. With slow, persistent vertical mixing in the oceans, some of the CO_2 eventually descends to the ocean depths, with estimates suggesting that on a global average, as of the early twenty-first century, about 30% of the human-generated CO_2 absorbed by the oceans over the past 200 years remains in the upper 200 meters of the ocean, the rest having descended.[45] Despite the depth of the ocean, even together the ocean and land do not constitute an infinite sink capable of absorbing all the CO_2 that the atmosphere might have to offer.

Combining all the human and natural processes that insert CO_2 into the atmosphere and take it out, the concentration of CO_2 rose from about 280 parts per million of the Earth's atmosphere at the start of the industrial revolution a little over 200 years ago to 386 parts per million in 2008.* This comes to an average rate of increase of about 0.5 parts per million every year. However, the rise has not been uniform, instead speeding up noticeably as industrialization increased. The average rate of increase over the first nine years of the twenty-first century, 2000–2008, was 2.1 parts per million each year, over four times the average rate for the previous 200 years.[46]

After carbon dioxide, methane accounts for the second-largest human-induced increase in the greenhouse effect,[47] with approximately 14% of greenhouse gas emissions from human activities in 2000 being from methane.[48] Like CO_2, methane is added to the atmosphere both through natural means and through human (or "anthropogenic") activities. By 2000, the natural sources were contributing about a third and the anthropogenic sources about two-thirds of the annual methane additions to the atmosphere. The largest natural source of atmospheric methane in the early twenty-first century is thought to be the methane gas produced by decomposing organic material in wetlands,[49] while volcanic emissions and hydro-thermal vents constitute important additional natural sources.[50] Major anthropogenic sources of methane derive from rice fields, coal mining, biomass burning,

---

* Atmospheric carbon dioxide has a seasonal cycle, with values decreasing during the Northern Hemisphere growing season, when large masses of continental vegetation in the Northern Hemisphere are inhaling carbon dioxide for photosynthesis. Exact numbers depend on location, but at Mauna Loa, Hawaii, which has the longest nearly continuous record, the contrast between the maximum and minimum amounts in an individual year is about 6 parts per million, with the values in 1980 falling from 341.7 parts per million in May to 335.7 parts per million in September and those in 2008 falling from 388.6 parts per million in May to 382.4 parts per million in October; that is, the seasonal cycle is clear, but the upward trend overwhelms it. (Mauna Loa CO_2 concentrations were obtained from Keeling et al. 2009.)

landfills, livestock rearing (and the accompanying methane in burps, belches, and flatulence), wastewater treatment, leaking pipelines, and the use of fossil fuels. Rice is a major component of the diet of a third of the world's human population, and the flooded rice fields common throughout much of Asia are releasing methane in significant amounts from the waterlogged soils. Essentially, the water in wetlands and in flooded rice fields cuts off the oxygen supply from the atmosphere to the soil, resulting in anaerobic (i.e., without oxygen) fermentation of the organic matter in the soil. This fermentation releases methane.

The anthropogenic methane sources have had a substantial impact on atmospheric methane amounts. In the years just prior to 1750, before the start of the industrial revolution, the global average atmospheric methane concentration is estimated to have been 730 parts per billion.[51] This was already slightly higher than the typical range during the past several ice age cycles, at least according to various ice core records that suggest that methane concentrations range from lows of about 400 parts per billion during glacial periods to highs of about 700 parts per billion during interglacial periods (chapter 2).[52] However by 1984, the atmospheric methane concentration had more than doubled from its preindustrial values, reaching about 1,650 parts per billion. It continued to increase until 1999, by which time it had reached about 1,770 to 1,780 parts per billion, but it leveled off after that, averaging about 1,775 parts per billion in 2006.[53] Although it is uncertain exactly what caused the leveling off, speculations include that 1) perhaps less methane is entering the atmosphere from human activities because of greater care in capturing some of the gas at coal mines and in lessening the leakage from pipelines and oil wells[54] and that 2) perhaps anthropogenic emissions have increased overall, especially with the greatly increased use of fossil fuels in Asia, but the increases have been masked by a drought-driven reduction in methane emissions from wetlands.[55] This is one of numerous instances with many remaining uncertainties.

Third in order of greenhouse gas emissions from human activities is nitrous oxide, coming in at 8% of total emissions in 2000.[56] Like CO_2 and methane, nitrous oxide is added to the atmosphere through both natural processes and human activities. Natural processes in soils and oceans release nitrous oxide to the atmosphere, as does chemical oxidation of ammonia within the atmosphere. Human activities causing nitrous oxide emissions include nitrogen-based fertilizer use, biomass burning, cattle raising, treatment of wastewater, nylon manufacturing, and catalytic reduction of fossil fuel emissions.[57] Prior to 1750, nitrous oxide concentrations in the atmosphere are estimated to have been about 270 parts per billion;[58] by the end of 2006, they had reached 320 parts per billion and were increasing at a yearly rate of about 0.77 parts per billion, with the human emissions by the early twenty-first century accounting for about 30% of the total emissions.[59]

As mentioned in chapter 1, the greenhouse gas ozone is desirable high in the atmosphere for its interference with ultraviolet radiation, although it is undesirable low in the atmosphere, where it not only contributes to the greenhouse effect but also is a pollutant. Unfortunately, human activities are decreasing ozone where we want it, in the upper atmosphere, and increasing ozone where we do not want it, in the lower atmosphere. In the upper atmosphere, ozone is continually produced and destroyed naturally through a variety of chemical reactions that humans have adversely affected, as described more fully later in this chapter. In the lower atmosphere, human land use changes and fuel combustion have contributed to an increase in undesired tropospheric ozone by about 38% over the past 200 years.[60]

The last of the greenhouse gases to be mentioned, atmospheric sulfur hexafluoride, is believed to come exclusively from human sources, with by far the most important source by the early twenty-first century deriving from its use in insulation for electric power distribution. Sulfur hexafluoride concentrations in the atmosphere, starting at zero prior to the industrial revolution, had reached 6 parts per thousand by the end of 2006 and were increasing at a rate of about 0.22 parts per thousand each year.[61]

All these trace gases—CO_2, methane, nitrous oxide, ozone, and sulfur hexafluoride—enhance the greenhouse effect and thereby exert a forcing toward warming the Earth's climate. Combined, they constitute the core of the reason why people are concerned that humans could be warming the Earth excessively, with cascading consequences.

However, not all the emissions to the atmosphere from human activities lead to warming, and in fact some human activities have the countering tendency toward cooling the atmosphere. In particular, humans insert sulfate particles into the atmosphere through fossil fuel combustion and, to a lesser extent, through biomass burning,[62] and these sulfate particles can act as cloud condensation nuclei and can increase cloud reflectivity, sending more solar radiation back to space and thereby producing a cooling tendency. Still, even though the sulfate particles might be beneficial in counteracting the enhanced greenhouse effect, limiting sulfur emissions to the atmosphere remains desirable, as these emissions also contribute to air pollution, smog, and acid rain, all with negative consequences to human health as well as to the environment.

Acid rain results when the pH of rainwater is lowered from its normal value of about 5.6 down to 5.5 or lower. This can occur when sulfur dioxide and nitrogen oxide emitted into the atmosphere from cars and other vehicles and from the burning of fossil fuels and other industrial activities react with water vapor to form sulfuric acid and nitric acid. The term "acid rain" arose in the 1800s, coined by British chemist Robert Angus Smith in the face of the London fogs produced by coal burning and made famous by novelists Charles Dickens and Sir Arthur Conan Doyle, the latter especially through the adventures of Sherlock Holmes.[63] The fogs seriously reduce visibility and clog the lungs, thereby noticeably impacting humans in the midst of them, but the sulfuric and nitric acids are also powerful corrosives that can damage bridges, buildings, sculptures, and other anthropogenic structures as well as many features of the natural environment. Acid rain has been identified as the culprit in damaging tens of thousands of acres of forests in Europe and North America and thousands of lakes around the world. Many of the lakes have become so acidic that fish and other species have disappeared from them.[64] With sulfur emissions an important contributor to the formation of acid rain, cutting back on them is a priority, even though the sulfur emissions might help to limit greenhouse warming.

Global sulfur emissions from human activities are estimated to have reached a peak of 74.1 teragrams, or 74.1 million metric tons, in 1989 but to have declined since then, in large part because of conscientious efforts and regulations to reduce the emissions. Even with the decreases, which averaged about 2.7% per year in the 1990s, the emissions remain high, estimated at 55.2 teragrams, or 55,200,000,000,000 grams, of sulfur per year as of 2000.[65]

The history of regional percentage contributions to the sulfur emission totals reflects the histories of industrialization and of environmental regulations. According to the substantial volume *The Economics of Climate Change* by Nicholas Stern, western Europe accounted for an estimated 87.6% of global nonshipping sulfur emissions to the atmosphere in 1850, but its percentage contribution has progressively declined as other regions have become more industrialized. North America in particular increased its percentage contribution in the late 1800s and early 1900s, and then eastern Europe and Asia increased their percentages in the mid- and late 1900s, with Asia becoming the largest contributor, at 33%, in the 1990s. Environmental regulations and other factors have led to a wave of reductions in emission totals in one region after another, with the emission totals from western Europe and North America peaking first in 1974, followed by those from eastern Europe and Africa in 1987, Asia in 1996, South America in 1999, and the Middle East and Oceania in 2000.[66]

Changing pollution levels can noticeably impact the amount of solar radiation reaching the ground (see chapter 1). Consequently, in the 1950s and 1960s, solar dimming was widespread throughout the world as pollution levels rose, whereas since the 1980s, many areas have experienced some solar brightening as aerosol pollutants have been reduced.[67] In addition to dimming, less solar radiation reaching the surface provides a cooling influence. Despite some warming from the absorption of sunlight by some aerosols (soot being particularly absorptive), the net atmospheric temperature effect of aerosols as a whole is widely felt to be a cooling.[68] Similarly, the net temperature effect of all the land use changes caused by humans, including fewer trees and more agricultural fields, roads, buildings, and other human constructions, is also thought to have been a cooling, through increasing, overall, the global albedo.[69]

GLOBAL WARMING

At least so far, the tendency for aerosols and land use changes to reduce temperatures appears to have been considerably outweighed over the past 150 years or so by the tendency of greenhouse gases to increase temperatures. Although temperature records prior to the satellite era are notoriously incomplete, with, for instance, no data for long periods for huge expanses of the oceans, extensive analyses of the records that do exist suggest that global surface temperatures have risen noticeably since the second half of the nineteenth century. Global temperature records based on the admittedly limited instrumental measurements have been constructed for the period since 1880. These records indicate that, overall, global temperatures by and large rose from 1880 until the 1940s, then fell somewhat until the 1970s, and then rose again. The 10 warmest years from 1880 through 2006 were all within the final 12 years of the 127-year period, with the rate of global temperature rise averaging about 0.06°C per decade over the past century, for a total warming of about 0.6°C (1.1°F) over 100 years. The rate is considerably higher, about 0.18°C per decade, when considering just the past 30 years.[70] Despite the incompleteness of the temperature data sets, they are probably correct in the conclusion of a long-term warming, which is verified by numerous additional measures, like lessened Arctic sea ice coverage and earlier spring blooming of various plants. There are also strong indications of twentieth-century warming from ice cores drilled on the Tibetan Plateau and in the tropical Andes of South America.[71]

No one knows for sure exactly how temperatures would have changed over the 1880–2006 period mentioned in the previous paragraph if humans had not been a factor; after all, temperatures have risen and fallen throughout Earth's history (see chapter 2). Certainly, it is possible that even without humans, temperatures might have risen between 1880 and 2006. Nonetheless, humans are widely believed to have contributed to the warming, with the temperature impact of our greenhouse gas emissions and other warming activities well outweighing the impact of our particulate emissions and other cooling activities.

In order to understand the basic changes and not get helplessly bogged down by counterexamples, it is essential to recognize that "global warming" refers to what is happening to temperatures averaged around the globe, not to what is happening to temperatures at each individual location on the globe. The Earth system is large and variable, and every year some regions are hotter than normal and others colder than normal. For instance, in January 2006, while the eastern United States was experiencing an unusually mild winter, much of eastern Europe and northern Asia were experiencing a cold spell, with temperatures in Moscow in particular sinking below any January Moscow temperatures since the 1920s.[72] Eurasia as a whole had its greatest snow extent on record in January 2006, while in the same month, North America had its second-lowest recorded January snow extent.[73] Similarly, in the winter of 2000–2001, despite global warming, a particularly cold spell in Scotland led to some Scottish lakes being frozen for the first time in decades, even allowing curling tournaments to be held on them.[74] Instances like this will continue to occur, with extremes in one direction or another arising despite the overall long-term trend. "Global warming" means only that, on average, the globe is getting warmer. Some locations have actually cooled in the long term, and all locations experience fluctuations, warming during some periods and cooling during others. Still, in line with overall global warming, the local and regional temperature records set in recent years are predominantly for high temperatures, not for low temperatures. Among the countries for which 2006 was warmer than any previous year with temperature records available were China, the Netherlands, Spain, and the United Kingdom.[75]

Although warming is discussed most frequently in connection with the atmosphere, especially near-surface air temperature (generally abbreviated "surface air temperature"), the global ocean is also believed to have warmed since the 1800s, at least in the upper layers, and, overall, the global land surface area is believed to have warmed as well. Measurements are incomplete, but over land the best temperature measurements for the period since the 1800s tend to be surface air temperatures, whereas the best records for the oceans tend to be sea surface temperatures. Combining the two for a global record, the global surface temperatures are estimated to have risen by approximately 0.74°C in the 100-year period from 1906 to 2005,[76] yielding a rate that is slightly higher than the 0.6°C-per-century rate mentioned by others for the longer record starting in 1880.[77] There are uncertainties in both numbers, but they provide good estimates of the magnitude of global warming.

Several attempts have been made to place the temperatures of the past century and a half in the context of the temperatures of the past millennium. One of the most noteworthy of such attempts was published in a paper by Michael Mann and Raymond Bradley of the University of Massachusetts and Malcolm Hughes of the University of Arizona in 1999. Mann, Bradley, and Hughes present a plot of Northern Hemisphere temperatures over the past 1,000 years based on multiple data records that shows a long period of gradual cooling for the first 900 years, followed by a sharp upturn in temperatures dominating much of the twentieth century (except 1940–1970).[78] Because of its shape, the plot has been labeled the "hockey-stick" plot. This hockey-stick plot depicts the late-twentieth-century temperatures to be well above the temperatures at the start of the curve, which lies in the midst of the medieval warm period. The plot was an ambitious and difficult attempt to merge a variety of data records into a coherent picture and of necessity incorporated records of differing quality. It has been criticized on both quantitative and qualitative levels,[79] with the debate even reaching the offices of the U.S. Congress. Among the criticisms, Stephen McIntyre and Ross McKitrick claim that obsolete data were used, that some of the data were unjustifiably truncated, and that the statistical methodology used by Mann and his colleagues artificially accentuates the twentieth-century warming.[80] In 2003, McIntyre and McKitrick presented a revised plot for the 1400–1980 subperiod, and the revised plot has higher temperatures in the early 1400s than at any time during the twentieth century up until 1980,[81] in marked contrast to the plot that Mann and his colleagues gave in 1999. Still, despite the criticisms, which have been examined extensively, many scientists believe that the basic aspects of the original 1999 curve are correct, that is, that a gradual cooling was followed by a steeper warming, with the temperatures at the end (in the late twentieth century) markedly greater than those at the start (in the medieval warm period). Others are less certain. For considerable further discussion and references on this controversy, the reader is referred to the 2007 report of the Intergovernmental Panel on Climate Change,[82] where it is concluded that the data are insufficient to establish a valid estimate of global temperatures during the medieval period.

CONSEQUENCES OF WARMING
AND INCREASED CO_2

If the only effect of our trace gas emissions to the atmosphere were to cause warming, with no further impacts, it is unlikely that there would be anywhere near as much concern as there is. Warming has the unpleasant effect of causing summers to become less comfortable in many regions, but on the other side of the ledger, it causes winters in many regions to become more comfortable. For the wealthy, an increased need for air-conditioning in the summer can be fully compensated for by a decreased need for heating in the winter. For the less fortunate, unable to retreat to the comfort of fully air-conditioned and heated buildings, the increased probability of dying from heat in summer in many localities is canceled by the decreased probability of dying from winter cold.

> Although we know approximately how much CO_2 and other trace gases humans are adding to the atmosphere, we do not know how much warming is caused by us or which impacts would have occurred in our absence.

However, higher temperatures and increased CO_2 concentrations have numerous propagating consequences. Some are immediately understandable, such as melting ice and snow, whereas others are less straightforward and often controversial. Among the propagating consequences causing the most concern are sea level rise, decaying permafrost, increased acidity of the oceans, and impacts on a vast multitude of species. Certainly, humans cannot reasonably be blamed as the sole cause of all the propagating impacts, as many of these impacts are entangled also with other changes, some of which are and some of which are not human induced. Although we know approximately how much CO_2 and other trace gases human activities are adding to the atmosphere, we do not have a good handle on how much of the climate warming is caused by us or which of the propagating impacts would have occurred even in the absence of our activities. Still, our role in adding greenhouse gases that foster warming makes us partly culpable.

The following sections describe a few of the propagating consequences of increased greenhouse gases and consequent warming and some of the reasons why they matter.

Sea Level Rise

As mentioned in chapter 1, increasing temperatures contribute to sea level rise both through thermal expansion of the water, causing it to take up more volume, and through the addition of mass to the oceans when land-based ice either melts or calves into the ocean. Both thermal expansion and land ice decay have been at play through much of the twentieth and into the twenty-first century, with ocean temperature increases[83] and glacial retreats (next paragraph). The combined result during the twentieth century was a sea level rise of about 1.7 millimeters (0.17 centimeters) per year on average, for a 100-year total of 17 centimeters.[84] The rise has not been uniform spatially or over time, but it has been enough to have caused problems for many low-lying coastal communities, as the rising sea has submerged some areas and in other areas has caused increased coastal erosion, flooding, and intrusion of salt water into underground and aboveground freshwater reservoirs and fields.[85] In the case of two uninhabited islands in Kiribati, an island nation in the central Pacific Ocean, sea level rise has totally eliminated the islands, as they disappeared into the ocean in 1999. Relatedly, the 100 inhabitants of Tégua, an island in the island nation of Vanuatu in the South Pacific, were forced to abandon their island in December 2005.[86] Other locations have not yet been eliminated but are being flooded repeatedly because of the rising seas, one such location being Saint Mark's Square in Venice. At the beginning of the twentieth century, the square flooded fewer than 10 times a year, by 1990 it flooded about 40 times a year, and in 1996 it flooded almost 100 times, and some predictions are that, without further protections, it will be flooding every day of the year by the end of the twenty-first century.[87]

Land Ice Decreases

For glaciers with summer temperatures near or above the melting point, warming increases melting on the glacier. However, the increased melt can under some circumstances be balanced or even outweighed by increased snowfall, as warming also tends to increase evaporation, getting more water into the air, most of which eventually falls back to the surface as either rain or snow. These countering tendencies for glacial growth versus decay, along with inherent variability, make it unsurprising that many glaciers have experienced fluctuating periods of advance and decay over the course of the twentieth and early twenty-first centuries. However, despite the fluctuations and marked contrasts from one glacier to another and from one time period to another, overall global land ice coverage appears on the decrease. In recent decades, glaciers have retreated in North America,[88] South America,[89] Asia,[90] and Africa[91] and along the Antarctic Peninsula in Antarctica.[92] The glaciers of Europe on average advanced over the period 1961–1990[93] although still retreated overall during most of the twentieth century[94] and in the late 1990s and early twenty-first century.[95]

Not all glaciers studied have retreated, and even in combination, the numerous relevant studies do not come close to covering all of the Earth's more than 160,000 glaciers, very few of which have much of a data record. Still, it does seem, from the uneven data available, that the global expanse of land ice around the world has decreased since the nineteenth century[96] and hence that the changes in land ice coverage have contributed to the sea level rise discussed in the previous section.

Arctic Sea Ice Decreases

Arctic sea ice has received a great deal of media attention starting in the late twentieth century, because of prominent decreases in areal coverage of the ice since the late 1970s that are visually quite clear from the satellite record, along with strong indications, from major compilations of ship reports and aerial reconnaissance, that the retreats began well earlier in the twentieth century[97] and, from submarine and ground-based measurements, that the ice has thinned along with its retreat.[98] The Arctic sea ice decreases since the late 1970s have averaged about 4% per decade on an annual average basis[99] but have been much greater, at nearly 9% per decade, at the end of summer, that is, in September,[100] the latter leading to much speculation about the possibility of late summer ice-free Arctic Ocean conditions within the next few decades.

As with global warming, the sea ice decreases are not uniform from year to year but instead are apparent in the longer-term trend. Demonstrating both the interannual variability and the downward trend in the Arctic ice, the late summer ice in September 2005 reached a record minimum for the period of the satellite observations, after which the ice recovered somewhat, with greater ice coverage in September 2006, followed in September 2007 by a substantially lower new record minimum.[101] The Arctic sea ice decreases have not been matched in the Antarctic, which instead has experienced increased sea ice coverage since the late 1970s,[102] but the Antarctic sea ice increases are quite a bit less than the Arctic sea ice decreases so that the globe as a whole has lost sea ice coverage. Chapter 11 includes more on the contrasts between the ice changes in the Arctic and Antarctic and the scientific and media coverage of them.

Although the lessened Arctic sea ice cover does not impact sea level (see chapter 1), it has raised concerns for other reasons. Among these are the following: the lessened ice cover is a highly visible indicator of our changing planet, in line with global warming; as the ice retreats, its highly reflective white surface is replaced by the far less reflective liquid surface of the ocean so that more solar radiation gets absorbed, further warming the system in a classic positive feedback; the ice reduction affects a wide range of species, including polar bears, which depend on the ice as a platform from which to hunt seals and other prey;[103] and the ice reduction affects the landscape of low-lying Arctic coastal regions, which are experiencing increased coastal erosion from the increased wave action due to reduced ice coverage.[104]

Decaying Permafrost

Permafrost underlies approximately 25% of the Earth's land area and introduces a new set of concerns resulting from warming, especially in the Northern Hemisphere, where permafrost underlies the surface not only in ice-covered glaciated regions but also in vast unglaciated regions, most extensively in Russia and Canada. The permafrost in these unglaciated regions, overall, has experienced significant warming and degradation since the mid-twentieth century.[105] Although few people live in these cold, harsh Arctic areas, many of those who do are suffering the consequences of the permafrost decay. Most dramatically, buildings, roads, and other infrastructure can (and sometimes do) crumble as the permafrost warms and decays beneath them, in substantial part because of the local disturbance to the surface energy balance caused by the infrastructure itself. Crumbled houses and other infrastructure are clearly of immediate practical concern for the residents and can provide visually alarming pictures of the impacts of the decaying permafrost, but there are also additional, larger-scale and potentially considerably more important long-term climate impacts of the permafrost decay. Specifically, because of the low temperatures in the polar regions, partially decomposed organic matter has resided for thousands of years in the permafrost, and some of the CO_2 and/or methane in this organic matter is released to the atmosphere as the permafrost thaws.[106] This release of greenhouse gases, particularly methane, to the atmosphere means that the decaying permafrost due to warming has substantial potential to cause yet further warming in another instance of a positive (or enhancing) feedback.

> Partially decomposed organic matter has resided for thousands of years in permafrost, ready to release CO_2 and/or methane to the atmosphere as the permafrost thaws.

Ocean Acidification and Affected Marine Life

One of the potentially most serious consequences of increased atmospheric CO_2 is that the oceans are becoming more acidic.* Essentially, with more CO_2 in the atmosphere, more gets absorbed into the oceans, and some of the absorbed CO_2 combines with water (H_2O) to produce carbonic acid (H_2CO_3), a weak acid that readily releases hydrogen ions (H+), central in the determination of acidity. Over the past 200 years, the pH of the global surface ocean waters has decreased from approximately 8.18 to approximately 8.07, and the concentration of hydrogen ions has increased 30%.[107] Of particular concern, the acidification impedes calcification,

---

* With an estimated average pH of approximately 8.07, the oceans are slightly alkaline, having a pH exceeding pure water's neutral value of 7.0. Water with a pH above 7.0 is "basic," or "alkaline"; water with a pH below 7.0 is "acidic." Nonetheless, as the pH value decreases, it is conventional to speak of the oceans as becoming "more acidic" (rather than the equivalent "less alkaline" or "less basic") and of the process as "ocean acidification."

whereby many marine organisms make shells and plates out of calcium carbonate ($CaCO_3$), and the calcium carbonate shells and exoskeletons that do form could dissolve back into the water if the concentration of carbonate ions in the surrounding waters becomes too low, a situation that can arise because some of the hydrogen ions react with carbonate ions (CO_3^{2-}) to become bicarbonate ions (HCO_3^-).[108] Consider for a moment the organisms affected. They face the possibility that their shells and exoskeletons might dissolve away because another species (i.e., humans) living elsewhere on the planet is pouring too much CO_2 into the atmosphere, and there is absolutely nothing that they can do to stop it. Life is certainly not fair.

Of the two common calcium carbonate forms, the aragonite form is more soluble than the calcite form so that organisms producing the calcite form for their shells and exoskeletons, such as coccolithophores and foraminifera, are considered less immediately vulnerable to ocean acidification than those producing the aragonite form, such as corals and pteropods.[109] Corals in particular have generated interest because of their beauty, the protection they provide to the nearby coasts, their importance to regional fisheries and the tourism industry, and the materials harvested from them.

Damaged Coral Reefs

Corals are being injured not only by ocean acidification but also by ocean warming. Although the reason that corals occur predominantly in low-latitude waters is that they prefer warm temperatures and substantial sunlight, too much warmth can lead to bleaching (whitening) and death of the corals. Bleaching and death can also occur from other stresses, including temperatures that are too low instead of too high, salinities that are too low, toxins, and infections, but warming has been a major culprit in recent decades. Reports of large-scale mass bleaching events in tropical and subtropical coral reefs began in the early 1980s, and the frequency and intensity of these events has increased from that time, at least into the early twenty-first century. Elevated ocean temperatures are considered a prime cause for the bleaching and for increased mortality rates of the corals,[110] with additional possible contributing factors being increased ultraviolet radiation, disease, sedimentation, pollution, and excessive shade.[111] An estimated, astounding nearly 30% of the Earth's warm-water corals disappeared between 1980 and the early twenty-first century, largely because of frequent periods of excessively high water temperatures.[112]

Additional Effects of Warming on Plant and Animal Life

Corals and shelled marine organisms are hardly the only life forms being affected by warming. Some animal ranges are shifting poleward, some plants are blooming earlier (by days in some cases and by weeks in others), some animals have shifted the timing of when they mate, and some grasses are now sprouting in higher latitudes than they had grown in previously. Among the signs of change are the following: in the late 1990s, robins were seen for the first time in some Inuit communities in northern Canada;[113] 22 species of European butterflies shifted their range north by 35 to 240 kilometers during the twentieth century, while only one species shifted south;[114] and frogs in upstate New York are reported to be mating 10 or more days earlier than in the past.[115] Species from mollusks to mammals have been found to exhibit phenological changes that are likely driven by recent warming.[116] Many of the changes are not inherently either good or bad, but even changes that seem to be innocuous can sometimes create major disruptions to local ecosystems by affecting the interactions between predators and prey,* between parasites and hosts, or between pollinating insects and the plants dependent on them. Many changes that involve the expansion of a species' range are favorable for that species but unfavorable for species into whose habitat they are expanding, like the poleward expansion in France of the pine processionary moth (*Thaumetopoea pityocampa*), an expansion that provides increased habitat range for the moth but new threats to the pine forests unfortunate enough to be along the path of the expansion.[117]

Warming has certainly contributed to changes in the distribution and abundance of species throughout the world,[118] but establishing the specifics of which particular changes were caused predominantly by warming is extremely difficult, as there is no way of knowing for sure what would have happened in the absence of the warming. Nonetheless, in an overview of ecological changes in marine, freshwater, and land-based plants and animals, biologist Camille Parmesan finds that "the direct impacts of anthropogenic climate change have been documented on every continent, in every ocean, and in most major taxonomic groups."[119] Proving cause and effect is rarely straightforward, but the changes observed are found to be "heavily biased" toward those that would be expected to occur with global warming.[120]

Species extinctions are among the most controversial claims attributed to recent warming, both because it is extremely difficult to know for sure that a species has

---

* For instance, a 32-year study of a Dutch population of the bird *Parus major* (the great tit) reveals that the timing of their breeding is becoming increasingly out of step with the emergence of the caterpillars that provide a core of their diet, as the caterpillars are adjusting more rapidly to the temperature increases than the birds are (Nussey et al. 2005). In contrast, a 47-year study of *Parus major* in England has found that there the species has adapted more closely in unison with the caterpillars (Charmantier et al. 2008). Lyon et al. (2008) compare the two cases, emphasizing that the reasons for the differences remain unknown.

become extinct (as survivors might exist in unchecked locations) and because there could well be a variety of causes for the extinctions that do occur. Among the animal groups thought to be undergoing a high rate of extinctions in which warming has been a contributing factor are the amphibians, thousands of species of which have declined in the past few decades, with hundreds of species said to be nearly or already extinct.[121] An estimated 74 of the approximately 110 species of *Atelopus*, an amphibian grouping endemic to the tropics of Central and South America, are believed to have gone extinct in the late twentieth and early twenty-first centuries. The immediate implicated cause is the pathogenic fungus *Batrachochytrium dendrobatidis*, but the reason for the fungal outbreaks appears linked to large-scale warming.[122] Another possible extinction linked in part to warming is the Monteverde golden toad. Likely because of the memorable name, the possibility that the golden toad has become extinct has received more attention than most of the other extinctions or possible extinctions. In 1987, 1,500 Monteverde golden toads were counted in Costa Rica's Monteverde Cloud Forest, but in 1988 only a few were seen, in 1989 only one was seen, and in the 1990s none were located despite intense searches.[123] With luck, some of the toads might have survived underground, where they typically spent much of their lives, and in that case the species might rebound and eventually thrive, although this is not considered likely.

Favorable Consequences of Warming

Like most changes, warming has favorable as well as unfavorable consequences. Warming has meant less need for artificial heat in winter, fewer deaths from cold, an expansion of the growing season in some midlatitude regions, a poleward expansion of where some crops might be able to grow, and fewer crop losses from frost. The Northern Hemisphere high-latitude oceans have witnessed an increase in coccolithophore blooms since the 1980s,[124] and with continued warming there could soon be an opening of regular shipping routes through the Arctic (good for shipping but bad for the Arctic environment). Taken together, the many favorable aspects of warming are quite significant, helping to explain why some people are not particularly concerned about the global warming phenomenon.

OTHER MODERN HUMAN IMPACTS

Increased greenhouse gases, aerosols, warming, and the propagating consequences constitute only a portion of the impacts humans are having on the planet. We hugely change the landscape by clearing the land for agriculture and other uses; by constructing buildings, roads, parking lots, and other structures; and by digging into the land to mine the many resources it has to offer. We siphon water away from rivers and lakes for our own purposes. We pollute water and land with our discarded garbage. We change ecosystems by our land use changes and by killing, growing, harvesting, and domestication. And even our emissions to the atmosphere have more impacts than the warming and pollution highlighted so far, a primary one being the damage done to the Earth's protective ozone layer. This section describes a few of these additional impacts.

Impacts of Humans on Other Species

All animals killed for food or sport are quite directly impacted by humans, as are all trees cut down for use of the wood or cut down or burned for clearing of the land. Just in the tropics alone, an estimated 142,000 square kilometers of forests were being destroyed by humans each year by the early 1990s.[125] On the positive side, humans are also planting trees, and some land vegetation is likely benefiting from the increased atmospheric CO_2, which provides more of this necessary ingredient for photosynthesis.

In terms of the seriousness of our impacts on individual species, nothing could top the instances where we are a prime cause for a species' extinction. Biologist Edward O. Wilson of Harvard University has suggested that humans have initiated the sixth great extinction episode in Earth's history, the first following the extinctions approximately 65 million years ago that included the last of the dinosaurs. He estimates that at least 27,000 species are going extinct each year and indicates that human activity is overwhelmingly responsible, increasing the background natural extinction rate by a factor of over 1,000.[126] Some of the extinctions are connected to climate change, but many are instead due to human hunting or to our cutting down or otherwise destroying trees and other plant life.

Among the most important causes of loss of species is the felling of primeval forests, destroying entire ecosystems that in some cases contained species that had existed only in that ecosystem. For instance, consider the Usambara Mountain forests in Tanzania, which were reduced to half their geographic expanse between 1954 and 1978 by human activities, largely through logging and clearing of land for agriculture. Wilson estimates that specifically because of the human activities in the Usambara Mountains, thousands of plant and animal species are now moving toward extinction in light of the number of species found exclusively in those mountain forests. On the other side of the Atlantic, about 25% of Colombia's Chocó forests were destroyed by timber companies and inhabitants clearing the forests for alternative land use in the 1970s and 1980s. These forests contain 3,500 known plant species and perhaps 6,500 unknown species, with perhaps 25% of the species existent only in these forests.[127] As the forests continue to be destroyed, hundreds or even thousands of species could go extinct.

Serious as the reductions of the Usambara Mountain forests and the Colombian Chocó forests are, percentage forest loss is much greater in other regions. When in 1832, during his famed voyage around the world on the *Beagle*, Charles Darwin encountered the coastal Brazilian forests to the southeast of the Amazon rain forests, these coastal forests spread over approximately 1 million square kilometers. By the early 1990s, they covered less than 5% of that original area, the other 95% having been converted by humans to agricultural and other land uses.[128]

While extinction is the most final of the consequences of human activities on other species, many other consequences can also be extremely damaging, sometimes in ways that might eventually lead to extinction or, less severely, to lower numbers and lesser quality of life. As humans populate regions, they necessarily disrupt and fragment the habitats of the species that had lived in the region, and although many times the small species, like ants and cockroaches and smaller animals and plants, can adapt quite well, often adaptation is much more difficult for the larger species.

Species that are migratory have special circumstances that can either aid or hinder them in adjusting to humans. Repeatedly, large migrations of large land-based animals have been totally disrupted by the increased presence of humans, in some cases because of outright killing of the animals and in other cases because of disruption along one or more key segments of their migratory route. In the case of bison that lived in North America at population levels in the tens of millions in the 1600s, their coming to the brink of extinction by the 1880s was due to intentional killing, initially by Native Americans for food but then in much greater numbers by white settlers and traders who killed the bison for at least three purposes: to obtain their hides, to protect settlements, and, worst of all, to destroy the bison-dependent nomadic lifestyles of the Native Americans. Since the 1880s, the bison have rebounded, in large part because of determined efforts, among the earliest being one started by Theodore Roosevelt in the 1880s. By the early twenty-first century, although the bison are no longer in danger of extinction, a large percentage of them (perhaps 96%) are not wandering free but are instead being raised for commercial purposes, including low-cholesterol bison meat; that is, the bison have been spared extinction, but this is perhaps as much for human benefit as for the benefit of the bison. With increasing numbers of farms and ranches in the American Great Plains facing bankruptcy and many of the local youth leaving the region, a movement is gaining adherents to turn large areas of the Great Plains into a buffalo commons, owned by the U.S. government and meant to restore the native wildlife as much as feasible, including bison, pronghorn, elk, prairie dog, and other species, while recognizing that the full past can never be restored. On other continents as well, large animal migrations have been severely impacted by humans, although some African migrations, though reduced in numbers, are still impressive, such as those of the million or more wildebeest and nearly 200,000 zebras migrating between southern Africa and northern Tanzania in response to seasonal rains and nutrients.[129]

Because of flying well above the normal human habitat, bird migrations have a better chance than land-based migrations of surviving human disruptions and climate change. Still, they too are not spared. Among the migrating birds that have gone extinct largely through overexploitation by humans (for sport and marketing) is the passenger pigeon, which had aggregated by the tens of millions across the Northeast and Midwest of the United States in the early nineteenth century, then became extinct in the wild by the end of that century and extinct altogether in 1914, when the last known elderly female died in the Cincinnati Zoo. A species that is still migrating, although in reduced numbers, is *Calidris canutus rufa*, a robin-sized red-knot shorebird that breeds in Arctic Canada and winters in southern South America, for a remarkable annual round-trip journey of 18,000 miles. Biologists studying these birds found about 53,000 making the migratory journey in the early 1980s and about 51,255 in 2000 but then recorded precipitous declines to only 27,242 birds in 2002 and 13,455 birds in 2006. Nothing substantial appeared to have changed at the end points of the journey, but a critical change had occurred at a refueling stop in Delaware Bay, along the East Coast of the United States. Fishermen in Delaware Bay were depleting the horseshoe crab population, and this deprived the red knots and other birds of the billions of horseshoe crab eggs that they had become accustomed to eating during their annual stopover in late May. Recognizing the impact on red knots and other songbirds in Delaware Bay, the states of New Jersey and Delaware in 2006 declared a 2-year moratorium on harvesting horseshoe crabs, hoping to reverse the damaging trends, an attempt that might or might not succeed.[130] At least the increased awareness of human culpability in the ongoing high rates of extinctions increases the chance that damaging trends can be reversed.

Impacts of Siphoning Water for Irrigation: The Case of the Aral Sea

In sharp contrast to the widespread rising water levels due to sea level rise, some water levels are actually falling. In fact, one of the most severe human-caused environmental disasters of the twentieth century was the damage done to the Aral Sea in central Asia because of water being siphoned away from the rivers flowing into it.

Between 1960 and 2000, the Aral Sea lost over 80% of its water, the salinity of the water increased by six times, and the populations of the formerly abundant fish and other life forms were reduced by orders of magnitude. Moreover, salt, sand, and dust from the newly exposed land, which had been part of the seabed before the waters retreated, have been transported throughout the region by vast salt/sand/dust storms, lessening the productivity of the surrounding lands[131] and sometimes totally engulfing houses in nearby villages unfortunate enough to be in the way of drifting sand dunes.[132] Fishing villages that originally were adjacent to the sea had the sea retreat further into the distance year by year until by the end of the twentieth century the sea was 30 kilometers or more away from some of these villages. The central cause of all this regional damage was the siphoning of water from the two main rivers feeding the Aral Sea, the Amu Dar'ya to the south and the Syr Dar'ya to the east, in order to provide water for irrigating crops, largely cotton and secondarily rice. Once the magnitude of the disaster became clear in the late twentieth century, efforts were made to limit further damage, with some success in the northernmost portion of the sea (see chapter 12). In the meantime, the tremendous reduction in the water volume of the Aral Sea produced additional concern because it resulted in Vozrozhdeniya Island, a former small island in the middle of the sea, becoming joined to the mainland as of 2001, at the end of a peninsula that formed as the waters retreated. This is a concern because Vozrozhdeniya was a testing ground for the former Soviet Union's biological weapons program. Various animals are believed to have been exposed in test cases on this island to anthrax, plague, typhus, Q fever, smallpox, botulinum toxin, and other pathogens. If the pathogens have survived, the land link to the mainland vastly increases the likelihood that they will be transported, in one way or another, to unprotected populations.[133]

Other bodies of water are also being deprived of their water supplies because of human activities, one of the largest being the Mediterranean Sea, which is now becoming increasingly salty because of freshwater siphoned off along the Nile and other inflowing rivers and also because of increased evaporation of Mediterranean waters from warming.[134] The siphoning of water away from rivers for agriculture and other human uses has led to many interstate and intercountry conflicts, as those downstream become deprived of water they otherwise would have.

Damage to the Stratospheric Ozone Layer

In addition to warming and pollution, another critically important impact of the emissions to the atmosphere from human activities is the damage done to the Earth's protective ozone layer. Because of the importance of stratospheric ozone in restricting the transmission of damaging ultraviolet radiation downward to humans and other life forms at the Earth's surface, great alarm arose in 1985 when a group from the British Antarctic Survey reported decreases in the October ozone amounts in the upper atmosphere above the British research station at Halley Bay, Antarctica.[135] This report generated considerable scientific and political interest, and the ozone decreases were soon confirmed through satellite observations to extend over a broad area, not just the Halley Bay location. Soon the "Antarctic ozone hole" was named, and extensive scientific research was undertaken regarding it, including ground-based studies, dedicated aircraft observations, and satellite studies.[136] (Just as the "ozone layer" is not a layer of pure ozone [chapter 1], the "ozone hole" is not devoid of ozone, instead being a region with depleted ozone amounts.)

By the end of the 1980s it was determined that the manmade addition of chlorofluorocarbons (CFCs) to the stratosphere's abundant chemical mixture was a major culprit in the ozone destruction, as forewarned by Mario Molina and F. Sherwood Rowland for the ozone layer in general.[137] Many chemical reactions take place in the stratosphere in the presence of sunlight, but one sequence in particular gives the gist of why the CFCs can be so destructive to the ozone layer. Specifically, ultraviolet radiation from the Sun can split a CFC molecule to release a chlorine atom (Cl), which then reacts with ozone (O_3) to produce chlorine monoxide (ClO) and a molecule of oxygen (O_2). On encountering an oxygen atom (O), the ClO molecule reacts with it to produce a single Cl atom and an O_2 molecule.[138] This means—and here is the crux of the matter—that the same Cl atom that has already destroyed one ozone molecule is now available to destroy another one. The process can continue through the destruction of untold numbers of ozone molecules: as long as the Cl atom is released into a region with ozone molecules, oxygen atoms, and sunlight, it can break down the ozone molecule by molecule. If, however, these reactions were the complete story, the ozone depletions would probably be comparable throughout the stratospheric ozone layer. The reason for the particularly large depletions over the Antarctic centers on the particularly cold temperatures in the

Antarctic stratosphere, approximately 10°C to 15°C colder than the temperatures in the Arctic stratosphere. These cold temperatures lead to a greater presence in the Antarctic of polar stratospheric clouds, which serve as breeding grounds for some of the chemical reactions leading to the formation of the atomic chlorine (Cl) that goes on to destroy the ozone.[139]

So far, the Antarctic ozone hole is largely a September/October phenomenon. Although the full mixture of relevant chemicals exists in the Antarctic stratosphere in the midst of the polar night, it is not until sunlight returns in the late austral winter and early spring (September and October) that the full array of chemical reactions starts up again and the ozone values decrease. Furthermore, during these months, the atmospheric circulation in the Antarctic stratosphere is dominated by a vortex that keeps the Antarctic stratosphere largely separated from the rest of the atmosphere so that the depleted ozone amounts remain over the Antarctic, forming the Antarctic ozone hole. In the October/November time frame, the vortex begins to break down, allowing substantial mixing of the air in the Antarctic stratosphere with the air in the stratosphere of lower latitudes. Hence, generally by November, the ozone hole is no longer prominent, as stratospheric ozone from lower latitudes has entered the Antarctic region.[140] Further, even in September and October, the ozone hole is not truly a hole, as some ozone remains, even though it is depleted. Technically, the ozone hole is defined as the region south of 40°S with ozone values less than 220 Dobson units,[141] a Dobson unit being a measure of the thickness of ozone in the atmospheric column, with 1 Dobson unit equaling a thickness of 0.001 centimeters. Prior to the development of the ozone hole, October ozone amounts over the Antarctic were typically only about 300 Dobson units, or 0.3 centimeters;[142] hence, even without the depletions, ozone, however important, constitutes an extremely small thickness of the atmosphere.

Fortunately, concerns over the Antarctic ozone depletions led to major international efforts to still the damage. A 1985 Vienna Convention for the Protection of the Ozone Layer had acknowledged potential problems, but, much more significantly, in September 1987 in Montreal, 27 countries signed a formal agreement to limit further production of ozone-depleting chemicals. This Montreal Protocol on Substances That Deplete the Ozone Layer was further strengthened during meetings in London in June 1990 and in Copenhagen in November 1992.[143] The Montreal Protocol and its follow-on agreements, all aimed at reducing emission of substances depleting the ozone layer, have appropriately been heralded as major instances when the international community has acted together to limit human-caused environmental damage. At least partially as a result of these agreements, emissions of CFCs into the atmosphere have been dramatically reduced, with the hope that eventually, sometime within the twenty-first century, the protective ozone layer will recover (see chapter 12 for more on the recovery). Former UN Secretary-General Kofi Annan has been quoted as referring to the Montreal Protocol as "perhaps the single most successful international agreement to date."[144]

Unfortunately, the problem of reduced stratospheric ozone has not yet been fully solved, both because a substantial fraction of the CFCs already in the atmosphere will remain for another several decades and because the CFC emissions have been replaced by other chemical emissions that themselves are not without impacts. Specifically, in many industrial processes, the CFCs have been replaced by hydrochlorofluorocarbons (HCFCs) and hydrofluorocarbons (HFCs). The HCFCs take part in chemical reactions that contribute to depleting the ozone (but not nearly as severely as the CFCs),[145] and both the HCFCs and the HFCs are also greenhouse gases, absorbing infrared radiation.[146] The net result is that reducing the CFC emissions through use of HFCs and HCFCs constituted progress but did not provide a full solution.

CHAPTER 5

The Future: Why Some People Are So Concerned while Others Aren't

Projected future changes on the planet, although not all bad, do include some very undesirable scenarios, generating considerable concern among the majority of scientists working on climate change issues and many people outside the science community. Still, not everyone shares the concerns or believes that the projections are valid.

CHAPTER 4 describes impacts that humans have already had on the environment. This chapter turns to the future, as it is the predictions of what might happen that raise the most concerns among scientists, the general public, and policymakers. Here, the projections of the Intergovernmental Panel on Climate Change (IPCC) serve as a focus, because the IPCC projections are widely used as indicators of the climate conditions that can be expected in the decades ahead under various scenarios of human behavior. Hence, in this chapter's discussion of why so many people are concerned about the current projected changes in global climate, I refer often to the IPCC documents.

CONCERNS HIGHLIGHTED BY THE IPCC AND OTHERS

The IPCC suggests that the present atmospheric CO_2, CH_4, and N_2O concentrations all now exceed any pre-1800 values indicated in the ice core record for the past 650,000 years.[1] In particular, after fluctuating between about 180 parts per million and 300 parts per million (i.e., between 0.018% and 0.030%) during the glacial/interglacial cycles of the past 650,000 years, atmospheric CO_2, faced by relentless inputs from human activities since the start of the industrial revolution, has risen well above 300 parts per million, reaching 386 parts per million in 2008. Similarly, atmospheric CH_4 concentrations, having varied between 320 parts per billion and 790 parts per billion for 650,000 years, have shot up over the past 200 years to reach 1,774 parts per billion by 2007.[2] (Like so much in paleoclimate, the pre-1800 numbers derive from proxy data that are not universally accepted. Harvard scientist Willie Soon, for instance, suspects that atmospheric CO_2 concentrations might have been higher than today's values as recently as A.D. 400 [personal communication, April 2007]. This view, however, is in the minority.)

Also recorded in ice core and other data is the rise in atmospheric sulfate aerosols during most of the industrial era of the past 200 years and the decline in these emissions in the late twentieth century and the early twenty-first century.[3] The recent decline in sulfate aerosols is due at least in part to government regulations and is hugely favorable from a health and pollution perspective, although it is not favorable from the perspective of global warming concerns, as the aerosols help restrict warming by reflecting solar radiation back to space.

The concern expressed most frequently is that human activities—largely through increasing greenhouse gases but also through limiting aerosols—will continue to warm the planet and bring about all the consequences that accompany the warming. This section summarizes the warming projections and some of the key concerns regarding future climate changes and their consequences.

Further Warming

The IPCC includes in its 2001 assessments a range of 1.4°C to 5.8°C for the projected warming from 1990 to the end of the twenty-first century.[4] The 2007 assessments include a range of 1.8°C to 4.0°C warming* for the best estimates from different scenarios of twenty-first-century population growth and energy usage.[5] For those unaccustomed to thinking about global averages and yearly averages, neither of these ranges might seem like much. However, with the average global temperature at the peak of the last ice age, when ice covered almost all of Canada

---

* The warming values for the 2007 IPCC assessments are for the period 2090–2099, i.e., the last decade of the twenty-first century, relative to the period 1980–1999, i.e., the last two decades of the twentieth century. The "best estimates" for the six scenarios tabulated in the IPCC Summary for Policymakers are 1.8°C, 2.4°C, 2.4°C, 2.8°C, 3.4°C, and 4.0°C. Each "best estimate" has associated with it a "likely range" of how much the warming will be. The likely ranges range from 1.1°C to 2.9°C on the low end to 2.4°C to 6.4°C on the high end (IPCC 2007).

and much of northern Eurasia, estimated to have been only 4°C to 7°C colder than now (chapter 2),[6] it is clear that 4°C to 7°C can make a phenomenal difference.

There are many reasons why the projections of significant future warming are a concern. Most directly, warming can certainly be expected to increase the frequency and intensity of heat waves (and decrease the frequency and intensity of severe cold). The seriousness of heat waves is demonstrated somewhere around the world probably every year, with, for instance, over 140 deaths in California blamed on a heat wave in July 2006[7] and, far more prominently, an estimated 35,000 heat-related deaths in Europe blamed on a severe heat wave in July–August 2003, the latter involving additionally agricultural losses estimated at $15 billion.[8] Nicholas Stern, adviser to the UK Government on the Economics of Climate Change and Development and author of the massive *Economics of Climate Change* (commonly referred to as the "Stern Review"), warns that heat waves like the one in Europe in the summer of 2003 will be "commonplace by the middle of the century."[9]

But warming does more than just warm, as it increases melting of ice and snow, changes environmental conditions for the world's ecosystems, and changes atmospheric and ocean circulation patterns, the latter with such potential consequences as increased water shortages in the western United States[10] and a less reliable Indian monsoon, conceivably leading to a return of the devastating droughts and horrendous death tolls from famine that occurred sporadically in India in the nineteenth century.[11] This section highlights a few of the serious potential problems, but there are many others, and for a far more complete presentation, the reader is referred to the IPCC volume *Climate Change 2007: Impacts, Adaptation and Vulnerability*[12] and the hundreds of references therein.

Enhanced Warming through Positive Feedbacks

The effects of the major positive feedbacks in the climate system could be quite substantial in further enhancing global warming. If the Arctic sea ice cover continues to retreat and if global land ice and snow coverage continue to decrease, each of these changes would encourage further warming through the ice-albedo and snow-albedo feedbacks. Similarly, decaying permafrost, brought about by warming, can release methane and carbon dioxide into the atmosphere, enhancing the greenhouse effect and thereby also enhancing the warming. Nicholas Stern indicates that the potential emission of methane and carbon dioxide from permafrost exceeds the total emissions from fossil fuels in the entire period since the beginning of the industrial revolution.[13]

Another huge potential source of methane exists in gas hydrates, trapped under the oceans in the seabed. If ocean warming reaches deep enough, destabilizing sizable portions of this methane and releasing it to the atmosphere could be a major additional factor in further warming.[14] Some of this release is already occurring in the early twenty-first century from the shallow seabed along the northern coast of Siberia, but in view of the depth of most of the hydrates, it is considered unlikely that methane release from hydrates will be a major factor in twenty-first-century climate. Still, because the reservoir of methane hydrates is huge, the long-term potential exists for a considerable climate impact from the hydrates.[15] Recall from chapter 2 that a leading candidate for what caused the prominent warming 55 million years ago is the release of methane from the ocean depths, providing a paleoclimatic analogue of what could happen with a major release of methane. Whether the effect is near term or long term, release of methane into the atmosphere as a result of warming provides a positive feedback toward further warming.

Another positive feedback associated with greenhouse gases concerns Earth's primary greenhouse gas, water vapor. With warming, more evaporation can be expected, sending more water vapor into the atmosphere, enhancing the greenhouse effect and thereby causing further warming.[16]

A very different positive feedback involving water derives from the greater capacity that cold water has than warm water for holding dissolved CO_2. Because of this contrast, as warming occurs, the uptake of CO_2 by the oceans from the atmosphere could decrease, thereby leaving more CO_2 in the atmosphere and providing another enhancing feedback toward further greenhouse warming.

The examples in this section are just a sampling of the many positive feedbacks within the climate system that could further compound continuing climate change. Sophisticated projections for upcoming climate changes factor in these and other known positive feedbacks, but it is quite likely that additional positive feedbacks exist that are not yet recognized and hence are not included in the models no matter how sophisticated the projections are.

Enhanced Warming from Pollution Reductions

In addition to positive feedbacks, another reason why warming might increase at an enhanced rate is that as localities make progress in lessening air pollution, thereby benefiting human and ecosystem health, a downside to this extremely positive development is the probable resultant increased local warming. Some of the aerosols constituting pollution are blocking sunlight from passing through the atmosphere, and as a result a reduction in these aerosols results in more sunlight reaching the surface, facilitating warming as well as solar brightening. The effect was visible in the late twentieth century in central Europe as some of the former Soviet-bloc countries made significant inroads to reducing rampant pollution from heavy industrial activity, pollution that had plagued such countries as Poland and the former East Germany and Czechoslovakia. As the air has cleared, central Europe has warmed at a considerably faster rate than the Earth as a whole. In fact, during the period 1990–2005, central Europe is estimated to have warmed at about three times the globally averaged rate.[17]

Other regions can be expected to compound their regional warming also as they reduce their air pollution. One country with noticeable potential in this arena is China, which by the early twenty-first century was experiencing 5% to 6% less sunshine in the most polluted regions of the south and east than they had in 1980.[18] The heavy pollution kept temperatures artificially subdued although at such great expense, healthwise, that the communities involved can be expected to set a far higher priority on reducing pollution than on avoiding the likely resultant warming.

Meinrat Andreae and colleagues have warned that because the computer models used to predict climate change fail to account fully for the aerosol effects, future global warming could considerably exceed the predictions of the IPCC and others. They calculate that warming by 2100 could, in an unlikely but possible eventuality, be several degrees Celsius higher than the predictions of the IPCC.[19]

Sea Level Rise

When in 1999 the two Kiribati islands mentioned in chapter 4 disappeared into the ocean, this was unfortunate although not a major event for humans, as, after all, the islands were uninhabited. As sea level rise continues, however, populated islands will also be submerged. Among the Pacific islands currently in danger are the remaining 33 islands of Kiribati, inhabited by 103,000 people,[20] and the islands of Tuvalu, inhabited by 10,000 people.[21] In fact, sea level rise could bring to an end the short-lived independence of the nation of Tuvalu, which originally received its independence from Great Britain in 1975. In 2005, the Tuvalu government, convinced that their nine inhabited islands are doomed to submergence by mid-century, signed an agreement with the government of New Zealand that the 10,000 Tuvalu citizens will be allowed to move to New Zealand as the situation worsens. Similarly, in 2005 the government of Papua New Guinea made the decision to move the 2,000 inhabitants of the Carteret Islands, suffering near continual coastal erosion since the 1960s and predicted to be totally submerged by 2015, to Bougainville, 4 hours by boat to the southwest. The migration is ongoing and is being made approximately 10 families at a time.[22]

> The 2007 IPCC and earlier estimates fail to factor in the potential lubricating effects of glacial meltwater from the top of ice sheets reaching the bottom, effects that could significantly speed the flow of ice into the oceans.

Mainstream expectations are that sea level will continue to rise at a rate comparable to what occurred in the twentieth century, estimated at approximately 0.17 meters per hundred years[23] or perhaps increased to two to four times that rate. More specifically, the 2007 IPCC assessment projects a twenty-first-century sea level rise of between 0.18 and 0.59 meters.[24] This would be quite serious, highlighted by the submergence of many more islands and coastal landscapes and communities. But it would be nowhere near as serious as what would happen with much more rapid sea level rise, something that many fear now is a real possibility. Most crucially, the 2007 IPCC and earlier estimates fail to factor in the potential lubricating effects of glacial meltwater from the top of ice sheets reaching the bottom, effects that could significantly speed the flow of ice into the oceans. These effects were revealed in the early twenty-first century, with the startling realization that surface melt can reach the bottom of the ice sheets within minutes rather than taking hundreds of years, doing so as water pours down through narrow chutes in the ice called "moulins."

Moulins had long been known to occur in glaciers. What had not been known was that some of these moulins can extend from the top surface of a major ice sheet clear to the bottom, with water sometimes pouring in torrents* down the chute, reaching the bottom, and from there lubricating the base and helping to speed the outward flow of the ice, something now recognized through observations on the Greenland ice sheet.[25] The new knowledge that the meltwater can reach the base and lubricate the ice sheet's bottom surface, with the possibility of marked acceleration of the outward flow of the ice, considerably changes the picture of how fast sea level might rise. The fact that the models used in the IPCC and other projections failed to include this mechanism means that those projections of sea level rise could

---

* Das et al. (2008) describe a case on July 29, 2006, when an entire lake covering 5.6 square kilometers of the Greenland ice sheet drained 980 meters vertically down one or more moulins and fractures within approximately 1.4 hours, the average drainage rate during this period being greater than the average rate of water flow over Niagara Falls.

be far too low. Furthermore, the models do not realistically incorporate the effects of ice streams and ice quakes, both of which could also speed the flow of ice to the sea. Hence, the sea level projections of the IPCC and others might considerably underestimate the amount of sea level rise likely in the twenty-first century.[26]

However, several studies examining the issue have concluded that the lubrication of the bottom of an ice sheet through the moulin mechanism will not have as severe an impact on ice sheet decay as many initially feared. For instance, one study concluded that the lubrication effect will likely be substantial but not catastrophic.[27] Using an ice sheet model enhanced to incorporate the lubrication mechanism, another study calculated that this mechanism might add anywhere from 0.6 to 6.5 centimeters to the projected 6-centimeter contribution of Greenland ice sheet decay to sea level rise by the end of the twenty-first century in the event of a global warming of about 2.5°C.[28] This is far from the catastrophic image sometimes portrayed of the entire Greenland ice sheet sliding into the ocean. Yet another study examined Antarctica as well as Greenland and concluded that it is not physically tenable that sea level will rise more than 2 meters by 2100 but that a rise of 2 meters is possible and that a rise of 0.8 meters is actually not only possible but also plausible.[29] However reassuring the 2-meter upper limit might be, even a 0.8-meter sea level rise would be serious for low-lying coastal areas, and a 2-meter sea level rise would have severe worldwide impacts. (A sea level rise of 1 meter would wipe out 21% of the land area of Bangladesh, eliminating the country's most productive agricultural land and the homes of 15 million people.[30])

Both the 0.18- to 0.59-meter IPCC sea level rise projections for the twenty-first century[31] and the 2-meter upper limit from the Antarctica/Greenland study mentioned in the previous paragraph are miniscule compared to the potential of what could happen under the most extreme conditions. Although there is almost no possibility that all the ice will enter the oceans within the next few hundred years, the total amount of land-based ice remaining on Earth in the early twenty-first century is sufficient to raise sea level around the world by about 70 meters, and more surprises could be in store as the understanding of ice sheet flow and collapse improves.

Of the total 70 meters of sea level rise potential, approximately 7 meters are locked in the Greenland ice sheet, approximately 5 to 7 meters are in West Antarctica, and the overwhelming majority of the rest is in the much bulkier East Antarctic ice sheet. In line with the traditional thought mentioned in chapter 1 that the massive East Antarctic ice sheet is safely stable, much of the East Antarctic, in marked contrast to portions of the smaller and perhaps unstable[32] West Antarctic and Greenland, appears to have grown rather than decayed in the 1990s and early years of the twenty-first century.[33] However, as of 2007, suggestions that some retreating ice streams in East Antarctica have similar geometries to ice streams in West Antarctica[34] raise the concern that even the East Antarctic ice sheet, with its much greater sea level rise potential, could be susceptible to an earlier collapse than had been thought.

Paleoclimate evidence suggests that in the mid-Pliocene about 3 million years ago, when the Earth is estimated to have been warmer than now by no more than 2°C to 3°C,* sea level was about 25 meters higher, providing a data-based rationale for the thought that if the Earth system warms another 2°C to 3°C—well within the range of IPCC predictions for the twenty-first century[35]—there is a possibility that sea level might rise another 25 meters,[36] although the time scale might be on the order of many hundreds of years if some analyses[37] of the tenable speed of ice sheet decay are correct. In any event, there is ample ice remaining on Earth to create a sea level rise of 25 meters (in fact, enough for about 70 meters), and such a rise would place a great many of today's coastal and near-coastal cities underwater.

Also from paleoclimate evidence, the fact that for the 400-year period from 14,500 to 14,100 years ago the rate of sea level rise averaged about 5 meters per century (chapter 2) is sobering. On the one hand, it shows that human impacts can still be swamped by natural variability; on the other hand, it also shows the troublesome fact that, regardless of whether the cause is human induced or natural, we could find ourselves subject to much larger and more rapid sea level changes than have occurred in recent centuries, which is a possibility that could cause quite severe disruptions to coastal populations.

Recognizing the failure of the computer models to incorporate well all the factors affecting sea level, including potentially important processes within the ice sheets, Stefan Rahmstorf of the Potsdam Institute for Climate Impact Research in Potsdam, Germany, decided alternatively to project sea level rise for the twenty-first century based on the relationship that the rate of sea level rise since the start of the industrial revolution has been roughly proportional to the rate of warming. Assuming that the relationship will continue and accepting the 1.4°C to 5.8°C range of projected global warming from 1990 to 2100 given in the IPCC's 2001 report, Rahmstorf projects that sea level will rise 0.5 to 1.4 meters above the 1990 level by

---

* Hansen et al. (2007) suggest that the global temperature in the mid-Pliocene was 2°C to 3°C warmer than today; Jansen et al. (2007) estimate that it was 2°C to 3°C warmer than preindustrial temperatures, that is, than in about 1800.

2100.[38] This is markedly higher than the IPCC's 0.18 to 0.59 meters, but like the IPCC's projection, it does not incorporate the possibility of a rapid ice sheet decay and hence is far from as high as the worst-scenario possibilities.

For balance, it is important to mention that not all human activities affecting sea level lead to its rise. In particular, humans siphoning off water for human uses and storage in reservoirs halts water from flowing into the ocean and thereby lessens sea level rise. Some estimates suggest that through impoundment of water in reservoirs, humans reduced the rate of sea level rise on average by 0.55 millimeters per year from the mid-twentieth to the early twenty-first century,[39] equating numerically to a sea level fall of 5.5 centimeters (0.055 meters) per century.

Reduced Availability of Freshwater

Although the sea level rise impacts have received most of the attention regarding land ice retreats, another extremely important eventual consequence of continued glacial loss is the loss of a formerly reliable freshwater supply for communities that depend on glacial meltwater. This directly affects a large number of communities and people, as more than a sixth of the world's population depends on glaciers and seasonal snow for freshwater.[40]

Even regions having no dependence on glaciers and snow could face increased freshwater difficulties with increased warming. If indeed warming continues, several pressures will likely increase the demand for freshwater, as more water is likely to be needed for cooling, bathing, drinking, watering gardens, and irrigating crops, even in the absence of expected population increases, which also would increase water demand. As humans continue to use freshwater, groundwater supplies in some regions are likely to decrease, adding further to the pressure for additional water sources. Other complications include the increased likelihood of saltwater intrusions (in the face of sea level rise) into near-coastal water supplies, such as the lower reaches of rivers.[41]

Ocean Acidification and Marine Biology

Estimates suggest that if current trends of CO_2 emissions from human activities continue, the resulting continued ocean acidification could decrease the average pH of the ocean surface waters by 0.5 units below the preindustrial level by 2100,[42] adding 0.4 pH units of decrease to the 0.1-pH-unit decrease that has already occurred (chapter 4). Not only is it thought that this pH level would be lower than any global level experienced during the past several hundred thousand years,[43] but the rate of change that it would imply is perhaps higher than has occurred in the past 20 million years.[44] The consequences to at least some marine ecosystems could be severe, especially to the species that produce calcified shells and plates of $CaCO_3$, such as foraminifera, mollusks, crustaceans, corals, echinoderms, and some phytoplankton. As explained in chapter 4, with lower pH, calcification becomes more difficult, and some of the $CaCO_3$ shells and plates that exist could dissolve. With a doubling of atmospheric CO_2 from the preindustrial 280 parts per million to 560 parts per million, calcification could decrease by anywhere from 5% to 25%.[45]

Although tropical corals and their prominent declines in the late twentieth century have garnered the most attention so far, cold-water corals may be even more sensitive to expected climate changes in the twenty-first century than warm-water corals. In fact, model calculations indicate that by 2100, acidification may make the Southern Ocean uninhabitable for corals.[46] Still, it is changes in the better-known, lower-latitude corals that have the more measurable economic impacts on humans. Published estimates suggest that climate-change–induced damage just to the Great Barrier Reef off the northeastern coast of Australia could cost the Australian tourist industry up to $8 billion in 19 years in the early twenty-first century[47] and that damage to coral reefs globally could cost global economies billions of dollars per year by 2100 because of the effects that such damage could have on the use of coral reefs for subsistence food gathering, tourism, and fisheries.[48]

Another concern regarding ocean life is that the increased acidification of body tissues and fluids in various fish and other marine animals can affect the ability of their blood to carry oxygen through their bodies and can lower respiratory activity, perhaps stunting growth and reproduction. Squid are viewed as potentially particularly sensitive because of the high energy demands for their jet-propulsive motions, but other marine species could be affected as well.[49] In one experimental study, the respiration rate in steelhead trout sperm was reduced 40% following a pH reduction from 8.5 to 7.5.[50]

Species Diversity and Extinctions

In their study concluding that warming has been a key factor in recent amphibian extinctions in the American tropics (referred to in chapter 4), J. Alan Pounds of Costa Rica's Golden Toad Laboratory for Conservation and 13 colleagues warn that we can expect the disappearance of many more species as temperatures rise in other highland regions to levels that are optimal for the growth of such fungi as the *Batrachochytrium* implicated in many of the amphibian extinctions.[51] Species living on mountains also have the additional complication that as conditions warm and they shift their habitat up the mountain, once they get to the top of the mountain, they have no further place to go, and so, unless the mountain is part of a chain that extends poleward, they arrive at considerable added risk for elimination. Basically, they will need to adjust to the changed climatic conditions or likely will die out. Elsewhere, other species have potential migration paths blocked because of the presence of cities or other human creations.

In a study examining the risk of extinctions from climate change for sample regions covering about 20% of the Earth's terrestrial surface, Chris Thomas of the Centre for Biodiversity and Conservation at the University of Leeds and numerous colleagues predict that, based on climate-warming scenarios for 2050 and projections of the distributions of 1,103 plant and animal species, 15% to 37% of the species in their sample would by 2050 be "committed to extinction."[52] There are many assumptions involved, including the assumption that species will not adjust at all to new climatic conditions, instead requiring the same range of temperature, precipitation, and other climate factors as exist in their current distributions. Still, the thought of 15% to 37% of species becoming "committed to extinction" is disturbing. Thomas and his colleagues are careful to note that "committed to extinction" by 2050 does not mean actually extinct by 2050. They make no prediction as to how soon the extinctions would occur, saying only that "decades might elapse."[53] Their assumption about species not adjusting to new climate conditions is surely overly stringent as a blanket assumption, as species have always had to adjust to changing climatic conditions in order to survive in the long term. On the other hand, the rapidity of possible climate changes and the many added complications that humans present (through direct killing, harvesting, destroying habitats, and so on) could make the actual extinction rates even worse than the predictions.

Beyond issues of species diversity and whether humans have a responsibility to be good stewards of the planet and its life, extinctions have significant further consequences. Loss in species diversity will be accompanied by a loss of whatever the lost species provided to the rest of the ecosystem, including to humans. When we lose species, we also lose both known and unknown benefits, such as any potential the lost species might have had for providing extracts that can help cure serious human diseases. There are numerous examples of relatively inconspicuous plants found to be hugely beneficial to humans. For instance, Madagascar's rosy periwinkle (*Catharanthus roseus*) was relatively inconspicuous before it rose to fame following the finding that two alkaloids it produces, vinblastine and vincristine, can cure at least some people suffering from Hodgkin's disease and acute lymphocytic leukemia, two of the deadliest cancers. Another species of Madagascar periwinkle, *Catharanthus coriaceus*, is approaching extinction as its habitat is being destroyed for agriculture.[54] With it, other potential cures of the same or other diseases could be extinguished. Edward O. Wilson and others wisely encourage greater "chemical prospecting," whereby wild species are examined for the potential that their chemicals might have in new medicines and other useful products.[55] The aim in such chemical prospecting is twofold: first, to encourage preservation of the wild species, and, second, to obtain the valuable new products.

Spread of Diseases

A biologically based concern in the opposite direction from the concern over extinctions is the concern that changing climatic conditions might expand the range of species and diseases that we would prefer to avoid, like the observed poleward expansion of the red fire ant, destroying native flora and fauna as they advance.[56] Warmer temperatures, increased rainfall, and the absence of subzero temperatures are all factors that can lead to extending the ranges of insects, rodents, and other organisms that carry diseases and otherwise cause problems for humans and other species. Among the most notable examples are the expansions to higher latitudes of diseases that often are associated largely with the tropics, such as malaria (see qualifiers later in this chapter), hookworm, schistosomiasis, dengue fever, leprosy, guinea worm, and West Nile disease. Other diseases likely to spread with warmer temperatures are Rocky Mountain spotted fever, Q fever, and Lyme disease, the latter already notably spreading in the United States and Europe,[57] and tick-borne encephalitis, already increasing substantially in Sweden since the mid-1980s.[58]

Health is also affected by warming through the influence that warming has on sea level rise and the consequent likelihood of increased saltwater intrusion into freshwater areas and increased saltwater flooding. More severe flooding raises not just the risk of flood damage but also the risk of such waterborne diseases as cholera, typhoid, and dysentery and such mosquito-borne diseases as yellow fever and malaria. Malaria, which is transmitted by mosquitoes, already results in a million deaths each year and has the potential of resulting in many more deaths, as higher temperatures in many regions will be more conducive to mosquito outbreaks.[59] Other factors are involved as well, however, as described later in this chapter.

Abrupt Climate Change and a Shutdown
of the Global Ocean Conveyor

When in the late twentieth century the standard climate change paradigm included the assumption that changes in the Earth's climate occur only very slowly (chapter 3), there was a comfortable sense that although the coming changes might be undesirable, at least they would develop slowly, giving humans a chance to adjust slowly as well. This comfort zone has vanished with the determination from Greenland ice cores and elsewhere that climate, at least regionally, not only can change abruptly but has frequently done so (chapter 3). In fact, one conclusion from the new results is that the fairly stable climate the Earth has experienced for the past several thousand years might be unusual. Another possibility is that periods of relative stability might be common enough; for instance, there might be long, relatively stable glacial states and long, relatively stable interglacial states, with the transitions between the two states fraught with multiple abrupt jumps. In any event, the evidence is now strong that abrupt shifts have occurred on many occasions in the past, prior to the past several thousand years, and hence could certainly do so in the future as well, whether triggered naturally or by human activities. This is cause for concern, as despite all our technological prowess, adjusting to abrupt climate change would probably be considerably more difficult for us now than it was many thousands of years ago, when the human population was much smaller, there was far less infrastructure and personal property to deal with, and the Earth had more unoccupied, unclaimed land to which people could migrate. If climate conditions worsened in one region in the distant past, bands of early humans could move to another region considerably more easily than communities could move today. They might have had to do it on foot, but even on foot, it was easier than moving a whole community under today's circumstances.

One possibility frequently mentioned as something that could trigger an abrupt climate change is a shutdown of the ocean conveyor circulation described in chapter 1. A shutdown of the ocean conveyor would cause major disruptions to climate systems throughout the globe. It could so alter warm-water flow into the polar regions, especially in the North Atlantic, that there could suddenly be a major cooling in Europe and the Arctic. For those most concerned about further warming, this could be seen as a favorable possibility, bringing cooling at least regionally, although adjusting to major climate changes would probably be extremely difficult regardless of their direction (warming or cooling). Some people have even suggested that a slowdown of the ocean conveyor could lead in the near future to an emergent ice age.

Recall from chapter 3 that massive iceberg discharges into the North Atlantic from the Laurentide ice sheet are thought to have reduced or even halted altogether North Atlantic deep convection and brought cooling to Europe and the North Atlantic vicinity during the Heinrich events, doing so by capping the regional ocean with a layer of freshwater that was less dense than the underlying salt water and therefore did not sink. Alternative possibilities of capping deep-convection regions with freshwater would be through a decay of the Greenland ice sheet, a massive inpouring of freshwater from rivers, or a significant regional increase in precipitation (which adds freshwater to the surface) without a corresponding increase in evaporation in that same region. Indeed, the latter mechanism, with a precipitation/evaporation imbalance capping the surface waters with freshwater and thereby greatly restricting regional deep convection, is exactly what was found to happen in a modeling study simulating future conditions in which atmospheric CO_2 is increased by 1% per year for 140 years, reaching four times its starting value, then is held constant thereafter.[60] In the model simulation, the ocean conveyor (or thermohaline circulation) almost stops altogether in the North Atlantic and also becomes weaker and shallower elsewhere. In other words, a slowdown and perhaps even a shutdown of the ocean conveyor could conceivably at some point be triggered by human activities through our trace gas emissions to the atmosphere.

For a brief period in the first decade of the twenty-first century, it looked as though the ocean conveyor might indeed be slowing.[61] Specifically, a report published in 2005 indicated a significant recent slowing of overturning circulation in the North Atlantic, raising concerns that maybe the feared slowdown of the ocean conveyor was under way.[62] However, the following year, new analyses indicated that the data examined in the 2005 study were very incomplete and that conceivably no slowdown has occurred.[63] It is now thought unlikely that a shutdown of the ocean conveyor will occur any time soon. Further, even if a shutdown or slowdown does occur, it will probably not have the severe impact that has often been associated with a conveyor shutdown, that is, triggering a new ice age by halting the northward heat transport in the Gulf Stream.[64] The 2007 IPCC assessment report states clearly that no realistic model simulation supports the speculation that a weakened ocean conveyor would cause a future ice age.[65] Still, although the 2005 study might have been a false alarm and despite the cautions expressed by the IPCC, concern remains that a slowdown with serious propagating impacts could conceivably occur. Although it might not trigger an ice age, a slowdown of the conveyor would have propagating impacts around the globe by changing heat, salt, and other transports within the world's oceans. Exactly what the impacts would be and which regions would be affected the most remain uncertain.

More Frequent or More Intense Hurricanes

In theory, since hurricanes are fueled from evaporation from warm waters and seem to form only over waters with temperatures of at least 26°C (78°F), an expansion of the area of waters this warm (and a lengthening of the period they remain this warm) should be favorable for more hurricanes and more intense hurricanes. This is a topic of tremendous interest, as more intense and/or more frequent hurricanes could be disastrous for communities in their direct path. Yet it is also a topic of considerable controversy, as the historical evidence connecting warmer conditions to more frequent hurricanes is not entirely convincing. This could simply be an issue of the incompleteness of the data, or it could be that the tendency for higher temperatures to encourage hurricane growth is outweighed in full or in part by the tendency for higher temperatures to cause other changes that discourage hurricane growth, one such possibility being changes in upper-atmosphere winds.

The North Atlantic hurricane season was fairly active in the 1940s, 1950s, and early 1960s, followed by a period of reduced hurricane activity preceding an upsurge of activity in the 1990s.[66] Without a much longer and more complete data set, it is impossible to be sure whether this sequence ties in well with global warming. Perhaps periods of more active and less active hurricane seasons continually alternate, almost regardless of warming or cooling. On the other hand, the 1930s and 1940s are noted for having warm atmospheric temperatures, and perhaps with the temperature lags between the atmosphere and ocean, the 1940s–1960s and 1990s might have had relatively warm ocean temperatures in the relevant low-latitude regions of the North Atlantic, accounting for the high hurricane activity. A good set of ocean temperature data throughout the twentieth century would help test the contrasting possibilities, if only such a data set existed.

In 1999, Eddie Smith found a slight increase in the number of North Atlantic hurricanes reaching the East Coast of the United States over the almost century-long period 1900–1998 but not an increase in their intensity.[67] However, by 2005, more support seemed to exist for the concept that hurricanes might become more intense with warming than that they might become more frequent.[68] In 2008 one study even suggested the possibility of a global reduction in the frequency of hurricanes in conjunction with global warming,[69] and another simulated a reduction more specifically in Atlantic hurricane frequency in the twenty-first century under warming conditions.[70]

Peter Webster led a team that examined the 1970–2005 period, which has the advantage of a far better data record, bolstered considerably by satellite observations, than exists for previous years but at the same time has the disadvantage of not including the previous decades with known more active hurricane seasons than the 1970s. The Webster-led team found, for the 1970–2005 period, an increase in hurricane intensity (more hurricanes of the most severe categories, 4 and 5) in the

North Pacific Ocean, southwestern Pacific Ocean, Indian Ocean, and, to a lesser extent, the North Atlantic Ocean but did not find, overall, an increase in the number of hurricanes despite the increasing sea surface temperatures.[71] Kerry Emanuel similarly found an increase in destructive potential of hurricanes since the mid-1970s, using an index of potential destructiveness based on the total dissipation of power over the lifetime of the hurricane.[72] Both the increase in destructiveness and the related increase in the number of hurricanes reaching the category 4 and 5 levels are tied to sea surface temperature increases,[73] in line with expectations. Still, evidence also exists that there was more storminess in the northeastern Caribbean in the second half of the Little Ice Age than in the early twenty-first century, despite the cooler temperatures, and that intervals of more hurricanes perhaps correspond better with periods of fewer El Niños than with warmth.[74] In their book *Hurricanes: Causes, Effects, and the Future*, Stephen Leatherman and Jack Williams also mention the hurricane/El Niño connection, specifically pointing out that upper-atmosphere winds blowing westward across the tropical Atlantic and Caribbean tend to increase during El Niños and that these winds can shear apart developing storms. The effect of shearing winds could also alter the hurricane/global warming connection, as these winds might well be enhanced under warmer conditions.[75]

Much uncertainty remains regarding the hurricane/global warming connection, and a 2005 review of the topic led by Roger Pielke Jr. concludes by saying that claimed linkages between the two are premature and that conclusive linkages are unlikely anytime soon.[76] However, despite the lack of conclusiveness, the sense that hurricanes might become more intense or more frequent with continued global warming can generate real fears.

Historically, we could be quite unusual in thinking of warming in a negative rather than a positive light. In fact, warmth has traditionally been viewed rather favorably, with periods of warmth in Earth history being given names like the "Medieval Climate Optimum" or the "mid-Pliocene Climate Optimum."

THE POSSIBILITY THAT THE PROJECTIONS ARE OVERLY DIRE

The previous section ("Concerns Highlighted by the IPCC and Others") highlights unfavorable changes feared as possible consequences of global warming. All these provide part of the rationale for why many people are so concerned and why some are considering massive schemes to counteract the projected climate changes. It is important to recognize, however, that the predicted substantially enhanced global warming might not materialize and that, even if it does, not all the resulting changes would be unfavorable. In fact, until the late twentieth century, warmth has traditionally been viewed rather favorably, with periods of warmth in Earth history being given positive names like the "Medieval Optimum" or the "Medieval Climate Optimum" for the warm period centered in the tenth through the twelfth centuries A.D.* and the "mid-Pliocene Climate Optimum" for the much earlier warm period approximately 3.3 million to 3.0 million years ago (chapter 2). We in our modern world could be quite unusual in thinking of warmth in a negative rather than positive light.

Projected Favorable Consequences of Warming

Among the most important advantages of a warming climate would be fewer deaths from cold. In fact, some people[77] feel confident that the reduction in cold-related deaths from the predicted warming would far outweigh the increase in heat-related deaths. Certainly, the estimated 35,000 heat-related deaths during the July–August 2003 heat wave in Europe are heartbreaking, and more intense heat waves could be expected in a warmer climate; however, according to Bjorn Lomborg, deaths from cold still far outnumber deaths from heat. For Europe specifically, as of the early twenty-first century, on average about 200,000 people are estimated to die each year from excessive heat, while about 1,500,000 are estimated to die each year from excessive cold.[78] Estimates for Great Britain suggest that a warming of 2°C (3.6°F) would increase the number of heat-related deaths by approximately 2,000 people, clearly unfortunate, but at the same time would decrease the number of cold-related deaths by approximately 20,000 people. This would produce a net favorable outcome of 18,000 fewer temperature-related deaths. Globally temperature-related deaths should substantially decrease with the projected global warming at least through the twenty-first century.[79] With warming continuing beyond 2100, the numbers would continue to shift toward more heat-related and fewer cold-related deaths so that eventually, if the temperature projections are correct, continued warming would result in more, not fewer, temperature-related deaths, but for the twenty-first century alone, temperature-related death projections do not support a picture of great horrors resulting from projected climate changes.

---

* As with all major climatic periods, the Medieval Warm Period (or "Medieval Optimum") had "winners" and "losers," had some regions that were not unusually warm, and had fluctuations in all regions. The warmth of this period, where it occurred, was, overall, beneficial to the people of Europe and the Arctic but not beneficial in many other areas. In fact, some regions suffered severe enough drought that it contributed to permanent declines in what had been thriving civilizations, including the Ancestral Pueblo in the American Southwest, the Maya in Central America, and the Angkorian Empire in Southeast Asia (Fagan 2008).

Warming should also reduce other cold-related problems, such as traffic complications from ice and snow, and floods resulting from ice jams on rivers. The problem of flooding from ice jams was severe in Europe during the Little Ice Age;[80] was demonstrated vividly in central Vermont in 1992 when an ice dam on the Winooski River suddenly broke, rapidly flooding all of downtown Montpelier, the state's capital city;[81] and in recent years has repeatedly caused major damage along the Lena River in Russia.[82]

Additional positive expectations due to warming include the following impacts on vegetation and other life forms: in some locations, warming will lengthen the growing season; in some regions, warming will likely extend the area of agricultural productivity poleward; and in some regions, particularly the Earth's coldest regions, species diversity is likely to increase. Regarding the latter point, the well-respected, mainstream Arctic Climate Impact Assessment projects increased species diversity and productivity in the Arctic as warming continues and more species expand northward into the polar region.[83]

Negative Feedbacks in the Climate System

Although much attention is given to positive (or self-enhancing) feedbacks, with the potential of significantly speeding up the continued warming of the planet, the negative (or dampening) feedbacks operating in the climate system are extremely important as well, and they help stabilize rather than destabilize the system.

One important negative feedback in the climate system involves the chemical weathering of rocks, during which rocks are eroded by reactions with carbonic acid, the latter formed from carbon dioxide and water. Increasing atmospheric CO_2 encourages increased chemical weathering because of the greater availability of CO_2 for formation of carbonic acid, but the process of chemical weathering consumes the CO_2, thereby providing a dampening effect on atmospheric CO_2 increases.[84]

On long time scales, weathering can also play a role in additional negative feedbacks. Specifically, consider a cooling climate in which the cooling leads to a significant increase in the land area covered by ice. Because of the ice, fewer rocks will be exposed, and hence less CO_2 will be removed from the atmosphere for chemical weathering, allowing the CO_2 to build up in the atmosphere, increasing the greenhouse effect and hence reversing or at least dampening the cooling.[85] This is another example of a negative feedback, in which the initial change (in this case a cooling) is dampened or reversed rather than enhanced.

In the case of both positive and negative feedbacks, some suggested feedbacks are controversial. (There really is quite a bit of uncertainty and controversy in our current understanding of the Earth system, a point made throughout this book and central to why we should be cautious about massive geoengineering projects.) One such controversial feedback is a negative feedback proposed in 2001 by Richard Lindzen of the Massachusetts Institute of Technology (MIT) and colleagues Ming-Dah Chou and Arthur Hou of NASA's Goddard Space Flight Center. This suggested feedback centers on what Lindzen and his colleagues have termed an "adaptive infrared iris," in analogy with the iris of a human eye and its beneficial opening and closing in response to changes in light levels. Essentially, they suggest that the cirrus cloud cover in the tropics might similarly open and close in a way that helps stabilize the Earth system, doing so as follows: perhaps when the surface temperature increases, the cirrus cloud cover decreases, thereby increasing the Earth's outgoing infrared radiation and helping the system to cool back down, and when the surface temperature decreases, the cirrus cloud cover could increase, decreasing the amount of radiation leaving the Earth system and thereby helping the system to warm back up.[86] They show supporting observational evidence and suggest that this negative feedback could cancel positive feedbacks within the system.[87] However, others have countered with evidence from 8 months of data from the Tropical Rainfall Measuring Mission's Clouds and the Earth's Radiant Energy System instrument that suggest that the cirrus clouds are contributing a slight positive feedback rather than the hypothesized negative feedback.[88]

Regardless of whether an eventual larger database confirms a negative or a positive feedback in the case of the proposed "infrared iris," the combination of the various negative feedbacks within the Earth system could conceivably be sufficient to keep the system from the out-of-control warming that is feared. This is certainly the view of some,[89] although it is not the majority viewpoint.

THE ISSUE OF THE SKEPTICS

As reflected in the IPCC reports, a strong consensus exists in the scientific community that 1) the globe as a whole has experienced warming since the start of the industrial revolution, 2) humans are at least partly responsible for the warming, and 3) we should do something (exactly what is wide open to discussion) to limit further human-induced warming because of the many potential negative consequences of the warming. Still, of these three points, the only one with close to unanimous agreement is the first one.

People disagreeing with any of the three points enumerated in the previous paragraph, especially either of the first two, tend to be labeled "skeptics." It is an unfortunate labeling, as through it the term "skeptic" has gained a very negative connotation for many people, even though all scientists should be at least somewhat skeptical (not necessarily on this issue) in the general course of their work. Nonetheless, the word is widely used in the context of people outside the global warming mainstream, and that usage will be adhered to here. Some of the best known of the skeptics in the United States are Richard Lindzen of MIT (mentioned in the previous section for his proposed adaptive infrared iris), Patrick Michaels of the University of Virginia, and, especially after the March 2008 publication of his book *Climate Confusion: How Global Warming Hysteria Leads to Bad Science, Pandering Politicians and Misguided Policies That Hurt the Poor*,[90] Roy Spencer of the University of Alabama in Huntsville. In Europe, one of the best known of the skeptics is Bjorn Lomborg of the Copenhagen Business School. Just as there exists a mixture of personalities and attitudes among the many individuals loosely grouped together in the "consensus," so there is also a mixture of personalities and attitudes among the skeptics, with some of the skeptics even agreeing with almost all the main points of the IPCC reports.

Regarding whether there has been warming since the start of the industrial revolution, I am not aware that any of the well-known skeptics has a strong disagreement with that point. Spencer states, "Every scientist-skeptic I know believes that global warming is real."[91] Some skepticism could legitimately exist, as the temperature records are far from complete, with nowhere close to global data coverage being feasible prior to the satellite record starting in the second half of the twentieth century. Still, there are enough records from a variety of sources that the fact of an overall warming over the past 200 years is not in serious dispute.

However, other issues are in dispute, including whether the warming that has occurred over the past two centuries and is projected to occur in the twenty-first century is necessarily bad. The skeptics often point out advantages of warming, including some of the advantages mentioned in the previous section. Spencer explicitly states that "global warming is unlikely to be a serious threat" and that "our present period of warmth might be more beneficial than harmful."[92]

Another issue in dispute is whether humans are responsible for the warming in the past two centuries and hence whether continued emissions from human activities will lead to further warming. We know definitively that over the past two centuries, we have poured considerable volumes of greenhouse gases into the atmosphere and that we are continuing to do so (see chapter 4), but we also know that the atmosphere is large. The consensus among knowledgeable scientists in this field is that our additions do matter, not only to the amount of greenhouse gases in the atmosphere but also to the temperature. For instance, the Technical Summary of the IPCC's *Climate Change 2007: The Physical Science Basis* states that it is "*very likely* that anthropogenic greenhouse gas increases caused most of the observed increase in global average temperatures since the mid-20th century."[93] The consensus is strong but not unanimous. Spencer doubts that human activities have made much of a difference, emphasizing that our CO_2 emissions amount on average to only one additional molecule of CO_2 for every 100,000 molecules of the atmosphere approximately every 5 years, hardly a statistic that sounds on the surface to be alarming.[94] However, all the trace gases exist only in trace amounts in the atmosphere, yet they do make a difference. Even Spencer agrees that "the extra carbon dioxide mankind is emitting is causing a slight enhancement of the Earth's natural greenhouse effect"[95]; the mainstream considers the enhancement more than "slight." Further, whether in the atmosphere or elsewhere, certain compounds can make a huge difference even in what seem to be miniscule amounts, as dramatically demonstrated, for instance, when a person dies from a miniscule amount of a very strong poison.

More broadly, Spencer feels that the Earth system is not as fragile or in such a delicate balance as some of the scientists in the mainstream suggest. Instead, according to Spencer, the system is quite resilient, with precipitation in particular playing a considerable role in the Earth system's beneficial self-regulation, as precipitation conveniently adjusts how much water vapor—Earth's major greenhouse gas—remains in the atmosphere.[96]

Spencer further believes that some of the policies proposed by the mainstream will be more damaging than beneficial both to the environment and to humans. He argues that the proposed policies "will have no measurable effect on future global temperatures," will hurt the economy, and could "delay the development of real solutions."[97] He feels that the economic recession possibly resulting from forcing industry to reduce CO_2 emissions could substantially delay the development of the technological advances that will lead to long-term solutions.[98] All of these are important considerations, and the concerns are shared in one way or another by others of the so-called skeptics.

In contrast to Spencer, Bjorn Lomborg accepts in full not only that global warming is real but also that humans have contributed significantly to it through their CO_2 emissions. Further, he even agrees with the consensus that the warming, if continued, will have serious impacts by the end of the twenty-first century. Where Lomborg disagrees with the consensus regards what should be done. He examines the issue from an economic standpoint, analyzing costs and benefits, and concludes that the mainstream suggested actions are not the best solutions.[99]

A basic tenet of Lomborg's argument is that, in the near term, many of the concerns feared from global warming can be much more cheaply and effectively alleviated by means other than reducing CO_2 emissions. For instance, damage from flooding would be more effectively reduced by appropriate investments in coastal protection (barriers, dikes, and levees), improved flood forecasts, and better warning systems; damage from hurricanes would be more effectively reduced by improving and enforcing building codes and not subsidizing insurance that makes living in high-risk areas more affordable; keeping cities cooler would be more effectively accomplished by planting trees, adding water features, and, to reflect away solar radiation, painting buildings and tarmac white; polar bears would be more effectively protected by instituting and enforcing stricter hunting regulations than those agreed to in 1973 with the signing of the International Agreement on the Conservation of Polar Bears and Their Habitat;[100] and malaria would be more effectively kept in check by insecticide-treated mosquito nets. Regarding the latter, Lomborg makes the highly relevant point that malaria was endemic in 36 states of the United States in the early twentieth century and was eliminated not by a change in climate but by spraying millions of American homes between 1947 and 1949.* He feels that insecticide-treated mosquito nets would have a much greater chance of diminishing malaria in the countries where it remains than climate change would; in other words, the malaria problem at this point is more wealth related than climate related.

Lomborg recognizes that for those problems that are climate related, tackling the symptoms one by one will not ultimately solve the problem, so he advocates

---

* Malaria in fact was the central focus of the U.S. Communicable Disease Center, later renamed the Centers for Disease Control and Prevention (CDC), in the first years after its founding in Atlanta, Georgia, in July 1946. This remained the case until in 1949 malaria was no longer a significant health problem in the United States (from the CDC website at http://www.cdc.gov/about/history.htm, December 28, 2007). Malaria has not returned as a wide-scale problem in the United States, at least through the first decade of the twenty-first century, despite global warming. Eradication of the disease in the United States involved many factors, including spraying more than 4.6 million homes with the pesticide dichloro-diphenyl-trichloroethane (DDT) (from http://www.cdc.gov/malaria/history), mosquito-control programs, and improved public health and nutrition (Lomborg 2007). In Africa in the early twenty-first century, progress is being made in combating malaria by using specially treated bed nets with insecticides bonded onto the fibers. Distributing millions of such nets for free has been one of many major activities of the Carter Center, founded in 1982 by former U.S. President Jimmy Carter and his wife Rosalynn Carter, in its efforts to combat and eradicate major preventable diseases (for more, see the Carter Center website at http://www.cartercenter.org).

substantial investment in research and development of alternative energy technologies that do not involve carbon emissions. He agrees with the mainstream that in the long run, it is necessary to reduce carbon emissions, but he feels that the most effective way of doing that is not to make the reductions now while doing so is still quite expensive but instead to place a heavy emphasis on research and development aimed at cutting the costs of the emission reductions later.[101]

Despite the many relevant points made by Spencer, Lomborg, and others of the skeptics, the consensus view on global warming is strong and is quite likely correct in its broad outline. This, however, is not certain. The models on which the consensus is based are imperfect (see chapter 10), and there are numerous pressures making the consensus appear to be stronger than it might actually be (see chapter 11). Furthermore, the uncertainties about future climate change are substantial enough that although the changes in the coming decades could be much less than the mainstream projections, as some of the skeptics believe, it is also conceivable that they could, alternatively, be much greater than the mainstream projections. In fact, Australian scientist Barrie Pittock has listed 10 reasons why climate changes might be greater than projected, starting with the possibility that climate sensitivity may be greater than estimated and including also the possibility that some positive feedbacks might accelerate changes faster than anticipated and that meltwater penetration to the bottom of the Greenland and Antarctic ice sheets could speed ice sheet disintegration.[102] A quite crucial point here is that we really are not sure.

Given the uncertainties and the major damage that we could conceivably be doing to the Earth system, I personally feel that we should be trying hard to limit our further damage, for instance by limiting (on a voluntary basis) further population growth, limiting our use of fossil fuels, reducing our materialistic excessive consumption, reusing items when feasible, and recycling items when reuse is not feasible. Chapter 12 includes more suggestions along those lines, but there are plenty of other resources available on how to reduce our environmental impacts, and hence that is not a central topic in this book. More central goals of the book are, first, to describe climate change in the Earth system and explain why there is so much disagreement on climate change issues and, second, to argue convincingly that despite whatever damage we might be doing to the climate system, we should be extremely cautious about undertaking large-scale geoengineering schemes to "fix" the system, as the damage we cause through the "fixes" could be far worse than the damage that we are already causing. So far, the book has concentrated on the first of those goals; the next three chapters concentrate on the second. All the proposed geoengineering schemes appear to be offered with the best of intentions, based on the expectations from the best models that we have so far. However, good intentions do not necessarily lead to good results, so before describing proposed geoengineering schemes (chapter 7), I turn to the topic of good intentions and how badly they can go awry.

PART III

Good Intentions and Geoengineering

CHAPTER 6

Good Intentions Gone Awry

Good intentions are all too frequently stymied by damaging unintended consequences.

THROUGH cooperation, humans for thousands of years have successfully carried out schemes that would be impossible by individuals alone or in small groups, from constructing Stonehenge and the pyramids and the Great Wall of China to constructing modern skyscrapers and interstate highways and telecommunication networks. Cooperation was essential for early human hunters to obtain food for themselves and their families, and the cooperative implementation of very "unnatural" schemes is a large part of why our species has succeeded so phenomenally well and why we have attained our current level of dominance. Nonetheless, when contemplating expanding our cooperative prowess to massive geoengineering projects, many dangerously within our reach, it is critically important that we be cautious. History is replete with instances of well-intended plans that go badly awry. Just as this happens routinely on small interpersonal scales (between a parent and child, a husband and wife, a student and teacher, or among siblings, colleagues, or friends), it also happens on much larger scales. Before turning to the far-reaching schemes being offered to geoengineer the Earth's climate, this chapter provides a few illustrative cautionary examples where purposeful activities have led to significant unintended and unwanted consequences.

THE 1665 LONDON PLAGUE

I begin with a specific historical instance, that of the Great London Plague of 1665. The London Plague started quietly enough, with one death in December 1664 and scattered deaths in the first 3 months of 1665. But by the end of those few months, the trend was clear, and after nine deaths in the first week of May 1665, King Charles II on May 12 appointed a committee to examine how best to halt the further spread of the infection. Houses with victims—and some houses without victims but suspected of being infected—were quickly quarantined, in many cases with all the residents prevented from leaving the house. Among those not quarantined, many with the means to do so exited the city altogether, including King Charles as of July 7, and, much more famously in the long run, the then young scientist Isaac Newton. Still, the death tolls rose, reaching 2,000 per week by late July and peaking at 7,165 plague deaths the week of September 12, 1665. Through 1665 as a whole, an estimated 70,000 Londoners died out of a population of about 500,000.[1]

Not surprisingly, attempts were made at the time to determine what might be spreading the disease and how this spreading could be stopped. Unfortunately, in the highly charged atmosphere as the death tolls climbed, the suspicion arose that dogs and cats were primary plague carriers, resulting in massive killings of these animals prior to the realization that the primary plague carriers were rats. Not only was this mistake most unfortunate for the slaughtered dogs and cats, but the problem was further compounded because the rats proliferated even more with the dogs and cats gone.[2] Good intentions in this instance clearly backfired horrendously.

THE 1810 RUNAWAY VERMONT POND

In the early 1800s, there was a sizable pond named Long Pond in Glover, Vermont. This pond was approximately a mile long and half a mile wide. To its north was a smaller pond, Mud Pond, from which a stream flowed that reached and powered a grist mill built by owner Aaron Willson. In the spring of 1810, residents of the region decided to cut a small channel from Long Pond to Mud Pond, with the admirable intention of increasing the flow of water in the stream powering the mill. So on June 6, 1810, approximately 30 to 40 men and boys proceeded to dig a trench connecting the two ponds. The unintended consequence: as the workers stepped aside after completing their work, to celebrate with a round of whiskey, Long Pond gave way. A huge torrent of water poured out, and within approximately 90 minutes the entire pond was emptied. The resulting flood spread throughout the Barton River valley, uprooting trees and affecting everything in its path, including destroying the Willson grist mill and a nearby sawmill. Despite all the destruction, amazingly no one was killed, in part because one of the workers sprinted ahead of the waters (mixed with trees, mud, and rocks) to the grist mill, arriving just in time to rush the one person at the mill uphill to safety. Ballads have been written about the event ("Beautiful lakelet with silvery wave, Prayers are now futile thy waters to save," from the ballad by Harry Alonzo Phillips), and it is still commemorated annually on Glover Day.[3] Clearly, this demonstrates another instance of good intentions with quite unintended negative consequences. At least the damage wrought by the runaway pond was local in extent, a favorable factor that is often not present when examining instead the consequences of introducing invasive species.

INTRODUCTION OF INVASIVE SPECIES

As humans migrate or just travel from place to place, they often bring with them—either intentionally or unintentionally—additional species that are nonnative (or "invasive") to the new location. Some of these introductions are benign, and some have actually been tremendously beneficial from the human perspective, a prime example of the latter involving the potato, which originated in the Andes Mountains but by now is a food staple for millions of people throughout the world, in large part because of purposeful introductions of the potato into regions where it was nonnative. However, of more relevance to this cautionary chapter, there have also been numerous instances when purposeful introductions of invasive species to rivers, lakes, or land areas have resulted in these species multiplying out of control and devastating the former in situ ecosystems. Habitat size is no limitation, as invasive species have proven quite capable of spreading across islands or continents and through intricate patterns of lakes and waterways. Here several examples are used to illustrate the all-too-common phenomenon, starting with various introductions of damaging invasive species to Australia.

Australia provides a unique case because of being an entire continent only sparsely inhabited by humans and fairly effectively isolated from other continents for thousands of years until the seventeenth century. Once Europeans started arriving, they brought with them animals and plants that had evolved quite separately from the Australian species. Among the animals noted for having been purposefully introduced to Australia with hugely damaging consequences are the following: rabbits, introduced in the nineteenth century for food and sport and subsequently multiplying out of control, causing enormous land degradation and consuming large volumes of plant life, especially fodder intended for sheep and cattle; foxes, also introduced in the nineteenth century (specifically for British fox hunting) and multiplying out of control, preying on and exterminating numerous small native Australian mammals; and the cane toad, introduced in the twentieth century to consume pests in Queensland sugarcane fields and subsequently expanding across the continent destroying native reptiles and amphibians.[4] Among the plants introduced purposely to Australia with devastating consequences is the rubber vine, brought to Queensland from Madagascar for its aesthetic quality (when occurring in small volumes) but becoming essentially a monster, multiplying out of control, forming impenetrable masses, smothering other vegetation, and, to top it off, being poisonous to various livestock.[5] The Australian example that I will detail more fully, however, is that of the prickly pear cactus.

The prickly pear cactus, indigenous to the Western Hemisphere and generally not a problem there, was intentionally introduced to Australia in 1787 when ships bringing European convicts and soldiers to the continent took on board prickly pear cuttings in Rio de Janeiro in the hopes that the prickly pear would form the basis of an Australian cochineal dye industry.[6] The attempted dye industry was unsuccessful, although later immigrants purposely imported more of the cacti as a hedge plant with edible fruit. Settlers who were impressed with its drought resistance and ability to provide food for cattle during dry spells contributed to the spread of the prickly pear northward from Botany Bay, where it had originally been introduced. Advantages of the plant seemed to outweigh disadvantages for decades, but in the 1870s serious concerns began to arise, as the prickly pear was multiplying uncontrollably in some locations, forming dense and unmanageable infestations. By 1913, an estimated 1,432,950 hectares in eastern Australia had dense pear infestation, and an additional 4,889,622 hectares had scattered infestation. By 1923, the numbers were 4,167,862 hectares with dense infestation and 5,504,021 hectares with scattered infestation. By 1925–1926, 12,000,000 hectares were virtually useless because of how dense the infestation was, and another 12,000,000 hectares were infested to a lesser extent. Photographs from the time show the prickly pear totally covering the landscape and abandoned homesteads helpless under the onslaught, as fighting the pest became a full-time and impossible task.[7]

Fortunately, despite the failed attempts for decades to bring the prickly pear infestations in eastern Australia under control through mechanical methods and cutting by hand the out-of-control plants, success did come eventually, through biological means. Limited initial success occurred in 1924 with the introduction of one particular cochineal insect (*Dactylopius*), and much greater success came 3 years later with the insect *Cactoblastis cactorum*. In the period September 1927–March 1929, more than 220 million eggs of *Cactoblastis cactorum* were distributed through the affected region. On hatching, the resulting caterpillars attacked the prickly pear voraciously, reducing it over wide areas to a pulpy mass. By the end of June 1930,

200,000 hectares of the prickly pear had been destroyed, and by the end of the following year, well over 10 million hectares of impenetrable thickets of prickly pear had been reduced to rotting vegetation, coming close to ending the prickly pear menace. A setback occurred in 1931–1932 as the insects died off and some prickly pear infestations reestablished themselves, but the insects resurged, and the setback proved only temporary. This is a story with a positive ending, although the devastating decades-long infestation would not have occurred in the first place without the introduction of the prickly pear from its native settings to the Australian landscape.[8] Of course, a new concern in any case like this is that the insects brought in to control one nuisance (in this case the prickly pear) could themselves become an even greater nuisance. Although this did not happen in the prickly pear case, it has happened in other cases.

A less successful instance in which one invasive species was imported to control another concerns the carnivorous snail *Euglandina rosea* and the giant African snail *Achatina fulica*. *Achatina fulica* was introduced from Africa to the Pacific island of Tahiti in 1967 with the intent of breeding it for food. Soon thereafter, it was introduced to the island of Moorea, 20 kilometers to the northwest of Tahiti, and to other islands in the same archipelago. It multiplied out of control, and, through its dietary habit of eating living plants, it became a considerable agricultural pest, damaging crops and gardens. The attempted solution on both Tahiti and Moorea was to introduce *Euglandina rosea*, native to Florida and central America, to control the *Achatina fulica* by consuming them. *Euglandina rosea* was introduced to Tahiti in 1974 and to Moorea in 1977, and indeed the number of *Achatina fulica* at least on Moorea decreased significantly in the 2 years following the introduction of *Euglandina rosea*, although apparently not entirely because of *Euglandina rosea*.[9] Sadly, the *Euglandina rosea* were much more effective in consuming the native snails (genus *Partula*) than they were in consuming the *Achatina fulica* and ended becoming a more serious problem than the problem they were brought in to solve.[10] The *Euglandina rosea* multiplied rapidly and advanced across Moorea at a rate of about 1.2 kilometers per year, consuming whole scale the endemic *Partula* snails as they went.[11] According to Edward O. Wilson, because of this wreckage, by 1988 every one of the wild tree snails of Moorea had become extinct, at least on that island, and those of Tahiti appeared headed in the same direction.[12]

The devastation from the *Euglandina rosea* is not confined to Tahiti and Moorea, instead occurring on numerous islands in the Pacific and Indian oceans. The *Achatina fulica* have had some value as a food item both for humans and for other animals and perhaps have some value as well medically. However, because of their various

negative roles, most prominently as agricultural nuisances and perhaps also as trans-
mitters of diseases to humans, along with their lesser annoying roles, such as being
unsightly in large numbers and being an occasional road hazard to cars, their control
has been a high priority. The attempted *Euglandina rosea* "cure" however, appears often to have been worse than the *Achatina fulica* "disease." In any event, both *Achatina
fulica* and *Euglandina rosea* were introduced purposely to numerous islands, only to
cause serious problems through their unintended consequences.[13]

Shifting from the Pacific islands to Africa, I turn to the Nile perch and its introduction into Lake Victoria for sport fishing in the twentieth century, an event
alternatively credited to Ugandan officials in the 1920s or to the British colonial
administration in 1954. The Nile perch is a gigantic fish, sometimes up to 2 meters
long and 180 kilograms. It quickly proved a dangerous predator, extinguishing some
of the native species in the lake and drastically reducing the populations of others.
Introduced in the northern portions of Lake Victoria, it systematically expanded to
the south, voraciously eating the native cichlid fishes. The propagating effects on the
rest of Lake Victoria's ecosystem included not just a serious decline of the cichlid fish
but also a proliferation of plant life formerly kept in check by the cichlids, depletion
of oxygen in the deeper waters of the lake, and further decline of many species.[14] On
the other hand, despite the damage to native species, the Nile perch introduction
was apparently successful economically.[15] As is often the case, whether something is
good or bad can vary greatly depending on one's perspective. In the case of the Nile
perch, this species introduction appears to have been favorable economically while
unfavorable ecologically. In many of the examples in this section, however, the species introductions are unfavorable both economically and ecologically, and this is
certainly the case of the next example, from the United States.

In the United States, the tamarisk plant, a large shrub native to Africa and
Eurasia, was purposely introduced to the American West in the early 1880s both
for wind and erosion control and for ornamental purposes. Over a century later,
it is now widespread throughout much of the western half of the continental
United States and is considered one of the most damaging invasive species in
the country. The reason: the plants voraciously soak up water through their long
roots, the result being a serious depletion of some of the underground aquifers.
A single large tamarisk plant can absorb up to 200 gallons of water a day, water that is often badly needed elsewhere in the drought-ridden western United
States. In 2006, the tamarisk was estimated to have infected over 3.3 million
acres in the West and to be spreading rapidly. If not eventually controlled, this
will be increasingly damaging to the region's water supply, at great expense to
the regional inhabitants.[16]

INTENTIONAL REMOVAL OF NATIVE SPECIES

Other cases of cascading unintended consequences in local or regional ecosystems involve the removal rather than the introduction of specific animals or plants. For instance, wolves and other large carnivores were purposefully removed from various regions of the western United States in the late nineteenth and early twentieth centuries in part to protect livestock. Documented unintended consequences include increased damage to vegetation caused by browsing elk (no longer kept in check by the wolves), reduced berry-producing shrubs, altered river channels and floodplains, disappearance of beavers, and reduced abundance and diversity of birds.[17] Wolves were so extensively hunted, trapped, and poisoned in the late nineteenth century that they almost completely disappeared from the contiguous 48 states in the United States by the early 1900s.[18] Because of the multiple undesired impacts on the local ecosystems, wolves were reintroduced into some regions in the 1990s, with very strong opinions on both sides regarding whether that should or should not be done. Here, the point is simply that when a major predator is removed or introduced or reintroduced to a region, it makes a difference well beyond that one predator, and it is important to recognize that we are not wise enough yet (and perhaps never will be) to predict accurately what all the consequences will be.

In the specific case of Yellowstone National Park, wolves were extirpated from the park in the 1920s, after which elk and other ungulates could browse on the aspen population and willow communities much more freely, no longer being hampered by the presence of the predatory wolves. The result included severe damage to the aspen and willow populations. Wolves were intentionally reintroduced to the park in 1995–1996, and by 2007 it was clear that at least part of the ecosystem was recovering in the desired ways, including the first significant growth of aspen in northern Yellowstone in over 50 years.[19] The aspen regeneration has come as the elk population has declined, and although there is no certainty regarding how the revised ecosystem will further evolve, certainly the prediction that reintroduction of the wolves would benefit the regeneration of aspen has proven correct.

FOREST MANAGEMENT

Forest management has become essential in many parts of the world, and indeed some forest management has been quite successful.* Nonetheless, unsuccessful forest

---

* Jared Diamond (2005a) specifically praises successful forest management policies in Germany and Japan since the 1500s.

management practices constitute yet another category of good intentions gone awry. As with invasive species introductions, there are varied forest management practices and consequences, and I include several illustrative examples.

In the first decade of the twentieth century, the U.S. Forest Service adopted a policy of putting out—or at least attempting to put out—all forest fires as quickly as feasible. Although this policy was adopted for the best of reasons, to protect trees, people, and property, it produced problems in the long run, as the extinguishing of small fires created a buildup of burnable materials and the increased possibility of very large, much more destructive fires.[20] Climate change has also had impacts on fire regimes, as warming has caused some regional fire seasons to lengthen.

Fires are an important natural feature of the forest environment. The ash from burning vegetation fertilizes the soil and keeps its acidity in check, thereby helping to maintain the vigor of the forest ecosystem. Furthermore, forest vegetation has evolved with the reality of natural fires so much so that, for an extreme example, some pinecone seeds cannot even be released until heated to fire-induced high temperatures.[21] Thus, a policy of putting out all forest fires as soon as possible not only is not necessarily conducive to the health of the forest ecosystem but also can be devastating to individual species. Such a policy also leads to accumulating plant material on the forest floor, increasing the chance of a catastrophic fire once any fire begins that is not brought under control quickly. All these factors, in addition to an intense regional drought, were involved in a catastrophic fire in Yellowstone National Park in the western United States in August 1988, making that devastating fire in part an unintended consequence of well-intended forest management policies.[22]

A very different misguided forest management policy was implemented in the early 1990s when President Suharto of Indonesia ordered that a large swamp forest in Borneo be drained and transformed into a giant rice paddy. As is so often the case, the intentions were admirable, the goal being to make Indonesia self-sufficient regarding rice, a primary staple in the diet of Indonesians and many other Asians. To accomplish the transformation from a swamp forest to a rice paddy, 2,500 miles of canals were dug, and 60,000 migrant farmers moved in. However, despite all the effort, almost no rice was grown, as the soils were found totally inadequate, and the project was abandoned. In the meantime, the drained swamps exposed peat that burns during the dry season, releasing major volumes of carbon into the atmosphere.[23] In fact, during the 1997–1998 El Niño, the carbon emissions from the smoldering Borneo swamps were significant on a global level. Estimates using satellite data and ground measurements in Borneo and extrapolating to Indonesia as a whole suggest that the amount of carbon released to the atmosphere as a result of the burning peat and vegetation in Indonesia in 1997 was 13% to 40% of the

1997 global carbon emissions from the burning of fossil fuels;[24] that is, burning peat and vegetation in just one country sent about one-quarter as much carbon into the atmosphere in 1997 as the burning of fossil fuels throughout the entire world.

The inadequacy of the Indonesian soils for routine agriculture is hardly unique for the tropical regions. In fact, one of the saddest aspects of the environmental tragedy of the large-scale destruction of the tropical rain forests in recent decades is that often portions of these rain forests have been cleared for agriculture only to find that the soil is extremely poor for agricultural purposes. Clearing of tropical rain forests for agriculture all too often results in a mere one or two years of crop growth before the soils are so leached of nutrients that the farms are abandoned and, tragically, additional forest is cleared and the failed sequence is repeated.[25]

On the island nation of Madagascar, 250 miles off the southeastern coast of mainland Africa, monsoon rains provide an additional complication. Over 92% of the forests covering Madagascar were destroyed between 1900 and 1996, burned down largely intentionally by residents wanting cleared land for rice and coffee farming and for cattle ranching. Without the trees, the soil becomes directly exposed to the heavy monsoon rains, which then wash the soil into the rivers and out toward the ocean. Sadly, instead of adjusting to the climatic conditions, once the soil gets depleted, the farmers and ranchers all too often simply clear more land and move their fields onto the newly cleared land.[26] This is a policy that clearly is not sustainable in the long run, as eventually—if the policy continues—the forest will be totally destroyed, with no more forested land remaining to clear.

At least Madagascar still has forests left. Easter Island, in contrast, does not. The island is thought to have been well forested when first colonized by humans, perhaps about A.D. 1200,[27] but human inhabitants, wanting to clear the land and use the wood for a variety of purposes, one by one destroyed the trees. Deforestation was nearly complete by Easter Day 1722, when the Dutch explorer Jacob Roggeveen became the first European to reach the island and saw no trees taller than 10 feet.[28] Jared Diamond, in an extended analysis of the collapse of societies throughout history, characterized Easter Island as "the clearest example of a society that destroyed itself by overexploiting its own resources."[29]

DAMS

The overwhelming majority of large dams ever built were almost certainly built with good intentions. But here, as elsewhere, the good intentions have not prevented many instances of unfortunate

> Large-scale technological projects are rarely (if ever) foolproof.
> —Stephen H. Schneider[30]

unintended consequences. For instance, the controversial Glen Canyon Dam along the Colorado River at Page, Arizona, constructed in the late 1950s and early 1960s to provide electricity and controlled water availability for the growing human population in the drought-prone southwestern United States, not only markedly altered a magnificent scenic view but also led to significant percentages of the water trapped behind the dam being soaked up by the porous stone of the Glen Canyon walls.[31] In writing of the Glen Canyon Dam, Steve Schneider commented that it provides yet "another example of how large-scale technological projects are rarely (if ever) foolproof,"[32] a comment as relevant today as when Schneider originally wrote it.

Elsewhere in the world, the much-heralded Aswan Dam along the Nile River, completed in 1971, has provided a sizable fraction of Egypt's electricity (approximately half in the mid-1990s), and the control of the dam and associated Lake Nasser reservoir have helped provide Egypt with a stable water supply. However, these intended favorable consequences of the dam have come at the cost of several unintended and unfavorable consequences. Among these are the following: freshwater snails that carry the visceral parasitic disease schistosomiasis are no longer flushed away by the Nile, instead accumulating and spreading the disease; water hyacinths, now no longer impeded by routine natural flooding of the Nile, are proliferating, clogging canals and waterways all along the course of the river from Lake Nasser to the Nile delta; and the dam has trapped sediment that otherwise would flow down the Nile, some to overflow along the edges and bring nutrients to agricultural land during the annual Nile floods and some to arrive at the delta and be deposited there. With so much sediment trapped at the dam, not only did the Nile delta cease to grow, but it began eroding away as well, in some locations by as much as 80 meters a year.[33] These various changes are serious consequences for a region that has been dependent on the Nile and its annual floods for thousands of years.

The Aswan Dam is far surpassed, at least in size, by the Three Gorges Dam in Yichang, China. Planned in large part for flood control, the Three Gorges Dam is also expected to generate 84.7 billion kilowatt-hours of electricity per year once operating at full capacity. These are admirable goals, but one wonders how long it will be before the advantages of the dam will be outweighed by the disadvantages. Even before construction began in 2003, there were objections in view of the likely environmental impacts, and some of the concerns were being realized by 2008. The rising waters in the reservoir have weakened sloping landscapes, increasing the risk of landslides, and species of fish that had been abundant in the region are having to adapt to severe disruptions in their migration routes, in some cases with precipitous declines in population.[34]

OTHER WATER MANAGEMENT

Dams are hardly the only water management schemes that have led to problems. Certainly the damage done to the Aral Sea during the twentieth century, described in chapter 4, fits into this category, with severe downstream environmental damage caused by the siphoning of water for irrigation.

For another example, in northern China and Mongolia excessive grazing, clearing of land for agriculture, and siphoning away of water for human uses has led to 4 million square kilometers of degraded grasslands, a massive dust bowl, and the blowing away of valuable topsoil by winds.[35] As in the case of the Aral Sea, agricultural irrigation in China and Mongolia dried up rivers and badly worsened regional environmental conditions. To combat the desertification, partly brought about by humans, over $1 billion have been spent in China since 1978 in the well-intentioned planting of trees. In portions of the region where rain is abundant, this resulted in a successful reforestation. However, a large part of the region is arid, and there the tree planting actually did further damage, for the trees—often poplars—have deep roots and have sucked up water, lowering the water table and further hindering the survival of native grasses and shrubs. Fortunately, the cascading unintended consequences have been reversed at least locally in Bayinhushu, where scientists and government officials have helped the residents live in a more sustainable fashion with the local conditions, with the result that the pastures have been restored and the dust storms have abated.[36]

LEADED GASOLINE AND PAINT

Industrial chemist Thomas Midgley Jr. of General Motors Research Corporation was heralded as a hero in the early twentieth century after determining that tetraethyl lead could be used effectively as an antiknock agent when added to the gasoline fueling airplane and car engines. The key discovery came in

> Disastrous unintended consequences typically have no guilty parties.

December 1921, after a long search by Midgley and others for anything that would reduce the knocking plaguing the internal combustion engines at the time.[37] Adding tetraethyl lead quite successfully raised the octane rating of the gasoline and helped General Motors' profits to soar in the 1920s. Leaded gasoline was used for decades before its negative environmental consequences were recognized. Lead was added also to paints and other products and for decades was viewed favorably. However, eventually its toxic qualities and the severe threats posed by them became widely recognized, after which leaded gasoline was banned in the United States, Canada, Japan, and many European countries. Expensive efforts have been made, one by one, to rid homes and other dwellings of the toxic lead paint, adding financial costs to the health costs of the well-intentioned lead paint.[38] The original finding by Midgley and its widespread application were appropriately heralded; disastrous unintended consequences frequently have no guilty parties.

THE CASE OF CHLOROFLUOROCARBONS

After his remarkable initial success with tetraethyl lead, Thomas Midgley Jr. was involved in a similar case with chlorofluorocarbons. In the 1920s, General Motors' Frigidaire Division had been using sulfur dioxide as the chief cooling agent in its refrigerators; but sulfur dioxide is toxic, as are methyl chloride and ammonia, the other two major coolants used at the time. Midgley was called on in the summer of 1928 to develop an alternative coolant for use in home refrigerators and air conditioners.[39]

In only three days, Midgley and his assistant Albert Henne developed and preliminarily tested an effective combination of chlorine, fluorine, and methane: dichlorodifluoromethane, later better known as freon.[40] Midgley then sent the compound to the U.S. Bureau of Mines Experiment Station in Pittsburgh for a sequence of toxicity tests carried out with dogs, monkeys, and guinea pigs. Although some of the tests were extreme, killing the animals involved, the sequence quite effectively established the nontoxicity of dichlorodifluoromethane under any concentrations that could reasonably be expected during accidental leaks or other mishaps.[41]

For its public unveiling, Midgley took his compound to a meeting of the American Chemical Association in Atlanta in April 1930 and dramatically demonstrated its nonpoisonous, nonflammable nature by personally inhaling it, with no ill effects, and using it to extinguish a burning candle. Before long, Midgley had developed other combinations of chlorine, fluorine, and carbon and called them collectively chlorofluorocarbons (CFCs). They were soon being manufactured on a large scale, some marketed under the name of Freons.[42]

CFCs were hailed as wonder compounds when initially introduced. Odorless, colorless, and effective coolants, CFCs seemed ideal for refrigeration and air-conditioning and became widely used for those purposes starting in the 1930s. In fact, they were a large part of the reason why between 1930 and 1935 the air-conditioning business in the United States boomed significantly, with sales multiplying by a factor of 16 despite the ongoing financial depression.[43] In the 1940s, CFCs became used also as an ingredient in Styrofoam,[44] and by the end of the 1950s, they were widely used additionally in insulation, packaging, aerosol propellants, and solvents.[45] As a coolant, the CFCs had tremendous safety advantages over the compounds previously in use, including methyl chloride, ammonia, and sulfur dioxide, all of which are toxic and could prove deadly if leaked into the surrounding air.[46] Hence, the high praise for CFCs was warranted.

The enthusiasm for CFCs abated in the 1970s after Mario Molina and F. Sherwood Rowland, both at the time employed at the University of California, Irvine, warned that the CFCs released into the atmosphere drift upward and take part in chemical reactions in the stratosphere that could lead to major reductions in the ozone layer.[47] In other words, the CFC compounds that seem so benign and favorable near the Earth's surface could potentially damage the protective ozone layer that shields human and other life at the Earth's surface from excessive ultraviolet radiation. With a few decades of widespread use, millions of tons of CFCs had reached the stratosphere.[48] As described in the last section of chapter 4 and in greater detail in scientific research articles,[49] when acted on by sunlight in the stratosphere, the CFCs can release free chlorine, which can act as a catalyst to facilitate the breakdown of ozone (O_3), thereby eating into the Earth's protective ozone layer. If the CFC emissions had continued unabated and the ozone layer had been destroyed throughout the global stratosphere, the resulting damage to humans and other life could have far outweighed the notable practical benefits that the CFCs provided; that is, the unintended consequences of the very well-intentioned use of CFCs could have been horrifying. The potential damage from CFCs is now considered so great that CFCs are banned for many uses.

ALBEDO CHANGES FROM REFORESTATION

Tree planting is strongly encouraged by environmentalists and others, in considerable part because the uptake of atmospheric CO_2 by trees should help to lessen the greenhouse effect and thereby lessen future warming. However, the CO_2 impact is not the only impact that trees have on climate. Among the other impacts, the low albedo (reflectivity) of most trees and forests makes it likely that in many areas planting trees in great numbers will lower the overall surface albedo and thereby cause increased absorption of solar radiation, encouraging, of all things, warming. Richard Betts from the Hadley Centre for Climate Prediction and Research in the United Kingdom has simulated the results of altered forest conditions and concludes that in many boreal forest areas, tree planting could result in enhanced warming rather than the intended cooling; that is, the warming from the resulting increased absorption of solar radiation due to the reduced reflectivity of the surface could well outweigh the cooling effect of the trees' uptake of atmospheric CO_2,[50] a very undesired unintended consequence.

FAVORABLE UNINTENDED CONSEQUENCES

Of course, not all unintended consequences are bad. Some are indeed decidedly good, like the tendency of traditional water-storing brass pitchers in India to release small amounts of copper that kill harmful bacteria, helping to ward off illness,[51] or the fact that the competitive nature of the Cold War between the Soviet Union and the United States (and others) through much of the mid-twentieth century spurred considerable scientific research in both countries. Separately, as a longtime NASA employee, I am reminded repeatedly of positive unintended consequences, through the huge variety of valuable technological spin-offs from the United States space program, well illustrated in annual NASA *Spinoff* volumes.[52] In fact, favorable unintended consequences might be every bit as common as unfavorable ones, and in our daily lives, to be happy and well-adjusted individuals, we should be at least as aware of and grateful for the favorable unintended consequences as we are dismayed by the unfavorable ones. However, in this book the emphasis is on the unfavorable unintended consequences because the potential unfavorable consequences of massive geoengineering schemes are simply too great to ignore without significant risk to ourselves and future generations. With that background, the next chapter describes a broad selection of some of the proposed geoengineering schemes.

CHAPTER 7

Geoengineering Schemes

In the effort to limit future global warming and other climate changes, various suggested geoengineering remedies have been proposed, some with quite scary potential unintended consequences.

RECOGNIZING the damage that humans have done to the environment and the likely human contributions to climate changes occurring now and expected in the next several decades, several proposals have been offered to undo or counteract some of these impacts. In general, trying to undo damage is a good thing. However, some of these schemes have the potential of inflicting even greater damage than the damage the scheme is trying to undo, and some of these unintended consequences are easily imaginable. Keeping the examples of the previous chapter in mind, there are likely even more serious potential consequences that we have not yet imagined. Our technological prowess is now sufficiently great that the unintended damage inflicted could be gigantic. This chapter describes some of the massive geoengineering schemes being contemplated. Some seem more reasonable and less potentially damaging than others; however, given all the uncertainties, incredible caution should be exercised before actually implementing any massive climate-altering geoengineering scheme.

Although definitions of exactly what constitutes "geoengineering" vary, here the term is used in the sense defined by David Keith of Carnegie Mellon University in a 2000 review article on geoengineering the climate. Specifically, geoengineering is defined as "the intentional large-scale manipulation of the environment," particularly for purposes of reducing undesired human-caused climate change.[1] Keith attributes the first use of the term in approximately this sense to Cesare Marchetti in a 1977 paper titled "On Geoengineering and the CO_2 Problem," where Marchetti suggests injecting excess CO_2 into the deep oceans.[2] For a further note on terminology, in line with Keith's 2000 article, suggestions of how to dispose of the CO_2 emitted by human activities are considered geoengineering, but suggestions for limiting the amount of carbon being emitted are not considered geoengineering, instead being considered carbon-emission management. As the use of the term "geoengineering" continues to evolve, Marchetti's original 1977 suggestion is on the borderline of what is or is not considered to be geoengineering, with some references categorizing it as geoengineering[3] and others categorizing it as outside of geoengineering.[4]

This chapter highlights several of the geoengineering schemes that have been proposed to counteract the damage humans have done and are projected to do to global climate. Notwithstanding the disagreement over whether it constitutes true "geoengineering," I begin with Marchetti's 1977 suggestion because it remains relevant regardless of how it is categorized.

DISPOSAL OF EXCESS CO_2 IN THE DEEP OCEAN

Concerned about the potential seriousness of allowing CO_2 to continue to accumulate in the atmosphere, with its high absorption of the infrared radiation headed outward from within the Earth system (termed "outgoing infrared radiation"), Marchetti considers the possibility of capturing some of the CO_2 at its industrial source, transporting it, and disposing it out of the way in the deep ocean.[5] He estimates that about 50% of the CO_2 normally injected into the atmosphere at large industrial plants could be sequestered, partly at marginal cost and partly at a somewhat higher but still "relatively low cost of a few dollars per ton of carbon."[6] After being sequestered, the CO_2 would be compressed to liquid form and transported overland in pipelines, then deposited in the ocean. The one location that Marchetti specifically advocates is the vicinity of the Strait of Gibraltar, where the injected CO_2 would presumably merge into the bottom current that flows out of the Mediterranean and gently sinks downward to a depth of about 1,500 meters prior to spreading out to cover much of the horizontal expanse of the Atlantic. This would place the CO_2 deep in the ocean (at midocean levels), likely not returning to the surface for 1,000 years, according to Marchetti's estimates. Marchetti leaves open the possibility of injecting the CO_2 in additional, yet-to-be-determined ocean locations and the possibility of transporting the CO_2 in solid rather than liquid form.

A paramount problem I have with this scheme is that while Marchetti recognizes that it is no longer appropriate to indiscriminately pour our waste products into the atmosphere, he continues to find it permissible to pour the waste (at least the CO_2) into the oceans. Just like the atmosphere, the oceans are a finite resource, and even if the 1,000-year estimate is close to correct (and it could instead be considerably too high), this scheme is postponing to future generations the need to solve the issue of what should be done with the CO_2 that current generations are expelling. Furthermore, injecting excess CO_2 into the deep oceans could exacerbate even further the acidification and other chemical changes that are already taking place in the oceans because of the propagating impacts of increased atmospheric CO_2. For these reasons, although I appreciate Marchetti's innovative suggestion for reducing anthropogenic atmospheric CO_2 increases, I reject disposing CO_2 in the deep ocean as a reasonable option for taking care of the atmospheric CO_2 problem.

CARBON CAPTURE AND STORAGE

Subsequent to Marchetti's suggestion, there have been additional proposals for CO_2 capture and sequestration (or storage). As an abstract concept, capturing the carbon emissions at the source of the emission and preventing them from being released to the atmosphere is highly attractive. Difficulties arise, however, in exactly how to accomplish the capturing and then in what to do with the captured carbon. Both issues are being studied from various angles.

One well-developed technology for CO_2 capture is amine scrubbing, in which aqueous amine is used to absorb and strip the CO_2 from natural gas and hydrogen. This technology has been in existence since 1930 and has been used successfully on a small scale but has not yet been tested on the large scale appropriate for capturing CO_2 from major coal-fired power plants. It is considered by some to be the most feasible technology in the near term for decreasing CO_2 emissions from coal-fired power plants.[7] However, even carbon capture—although helpful—would not cancel the environmental damage being done by coal-fired plants.

Another proposal is to capture CO_2 not from the source of its emission but from the ambient air. No cost-effective technique yet exists for doing this, but if one is devised, this capability would have several advantages. For instance, it would allow capturing CO_2 regardless of the source of the CO_2 emissions, it would allow capturing past emissions, and it would allow the capturing to take place at a location convenient to the selected storage location, thereby limiting the transportation necessary from capture to storage.[8]

Storage locations being considered for the captured CO_2 include underground locations both beneath dry land[9] and beneath oceans.[10] A key consideration is the need for low-permeability rocks above the storage region to prevent the CO_2 from flowing upward and escaping to the atmosphere, where it not only would add to the greenhouse effect but also could cause a deadly human health hazard in the immediate vicinity of its release because of its high concentration. Probably the most obvious potentially safe locations for CO_2 storage would be depleted oil and gas reservoirs, which are known to have kept oil and gas confined for many millennia.[11] If it can be done without leaks, putting something into these reservoirs has the added advantage of lessening the chance of collapse of the rock formations above.

Although carbon capture and sequestration or storage does not yet provide a viable solution to the anticipated climate problems, there is considerable ongoing work that could conceivably lead to partial solutions.[12] At least carbon capture and storage are not pouring additional materials into the atmosphere, as the next proposed scheme does.

AEROSOL INJECTION INTO THE ATMOSPHERE

Both of the next two schemes are aimed at counteracting global warming by reflecting away incoming solar radiation before it reaches the Earth's surface. The most widely publicized of the two suggests artificially increasing the Earth's reflectivity (or albedo) through the purposeful addition of more aerosols into the stratosphere. Although earlier mentioned by Soviet climatologist Mikhail Budyko[13] and others, this suggestion gained considerably increased attention in 2006 when Nobel Laureate Paul Crutzen spoke out in support of further studying the possibility.[14]

To understand why anyone might suggest purposefully injecting aerosols into the atmosphere, it is relevant to recall the discussion in chapter 1 of the temperature impact of volcanic eruptions, with tropospheric cooling resulting naturally from eruptions that send large volumes of volcanic materials into the stratosphere. As mentioned in chapter 1, the April 1815 eruption of Mount Tambora led to an estimated 0.4°C to 0.7°C cooling of the Northern Hemisphere, lasting approximately 2 years. Placing this in perspective, the high end of the estimate is the same magnitude as the estimated approximately 0.7°C global warming for the entire twentieth century[15] from greenhouse gases and all other sources combined. More recently and with better data sets* available, the effect of the June 1991 eruption of Mount Pinatubo on lower tropospheric temperatures in 1992 is estimated, based on satellite observations, to be a cooling peaking at about 0.5°C globally and 0.7°C for

---

* In the case of the Mount Pinatubo eruption, the emission plume was tracked by satellite data as it traveled westward and diffused outward, making the trek completely around the globe in the following 22 days (Bluth et al. 1992; Parkinson 1997).

the Northern Hemisphere.[16] Among the prime "culprits" in this cooling were the estimated 20 million tons of erupted sulfur dioxide that reached the stratosphere.[17] As with other eruptions, the temperature impacts of Mount Tambora and Mount Pinatubo were temporary, generally peaking in the year or two after the eruption, before the settling out of the bulk of the erupted materials from the stratosphere.

Recognizing the impact of materials in the stratosphere on temperatures and their likely longevity there, Budyko calculated that it would take approximately 10 million tons (10 teragrams) per year of additional sulfur dioxide (SO_2) in the stratosphere to counteract the global radiative effect of doubled CO_2.[18] A 1992 report of the U.S. National Academy of Sciences indicates the possibility of injecting these amounts into the stratosphere at low cost,[19] and Keith confirms that the costs involved in any of the schemes to deploy stratospheric scatterers would be "trivial compared to the cost of other mitigation options."[20]

In his 2006 study, Crutzen indicates that the necessary artificial aerosol injection into the stratosphere to compensate for enhanced climate warming would be about 1.9 teragrams of sulfur each year, at a cost of perhaps $25 billion to $60 billion. He unambiguously asserts that the preferred solution to human-induced global warming would be sufficient reduction in greenhouse gas emissions to halt the warming, but he is pessimistic regarding the possibility that will happen. As a result, he advocates further consideration of the option of intentional aerosol injections into the stratosphere.[21] Crutzen specifies that the desired increased albedo "can be achieved by burning S_2 or H_2S, carried into the stratosphere on balloons and by artillery guns to produce SO_2."[22] Natural chemical and microphysical processes would then produce submicrometer sulfate particles in the stratosphere, mimicking the natural processes that occurred, for instance, following the June 1991 Mount Pinatubo eruption. Crutzen explicitly recommends not only intensifying research into such schemes but perhaps also proceeding to small-scale atmospheric tests. Much as I respect Crutzen's Nobel Prize and his scientific credentials, I strongly disagree with this suggestion regarding intentional aerosol injections.

Crutzen was involved in a further study led by Philip Rasch of the National Center for Atmospheric Research (NCAR) that examines the effect that the size of the aerosol particles might have on their effectiveness in counteracting global warming, concluding that smaller particles would be more effective. With small particles, they estimate that 1.5 teragrams of sulfur additions each year would suffice to balance the warming from a doubling of atmospheric CO_2.[23] This is reduced from Crutzen's earlier estimate but still a very large amount of added aerosols.

The massive injection of aerosols into the stratosphere would have substantial potential negative consequences not only to climate but also to the efficiency of systems generating power from solar energy and to human health. Both Michael Mac-Cracken and Daniel Murphy warn, in separate studies, about the significant reduction in solar power generation efficiency expected in the event of proposed aerosol additions to the stratosphere, with Murphy offering the historical example that in the year after the eruption of Mount Pinatubo, peak power output from systems of concentrating solar collectors was reduced by about 20%.[24] Crutzen himself mentions that the World Health Organization estimates[25] that the sulfate particles that already help lessen global warming lead to more than 500,000 premature deaths each year.[26] Of course, no one (to my knowledge) is recommending pouring particulate matter into the ground-level air, but particles do eventually descend, and yearly injections of matter into the stratosphere will eventually make a difference at the ground level as well.

By now, humans have severely polluted the lower atmosphere, the upper ocean, and much else. It is not appropriate to expand this pollution to the upper atmosphere and other relatively uncontaminated regions of the Earth system. Furthermore, even if the scheme of aerosol injections into the stratosphere had the desired effect of cooling the lower atmosphere, it is possible that this cooling effect would apply to the lower stratosphere as well, thereby likely having the additional significant negative effect of further depleting the stratospheric ozone layer, a point made also by Keith[27] and others.[28] A cooler stratosphere would encourage more of the polar stratospheric clouds that serve as breeding grounds for the chemical reactions leading to stratospheric ozone destruction (chapter 4). An examination of the sensitivity of polar stratospheric ozone to proposed geoengineering injections of sulfur into the stratosphere concludes that the Arctic ozone layer would be thinned "drastically" and that the sought-after recovery of the Antarctic ozone hole would be delayed by between 30 and 70 years.[29]

INJECTION OF OTHER SMALL
REFLECTORS INTO THE ATMOSPHERE

The second scheme to counteract global warming by reflecting away solar radiation before it reaches the Earth's surface involves miniature balloons rather than aerosols. Keith, in his 2000 review, describes work by Edward Teller and colleagues proposing thousands of small hydrogen-filled metal balloons floating at altitudes of about 25 kilometers. Individually, the balloons would have diameters of only about

4 millimeters, but in total the estimated required mass would be 1 million tons. The thought is that manufacturing such balloons with aluminum would minimize the impact on ozone chemistry and would allow a long lifetime in the stratosphere, with the added advantage of rapid oxidation when the material descended into the troposphere, limiting the volume of material eventually deposited on the Earth's surface.[30]

Although this scheme might have some important advantages over the insertion of aerosols, the thought of a million tons of hydrogen-filled metal balloons inserted into the atmosphere is not a great deal more appealing than the thought of the proposed additional aerosols. Consider the damage they might cause to very high flying aircraft or to spacecraft lifting off or returning through them, plus the damage to the clarity of the night sky for astronomical observations, the damage to daylight clarity, and any potential effects on the rest of atmospheric chemistry as the balloons become oxidized when descending. Schemes of pouring massive amounts of material into the Earth's atmosphere, whether aerosols or hydrogen-filled balloons, are simply too fraught with potential unintended consequences to warrant serious consideration, at least until the Earth's condition becomes far more definitively dire than it is now.

Columnist George Monbiot places the potential damage from this scheme into a more dramatic context in his book *Heat: How to Stop the Planet from Burning*. He reminds his readers that Edward Teller developed the hydrogen bomb and mentions that Teller's disciples work both on nuclear weapons and on further developing the idea of launching a million tons of hydrogen-filled balloons. He then declares, "It is hard to decide which of their activities is more dangerous."[31]

SHIELDING THE EARTH FROM AFAR

A proposed scheme that is overwhelmingly less inherently damaging than the previous two is to place a massive reflector far out in space in order to lessen warming by shielding the Earth from a portion of the Sun's rays.[32] This would be expensive, but it would have two decided advantages over the aerosol and small reflector schemes: 1) it would not further pollute either the atmosphere or the oceans, and 2) if it proved either ineffective or overly effective in its intended goal, it could presumably be undone with relative ease, for instance by sending a signal to the spacecraft to retract the reflector or to modify its size. Many satellites have solar arrays (to collect solar energy to run the satellite instruments) that have an accordion-style construction for compaction during launch and unfolding after launch. The massive reflector for shielding the Earth could be constructed similarly, providing an easy option for size adjustments by adjusting the number of panels unfurled.

Bala Govindasamy and Ken Caldeira from the Lawrence Livermore National Laboratory estimate, from calculations with a global climate model (specifically the NCAR Community Climate Model 3), that a shielding of about 1.7% of the incoming solar radiation would be the amount required to offset the 1.75°C warming simulated in the same model for a doubling of atmospheric carbon dioxide.[33]

The scheme, however, has imperfections, a primary one being that the shielding would tend to reduce temperatures throughout the atmosphere, whereas the CO_2 increases tend to warm the lower atmosphere but cool the stratosphere. The additional stratospheric cooling from the shielding could have the damaging consequence of further depleting stratospheric ozone. Still, Govindasamy and Caldeira conclude from their simulations that "geoengineering may be a promising strategy" although sensibly add at the end that the most prudent policy may instead be to cut back on greenhouse gas emissions.[34]

Walter Seifritz performed some rough calculations of what would be required for the shield—in his case a mirror—to be effective. To compensate for a 2.5°C warming, the shield would need to block about 3.5% of the incoming solar radiation. With the shield positioned at a location where its orbit would keep it constantly between the Earth and the Sun, this would require a shield area of at least 4.5 million square kilometers, which Seifritz estimates as requiring at least 4.5 million tons of material, all needing to be transported from the Earth to the shield location. Seifritz offers a very rough estimate that the annual cost of this endeavor would come to about 6% of the annual gross national product of the entire world, making it quite prohibitive from a fiscal standpoint. Moreover, the shield would be in an unstable orbit and hence would require additional effort to keep it in place.[35]

The solar shield suggestion has been enhanced by a proposal to make the shield steerable, enabling it to be used not only to block sunlight but also to direct the sunlight toward specific regions of the Earth.[36] While adding potentially valuable capabilities, this would also add another major concern to the items mentioned in the previous paragraph, as the steerable shield would have terrifying possibilities if the steering controls were in the wrong hands. One could easily imagine well-positioned terrorists concentrating solar energy and beaming it in to burn down entire cities or nations. Even without criminal intents, multiple legal problems could arise, as the climate and weather changes brought about (or believed to have been brought about) by a steerable solar shield would almost certainly be damaging to some people and communities even if successful in being beneficial to others.

The steerable solar shield is reminiscent of the suggestion in the 1970s to warm the Earth by putting huge mirrors in orbit around the Earth,[37] this time not to block solar radiation from getting in but instead to redirect toward the Earth solar rays that would otherwise have bypassed it. At that time, a concern was a possible approaching ice age, so a desire to warm the Earth made sense. In both cases, until our understanding of climate and climate change improves significantly, the shield and mirror schemes are more appropriate for science fiction than for policy implementation.

PLACING REFLECTIVE WHITE PLASTIC OVER DESERT AREAS

Instead of blocking some of the Sun's radiation from getting to the Earth/atmosphere system (e.g., with a shield) or reflecting some of the radiation while within the atmosphere (e.g., with miniature balloons or added aerosols), an alternative means of reducing the Earth's absorption of solar radiation would be to allow the radiation to reach the surface but reflect it once it gets to the surface. Along that line, as part of the Global Albedo Enhancement Project, Alvia Gaskill examined the potential impact of increasing the reflection of sunlight back to space by increasing the albedo of one or more of the Earth's deserts. The proposed scheme would be a 60-year effort starting around 2010 and placing white plastic polyethylene film over a new 67,000 square miles of desert each year. By the end of the 60 years, over 4 million square miles would be covered, which would be somewhat more than the size of the entire Sahara Desert. This is suggested by the proposers as a temporary measure, essentially to buy time for more permanent strategies to be developed, with the thought being that 60 years should be sufficient time to plan and be ready to implement other, better strategies for keeping the Earth from excess warming.[38] Gaskill identifies the Sahara, Arabian, and Gobi deserts as the prime targeted locations, perhaps unfairly sparing the Australians and Americans from this intrusion on their desert environments, although indeed the southwestern United States is at least mentioned.

Gaskill identifies 11 advantages of this desert-based scheme for reducing global warming. Among them are the following: the technology exists; the cost, estimated at $15 per metric ton of carbon equivalent, is deemed to be relatively low compared to other strategies; the reflective covering would not necessarily have to be a single continuous sheet; surrounding communities might benefit in particular from the anticipated resulting regional cooling (or at least lessened warming); and knowledge could be gained about land use changes and how to mitigate dust storms.[39]

Despite the 11 advantages identified by Gaskill, the concept of covering deserts to make them more reflective is frightening. It treats the deserts as totally expendable, essentially ignoring that deserts can be stunning environments with an abundance of well-adapted life forms and often vibrant ecosystems. It also discounts the vital roles the deserts play in the rest of the global system. Among these roles, desert dust affects phytoplankton growth and hurricane formation, often far from the desert itself, through the transport of the desert dust by atmospheric circulation.

Gaskill mentions some of the disadvantages of the desert-covering scheme, including that the proposed covering would require significant maintenance, would alter dust flow patterns and hence the transport of nutrients to locations sometimes thousands of miles away, and, according to Gaskill, would kill all species of plants and animals throughout the covered areas. In particular, if the white plastic film covered sizable areas of the Sahara, the iron and phosphorus supply typically carried westward in dust storms could be significantly reduced, perhaps seriously depleting the North Atlantic and Amazon of these critical nutrients. Gaskill also mentions that some of the countries containing the targeted desert areas might not be agreeable to having the reflective white covering placed over their land (no wonder) and then proceeds to make the valid but troubling comment that this reluctance might be overcome by offering sufficient debt forgiveness in compensation for permission to cover the land, as many of these countries are impoverished and ridden with debt.[40] However successful such a strategy might be in terms of getting agreement to desecrate the land, I find it chilling. It would be comparable to an aggressive debt collector or loan shark stripping an individual of his or her most prized possession or inheritance, only scaled up from the individual level to the national level.

MODIFICATION OF OCEAN REFLECTIVITY

In addition to the criticisms mentioned in the previous paragraphs, another problem with the desert-covering scheme to reduce warming is that the deserts already have fairly high solar reflectivities and hence are already reflecting much of the solar radiation incident on them. This is not true of the oceans, which instead absorb by far the majority of the solar radiation reaching them. Furthermore, the oceans in general are not owned by anyone, removing that complication as well. Hence, it is understandable that people inclined to think about making surfaces more reflective in order to decrease future warming would at least consider the oceans. And, sure enough, the concept was under consideration as early as the mid-1960s.

In 1965, the U.S. President's Science Advisory Committee wrote a report, *Restoring the Quality of Our Environment*, in which they analyze a scheme to disperse buoyant reflective particles over the ocean in order to modify the ocean albedo. They conclude that at a cost of $500 million annually, the albedo could be changed by 1%.[41] Unfortunately, although the reflection of sunlight back to space would help cool the planet, the costs would extend far beyond the direct costs of $500 million per year in terms of the damage done to the ocean habitat. The buoyant reflective particles would certainly restrict the amount of sunlight penetrating into the oceans, would likely become a choking hazard for marine life, and could also become a hazard on land after washing up onshore. Such damages in total would almost certainly outweigh any benefit of the particles in modifying the ocean albedo.

An alternative scheme proposed to increase the ocean reflectivity is to cover large areas of the oceans with a single gigantic floating reflector. This suggestion is even more horrifying than the suggestion of huge numbers of smaller reflectors, as, similarly to the desert-covering scheme, it would put a cap over the ocean, even more thoroughly destroying the ocean habitat. MacCracken comments that the single floating reflector would "ultimately need to grow to roughly the size of a continent."[42] Schemes like this are so destructive to major portions of the global ecosystem that if the rest of the Earth system had a say, it would be understandable if it voted to eliminate mankind and let the remaining system evolve without us.

> Schemes like this are so destructive that if the rest of the Earth system had a say, it might well vote to eliminate mankind and let the remaining system evolve without us.

DAMMING THE BERING STRAIT

An environmentally gentler way to increase ocean reflectivity would be to take advantage of natural processes and contrasts, in particular the markedly higher reflectivity of ice versus liquid water. The Arctic sea ice reductions of the late twentieth and early twenty-first centuries have lessened the area of the highly reflective white covering that the ice provides, thereby retaining more solar radiation within the Earth system and contributing to additional warming. If the continued loss of the reflective Arctic ice cover could be slowed, halted, or even reversed, this could have the desired result of reflecting more radiation back to space without the need for an environmentally destructive artificial covering. However, no one has yet devised a scheme that would do this without being environmentally destructive in some other way.

The scheme most often mentioned as a possibility for retaining more of the Arctic ice is to dam the Bering Strait.[43] The object would be to prevent relatively warm Pacific waters from entering the Arctic and further melting the remaining sea ice. Considerably more warm water enters the Arctic from the Atlantic Ocean through the passages between Greenland and Europe than enters from the Pacific Ocean through the Bering Strait, but the passages between Greenland and Europe are far too broad and deep to make damming them feasible. Being narrow and shallow makes the Bering Strait considerably more amenable to damming.

Damming the Bering Strait, if successful, would indeed prevent Pacific waters from flowing into the Arctic and hence likely would result in retaining more of the Arctic Ocean ice cover, at least temporarily. However, damming the Bering Strait would also prevent the cold waters from the Arctic Ocean from flowing into the Bering Sea, with a consequent reduction in wintertime ice in the Bering Sea. This loss of ice in the Bering Sea might become only a minor consideration because with continued warming there might not be much wintertime ice in the Bering Sea anyway. However, damming the Bering Strait would also mean that all the many animal species that currently migrate through the strait would no longer be able to do so. This would tremendously disrupt the Arctic ecosystem and the Bering Sea ecosystem and would quite likely have major effects on lower-latitude marine ecosystems as well. For instance, it would necessarily mean adjustments for animals such as the beluga whales that annually migrate between the coastal waters off California and the coastal waters off northern Alaska. All these consequences to numerous ecosystems make the idea of damming the Bering Strait highly unappealing.

EARLIER SUGGESTIONS
FOR GEOENGINEERING THE ARCTIC

As opposed to trying to preserve the Arctic sea ice cover, several suggestions have been made in the past regarding how to decrease the ice cover. Prior to significant concerns about global warming, the possibility of a reduced Arctic ice cover was viewed quite favorably by many, as it would help to warm the very cold Arctic region and would be a boon for shipping.

Interestingly, in view of the previous section, with suggestions to dam the Bering Strait to increase the ice cover, one of the schemes to decrease the Arctic ice cover also involved damming the Bering Strait. This time, the idea was not to block warm water from entering the Arctic from the Pacific but rather to supplement the dam with a mechanism to pump water from the Arctic to the Pacific and thereby draw more of the warmer water in from the Atlantic, helping to melt the ice.[44]

Another scheme, repeated frequently as a major suggestion for climate modification on a large scale, was to cover the Arctic sea ice with soot or an alternative absorbing material, which would reduce rather than increase the region's reflectivity. The intention in that case was to melt the Arctic ice through the increased absorption of solar energy, for the dual purposes of creating an open passageway for shipping and warming the frigid bordering land areas. In the mid-twentieth century, the Soviets in particular considered such schemes to remove the Arctic sea ice,[45] hoping to lessen the severity of their bitterly cold winters and perhaps also to lessen drought conditions. Lessening drought was perceived to be a likely consequence of the increased evaporation expected from the less ice-cluttered Arctic and the subsequent increased regional precipitation.[46] Among the Soviet scientists considering such schemes was renowned climatologist Mikhail Budyko, who proposed using airplanes to dust the Arctic ice with soot, thereby encouraging melting.[47]

In the United States, Harry Wexler, the director of meteorological research at the U.S. Weather Bureau, also considered schemes to warm the Arctic and proceeded to quantify how much effort this would require. In a 1958 paper, Wexler estimated that blackening the Arctic ice (to decrease its reflectivity and retain more solar radiation) by covering it with black carbon to a thickness of 0.1 millimeters would require 1.5 billion tons of carbon and approximately 150 million airplane flights.[48] Not only would this be time consuming and expensive, but the amount of blackening would be reduced by wind erosion and any snowfalls subsequent to the laying down of the black carbon. As a result, Wexler turned to examining an alternative means of warming the Arctic, namely, setting off nuclear explosions within the Arctic Ocean to create infrared reflecting ice clouds in the Arctic atmosphere.[49] The reflection in this case would be reflecting infrared radiation from the Earth system back toward the Earth surface, thereby reducing the amount of infrared radiation escaping to space. Thus, the Arctic would be warmed through cloud modification rather than surface modification. In 1974, Will Kellogg and Steve Schneider alternatively mentioned detonating thermonuclear devices within the ocean to fragment the ice and to stir up and bring to the surface some of the warmer waters underneath.[50] (Throughout much of the global ocean, the warmest waters are in the top layer, but in the Arctic, the top layer is underlain by warmer waters advected into the Arctic from the North Atlantic.)

In a footnote to his 1958 paper, Wexler mentions that A. B. Markin of the Moscow Institute of Power suggested melting the Arctic ice by pumping water across the Bering Strait and that R. A. McCormick of the U.S. Weather Bureau calculated that it would take at least 100 years to succeed fully in eliminating the Arctic ice through this means. In contrast, Wexler estimated that the ice cloud he proposed could accomplish this feat in fewer than 10 years. Wexler wisely concluded his article by indicating the importance of more comprehensively predicting the results before acting on any of the proposed large-scale weather modifications "so as to avoid the unhappy situation of the cure being worse than the ailment."[51] I can heartily agree with him there.

Others estimated that an effective soot covering of the Arctic ice might be possible at a lesser effort than Wexler indicated. In particular, Halstead Harrison at the University of Washington calculated that 500 Boeing 747 aircraft could do the job in 50 days, with two flights a day by each aircraft. During these 50 days, 10 million metric tons of soot would be spread over 10 million square kilometers of ice, with a total cost of $2 billion[52] (still not cheap—or wise).

Turning to how effective an albedo (reflectivity) reduction might be, Gary Maykut and Norbert Untersteiner of the University of Washington in Seattle used a one-dimensional sea ice model to calculate that by lowering the albedo of the ice by 20%, the dusting of the ice cover could lead to the disappearance of the ice in just 3 years.[53] Maykut and Untersteiner also mention that some small-scale experiments were actually carried out on the ice in the mid-twentieth century, testing different dusting materials and reported by K. C. Arnold in 1961.*

Another suggested scheme to reduce the Arctic sea ice was to divert major rivers from flowing into the Arctic from either the Eurasian or the North American side. This would deprive the Arctic of a major source of freshwater, which not only freezes more readily than salt water[54] but also contributes to the strong "stratification" of the Arctic Ocean, in which the top layer is less salty and dense.[55] Stopping the freshwater inflow to the Arctic would lessen the salinity and density stratifi-

---

* Separately, a small-scale and very positive use of absorbing materials to reduce ice coverage was successfully carried out in 1903 for practical rather than climate purposes and in the Antarctic rather than the Arctic. Specifically, in 1903, German scientist and Antarctic expedition leader Professor Erich von Drygalski faced potentially dire conditions for himself and his crew as his ship the *Gauss* was beset in the ice off the coast of East Antarctica and all attempts to free it by blasting or sawing through the ice had failed. Fortunately, von Drygalski understood the effects of albedo and had his crew spread a layer of dark-colored trash over the ice along a 610-meter strip to the open ocean. Sure enough, the dark material absorbed the incoming solar radiation and resulted in the melting of the ice along the debris-covered strip, creating a convenient channel for the ship to pass through and safely emerge from the ice (Fogg 1992).

cations of the waters, thereby increasing the likelihood of vertical mixing of the waters. This would bring up to the surface some of the heat routinely brought into the deeper waters of the Arctic from the Atlantic. This heat would be available to warm existing ice and to hinder formation of new ice. I am not aware of any serious suggestions to follow through and divert the northward-flowing North American rivers from flowing into the Arctic, but suggestions for diverting some of the rivers in the former Soviet Union were seriously considered, even to the point of being included in the Draft Guidelines for the Soviet Union's Twelfth Five-Year Plan, covering the years 1986–1990.[56] Reducing the Arctic sea ice coverage was by no means the only goal; in fact, the primary goal was probably to use the water elsewhere, with an additional goal being to lessen the chronic flooding in western Siberia. Of most relevance here, however, are the potential impacts on the Arctic sea ice.

Long before the Twelfth Five-Year Plan, Knut Aagaard and Lawrence Coachman of the University of Washington concluded in 1975 that if the Russians were to carry through on suggestions to divert the Yenisei and Ob rivers so that they no longer flowed into the Arctic, the consequences could include prolonged Arctic ice-free conditions,[57] which, at the time, some people viewed favorably. (Notably, Budyko had already, in 1972, projected an ice-free Arctic by 2050 even without river diversions, based on CO_2-induced warming,[58] similarly to numerous projections in the early twenty-first century of an ice-free Arctic within the century.) However, several Soviet analyses reached very different conclusions, indicating that the river diversions would instead increase the ice coverage, and warned against the river diversions partly on that basis.[59] American Philip Micklin from Western Michigan University analyzed possible consequences of the river diversions with a conceptual systems analysis model and concluded that the river diversions likely would, overall, increase the ice cover but that the Arctic system is too complex to be certain.[60] In line with the complexity of the climate system, this instance is hardly unique in terms of having different researchers predict opposite results (more versus less Arctic sea ice) from the same cause (reduced river inflow), making the need for caution before implementing a climate-altering scheme all the more clear.

Although the river diversions might have been proposed largely to redirect water to where it was needed, the Soviets did additionally consider schemes aimed exclusively at climate amelioration. For instance, in an effort to warm Russia, they considered injecting metallic aerosols into near-Earth orbit, with the thought that aerosol rings around the planet would heat and illuminate the high latitudes[61] rather than reflect solar radiation and cool the planet, as in the schemes to counter global warming with aerosol injections.

The various schemes of the Soviets, Wexler, and others to decrease the Arctic ice cover found adherents in the 1970s, when the possibility of a coming ice age loomed as a concern in some people's minds and others unconcerned by an ice age could still see the advantages of using technology to lessen the bitter cold of the far north. Fortunately, no one successfully followed through on the schemes to decrease the ice cover, for if they had done so, global warming now would almost certainly be even worse than it already is. Correspondingly, now in the early twenty-first century, when the greater concern is continued warming, it will also be fortunate for future generations if no one follows through on any other inadequately thought-through climate-altering scheme.

DAMMING THE STRAIT OF GIBRALTAR

Like the Bering Strait, the Strait of Gibraltar is narrow enough and shallow enough for people to consider the technological possibilities of damming it and the resulting environmental consequences. In the case of the Strait of Gibraltar, Robert Johnson of the University of Minnesota in 1997 advocated not just studying the feasibility and consequences but actually proceeding to construct the dam.[62] Most surprisingly, the ultimate purpose of the proposed dam is to stave off the next ice age, despite the advocacy coming well after the scientific consensus had shifted to a major concern regarding continued warming. Although others besides Johnson have also said that Europe might cool while the globe as a whole continues to warm, these others generally do not project an imminent cooling so great that it brings on full-scale ice age conditions. However, Johnson does at least entertain that possibility.

Johnson expresses concern that continued loss of freshwater river inflow to the Mediterranean Sea because of human activities, combined with evaporation increases due to rising temperatures, will increase the salinity of the Mediterranean waters and thereby lead to greater outflow at the Strait of Gibraltar. In his scenario, the outflowing waters will sink to lower levels in the Atlantic; mix with the cold, deep Atlantic waters; move northward; and upwell in the vicinity of Ireland, after which they will continue northward and, more important, divert the water of the North Atlantic Drift extension of the Gulf Stream in such a way that the waters of the North Atlantic Drift will flow westward toward the Labrador Sea rather than northward into the waters west of Norway. The result will be a cooling both of the waters west of Norway and of northern Europe and a warming of the Labrador Sea and Baffin Bay to the south and west of Greenland. With the concomitant changes in atmospheric circulation, Johnson foresees moisture advection from the Labrador Sea and Baffin Bay region triggering ice sheet growth in Canada. Continued CO_2-induced warming of lower latitudes would cause increased moisture advection, feeding rapidly expanding ice sheets in both northern North America and Eurasia[63] (quite a proposed sequence).

To forestall the potential approaching ice age that he envisions, Johnson recommends halting the sequence of events at the critical adjustable point at the Strait of Gibraltar. The proposed dam would extend most of the way from Spain to Morocco, with a full-depth gap near the middle to allow an outward flow at approximately 20% of the late twentieth century's rate. The dam would be constructed with large rocks with a total volume of approximately 1.27 cubic kilometers, about 420 times the volume of the Great Pyramid.[64]

Issues regarding the proposed Strait of Gibraltar dam begin with the validity or lack of validity of the two premises that a Northern Hemisphere ice age is imminent and that damming the Strait of Gibraltar would prevent it. It is conceivable that both premises are correct, but with so much concern about global warming it would be very difficult to convince the scientific community (including me), the general public, or governmental authorities of the need to undertake an expensive construction project in the waters between Spain and Morocco in order to prevent the imminent arrival of an ice age farther north. Hence, until some substantial new glaciation begins, I have no fears that this particular project will go forward. If it did, other issues would center around what it might do to the ecosystems on both sides of the divide and how well the dam could withstand the various likely storm-surge and other pressures to which it might be subjected.

IRON FERTILIZATION OF THE OCEANS

Most of the schemes mentioned so far for counteracting global warming center on reducing the amount of solar radiation retained within the Earth system (by shielding the Earth or by increasing surface or atmospheric reflectivity) or on selectively changing ocean circulation (by damming the Bering Strait or the Strait of Gibraltar). A very different proposed tactic is to remove some of the atmospheric carbon dioxide that is helping to cause the warming. In particular, a leading plan along these lines is to increase the ocean uptake of CO_2 from the atmosphere through facilitating increased phytoplankton growth. More phytoplankton would mean more conversion of CO_2 by phytoplankton during photosynthesis. Hence, warming would be reduced by decreasing the concentration of the greenhouse gas CO_2 in the atmosphere.

In some regions, notably the Southern Ocean, a lack of iron is thought to be the current limiting factor in phytoplankton growth.[65] In these regions, adding iron to the ocean—a process called "ocean fertilization" or "iron fertilization"—would reduce the iron deficiency and thereby presumably increase phytoplankton growth and the uptake by the oceans of atmospheric CO_2.

John Martin of Moss Landing Marine Laboratories in Moss Landing, California, was, until his death in 1993, a primary leader in examining and advocating the potential value and feasibility of the iron fertilization possibility. Martin and his colleagues studied the phytoplankton/iron connection in various locations of the global ocean[66] and in particular tested the hypothesis that in the Southern Ocean, the lack of iron is the limiting factor restraining phytoplankton growth.[67] The Southern Ocean work involved sampling water at four stations in the Ross Sea, adding iron to some samples and not others at each station, and tracking the changes in the bottled water samples. After analyzing the results, they concluded that in the Southern Ocean, iron is indeed the limiting factor in phytoplankton growth. Further, they estimate that the amount of iron required for effective large-scale iron fertilization of the Southern Ocean would be 100,000 to 500,000 tons annually, an amount they consider feasible for large-scale implementation. Still, they caution that although the amount of iron required is feasible, distributing it appropriately in the ocean might not be.[68] Elsewhere, Martin estimated that fertilizing the Southern Ocean with 430,000 tons of iron annually could lead to the annual removal of 3 gigatons of carbon from the atmosphere.[69]

Others question both the amount of iron that would be required, arguing that Martin might have greatly underestimated it, and the potential impact on atmospheric CO_2 levels.[70] For success, the carbon not only needs to be taken out of the atmosphere by the photosynthesizing phytoplankton but also needs to get transported downward in the ocean, away from the surface layer,[71] and this is not something that would necessarily happen. Further, Wally Broecker points out that one of the unintended consequences of iron fertilization could be that portions of the ocean underlying the fertilized region would become anaerobic, which would be extremely damaging to some of the marine life inhabiting these regions.[72] Another potential complication, pointed out in a 2007 review in *Science*, is that the added iron could induce increased ocean emissions of methane and/or nitrous oxide. As both of these are greenhouse gases as well, there is a possibility that the iron additions would result in an increase rather than a decrease of the greenhouse effect, even if they succeeded in the primary objective of reducing atmospheric CO_2.[73] Summarizing the broad picture, George Monbiot concludes that not only would iron fertilization not accomplish its intended goal, but it would foul up ocean ecology and cause increased rather than decreased global warming.[74]

Despite the issues raised, the concept of countering global warming by iron fertilization of ocean regions with iron deficiencies has found sufficient favor in the international oceanographic community that 12 major field programs carried out intentional (although geographically limited) iron fertilization experiments in the period 1993–2007. These experiments make iron fertilization unusual in being a geoengineering scheme that has actually been tested, though only on a small scale. Results from these field programs have established that iron supply does appear to limit biological productivity in perhaps a third of the global ocean[75] but have also left open many uncertainties regarding both the level of impact that iron fertilization could have on atmospheric CO_2 amounts and the unintended changes that would take place in ocean ecosystems.[76]

The uncertainties and potential unintended consequences make it all the more scary that commercial companies are now looking toward the possibility of seeding portions of the ocean with iron and selling "carbon credits" for the amount of carbon calculated to have been sequestered from the atmosphere as a consequence of the iron fertilization.[77] Once commercial companies feel that they can make money doing this, there could be a great deal more iron fertilization taking place than is warranted, with potentially very negative impacts on the ocean. The same comment applies also to other ocean fertilization schemes, such as the proposed fertilization with nitrogen-rich urea under development at the Australia-based Ocean Nourishment Corporation[78] or earlier proposals to fertilize with phosphate or nitrate.[79] It is extremely important that the results of small-scale studies be analyzed thoroughly before any large-scale efforts are implemented. Until proved otherwise, pouring iron, urea, phosphate, nitrate, or anything else into the ocean should be regarded primarily as polluting the ocean environment, with serious potential additional consequences, and should be avoided.

> The potential unintended consequences make it all the more scary that commercial companies are now looking toward the possibility of seeding portions of the ocean with iron and selling "carbon credits" for doing this.

INJECTION OF ADDITIONAL OZONE INTO THE STRATOSPHERE

Geoengineering goals are not confined to adjusting temperatures. Another major articulated goal is to use geoengineering to suppress the destruction of the Earth's protective stratospheric ozone layer. While most discussions of stratospheric ozone reductions focus on the Antarctic stratosphere and the Antarctic ozone hole (e.g., chapter 4), comparable ozone reductions might eventually occur elsewhere as well, especially in the Arctic stratosphere. Most crucially, if stratospheric temperatures continue to decrease (while tropospheric temperatures increase), the stratospheric temperatures in the Arctic could get cold enough to form more of the polar stratospheric clouds that serve as breeding grounds for the chemical reactions leading to ozone depletion.

One way to attempt to suppress the destruction of the ozone layer would be by replenishing the lost ozone through producing vast amounts of ozone and transporting them to the stratosphere, a suggestion made repeatedly according to the president of the U.S. National Academy of Sciences, Ralph Cicerone, who explicitly opposes the idea.[80] Without a further scheme to limit the destructive chemical reactions, this scheme would require continual replenishment of the ozone, requiring continual funding as well.

Although adding a gas to the atmosphere to replenish losses of that gas from the atmosphere seems substantially less objectionable than adding aerosols, the scheme to add ozone to the stratosphere would still be expensive and would still have the potential for unintended consequences, especially in view of all the chemical reactions that occur in the stratosphere, powered by the Sun's radiation. Further, any of the ozone that made its way downward to near-ground level would be an undesired pollutant. From my perspective, the preferred solution to stopping the stratospheric ozone destruction remains the route taken in the Montreal Protocol and follow-on agreements, that is, to halt the insertion into the atmosphere of the chemical compounds that are instigating the ozone destruction. That should, if the consensus reasoning is correct, solve not only the existent problem in the Antarctic stratosphere but also the potential problem throughout the global stratosphere.

INJECTION OF ETHANE OR PROPANE INTO THE ANTARCTIC STRATOSPHERE

Rather than injecting ozone itself into the stratosphere to replenish the Earth's protective ozone layer, in 1991 Ralph Cicerone of the University of California, Irvine, Scott Elliott of the Los Alamos National Laboratory, and Richard Turco of the University of California, Los Angeles (UCLA), considered injecting ethane or propane to trigger a sequence of chemical reactions that would prevent the damaging ozone-destroying reactions. Specifically, the injection would be into the Antarctic stratosphere, and the focus would be on the Antarctic ozone hole. Using a detailed numerical model of stratospheric photochemistry and multiple scenarios, the group simulated some situations in which ozone loss is greatly reduced, as desired. But although they indicate in their resulting journal article that the calculations are encouraging, they are rightly cautious and explicitly state that many questions need to be addressed before attempting the ethane or propane injections.[81] Among the technical issues, the numerical model assumes an even distribution of the ethane or propane through the selected portion of the atmosphere, possibly an insurmountable practical challenge. The researchers estimate that in the case of ethane, the effort would require approximately 50,000 tons of ethane, distributed by several hundred large airplanes, with no certainty that the mixing would prove adequate. Furthermore, as in the case of other model predictions, the possibility exists that the model may be incomplete to a degree that would invalidate the projections. Although the model used by Cicerone and his colleagues included approximately 130 chemical reactions, the authors recognized that there could be other reactions also critical to the formation of the ozone hole.[82]

Fortunately, in the case of the ethane/propane scheme for reducing ozone depletion, the team that introduced and worked on this scheme, consisting of Cicerone, Elliott, and Turco, recognized the incompleteness of the numerical model and cautioned against any real-world testing of the idea as they proceeded to enhance the modeling effort, joined by colleagues Katja Drdla and Azadeh Tabazadeh of UCLA. In the expanded group's enhanced study, they found that by adding to the calculations one particular additional chemical reaction, known also to occur in the polar stratosphere, the net effect of the ethane injections, at least in some simulations, was to increase rather than decrease the undesired ozone depletions. They wisely concluded that since adding one additional photochemical reaction to the 130 or so that they had already included resulted in such a major revision to the earlier conclusions, their model and other state-of-the-art models are simply not adequate for reliably predicting the ramifications of attempted interventions.[83] This remains true not only for the ozone depletion studies but for many others as well and is an important caution to keep in mind; the complexity of the Earth system, including its entangling profusion of chemical reactions, is far beyond anything anyone has yet come close to modeling in full.

Regarding the potential effectiveness of the ethane/propane scheme, I am also struck by the result[84] that in one set of simulations, adding 1.8 parts per billion of hydrocarbon had the undesired effect of increasing the ozone depletion, but adding 3.6 parts per billion of the same hydrocarbon significantly reduced the ozone depletion. The sensitivity of the results to the amounts injected suggests that the appropriate amount probably also depends on the precise chemical composition of the stratospheric layer at the time of the injection, and this is likely not known to the required accuracy. Hence, the expensive, time-consuming ethane or propane injections might produce the opposite result—increasing ozone depletion—to the desired one. Further, it could precipitate additional unforeseen and undesired chemical reactions.

MODIFYING THE EARTH'S ORBIT

Considering the much longer term, in 2001 Donald Korycansky of the University of California Santa Cruz and colleagues Gregory Laughlin of NASA Ames Research Center and Fred Adams of the University of Michigan proposed a novel solution for adjusting for the Sun's increasing energy output, which is calculated from solar physics to be very gradually rising, leading to the Sun's eventual fate, billions of years from now, of bursting into a red giant star and engulfing the Earth and other inner planets before shrinking into a white dwarf, a very small and extremely dense star. Projections suggest that in comparison with today, the Sun will be 11% more luminous about 1.1 billion years from now, 40% more luminous about 3.5 billion years from now, and 2.2 times as luminous by about 6.3 billion years from now. Korycansky and his colleagues feel that an 11% increase would have a catastrophic impact on life on Earth and that a 40% increase would likely bring an end to life on Earth altogether.[85]

To forestall the eventual disaster caused by the ever more luminous Sun, the Korycansky group suggests harnessing a large asteroid, comet, or other object to nudge the Earth outward, through the gravitational attraction between the Earth and the chosen object, and thereby move the Earth into an orbit somewhat farther from the Sun. Performing such an orbit correction approximately every 6,000 years could extend for a considerable length of time the viable life span of the Earth's surface biosphere, perhaps by as many as 5 billion years. By 6.3 billion years from now, the proposed incremental outward nudging would place the Earth at 1.5 times its current distance from the Sun, roughly at the same distance as the current orbit of Mars. The expanded orbit would cancel the Sun's luminosity increase so that the Earth would still be intercepting approximately the same flux of solar energy as it does now.[86]

As explained by Korycansky and his colleagues, the nudging object or objects would likely come from the Oort Cloud or Kuiper Belt at the outer regions of the solar system, and the object would serve as a catalyst, transferring to the Earth orbital angular momentum and energy from Jupiter. Korycansky and colleagues include detailed calculations, mention ways in which the scheme might be improved, and conclude not only that the scheme is feasible but also that we are close to already having the technical expertise to carry it out. At the same time, they recognize that the calculations incorporate simplifications that would need to be overcome and that there are some quite serious potential risks, like having the nudging object instead collide into the Earth, losing the Moon as a satellite of the Earth, or seriously speeding up the Earth's rotation rate.[87] Still, when considering the very long term (billions of years) and knowing with fair certainty that the Earth will become uninhabitable at its current distance from the Sun well before it would become uninhabitable farther out, the concept of eventually somehow adjusting the orbit outward does make sense, leaving it to future, much more advanced generations, millions of years from now, to devise the exact means of doing so safely.

Although not mentioned in the 2001 article by the Korycansky group, the concept of enlarging the Earth's orbit could equally be proposed to adjust for the warming from human activities: bump the Earth outward, by whatever means, and thereby reduce the Sun's energy supply to the Earth/atmosphere system. Such a scheme would definitely help cool the Earth. However, despite supporting the concept of systematically adjusting the Earth's orbit outward in the extremely distant future, to adjust for the apparently inevitable increase in solar luminosity, I would strongly oppose a similar adjustment to correct for human-facilitated warming. There is a huge contrast between the two situations. If the current understandings of the Sun's future evolution are correct and no method is devised to change that evolution, then in the very long term, if not earlier destroyed by war, asteroids, or other means, the Earth is unalterably doomed by the relentlessly increasing luminosity of the Sun and the natural processes that will lead to the Sun's eventual dramatic expansion and subsequent contraction. Moving the Earth outward could be a way to extend the habitability of the planet by billions of years. In contrast, in the much nearer term (on the order of hundreds rather than billions of years), the Earth is perfectly habitable where it is. Furthermore, with the Sun's current energy output and with human societies (and the rest of life) so dependent on energy supplies, it would be far wiser to stay in the orbit that has served the Earth well for billions of

> Instead of devising means for getting away from the Sun, we should be devising more effective means of capturing and making use of more of the Sun's energy, which is so freely given to us, with no charge either for the energy or for its transport across 93 million miles of space.

years, keeping exactly the distance we already are from our main energy source. Instead of devising means for getting away from the Sun, we should devise more effective means of capturing and making use of more of the Sun's energy, which is so freely given to us, with no charge either for the energy or for its transport across 93 million miles of space. The Sun is by far our biggest and most reliable energy source as well as being our biggest and most reliable source of visible light, and we should embrace what it provides. We need more creativity in capturing and using the energy the Sun sends our way, not more creativity in moving away from that energy. Finally, consider what a mistake it would be if we opt for an orbit adjustment and an error is made, sending spaceship Earth careening too far in the outward direction.

FURTHER READING

Readers interested in learning more about the many geoengineering schemes being proposed have a wide range of options for where they can find additional information. A simple Internet search on "geoengineering" and "climate" yields many thousands of links, including to video and audio as well as written sources. Of the many traditional sources, Keith's 2000 review article mentioned earlier provides a scholarly scientific review, with many references where further details can be found.[88] Keith had earlier, with Hadi Dowlatabadi, provided a shorter review titled "A Serious Look at Geoengineering" and summarizing many of the schemes proposed as of the early 1990s.[89] James Fleming in 2007 published in the *Wilson Quarterly* an engaging and illustrated article for the general public titled "The Climate Engineers" that gives a historian's view of the unwarranted confidence of some of the most assured of the geoengineering proponents.[90] David Victor delves into issues surrounding the need to establish appropriate norms to govern the implementation of geoengineering in an article titled "On the Regulation of Geoengineering" published in 2008 in the *Oxford Review of Economic Policy*,[91] and Alan Robock offers a succinct list, "20 Reasons Why Geoengineering May Be a Bad Idea," published in 2008 in the *Bulletin of the Atomic Scientists*.[92] Terry Barker and others include only two pages (pp. 624–625) specifically on geoengineering in a much longer article, but those two pages warrant mention because they provide the 2007 published perspective of the highly esteemed Intergovernmental Panel on Climate Change (IPCC).[93] The IPCC reference describes iron fertilization, a solar shield, aerosol injection into the stratosphere, and albedo enhancement of clouds all as "apparently promising techniques" although cautions against the inherent risks.[94]

SLIPPERY SLOPE OR
REASONABLE ACCOMMODATION?

Contemplation of geoengineering has a long history, going back at least to 1901, when Nils Ekholm suggested at the end of a 62-page article on past climate variations that humans may be able to forestall a new ice age by judiciously maintaining sufficient carbon dioxide in the Earth's atmosphere.[95] Both he and his fellow Swede Svante Arrhenius (about whom more appears in chapter 9) were openly appreciative of the role played by anthropogenic atmospheric CO_2 increases in warming the planet. By 1974, Will Kellogg and Steve Schneider of NCAR were able to list the following schemes that had been proposed by that time: spread black particles over the Arctic ice, dam the Bering Strait, detonate thermonuclear devices in the Arctic Ocean, or divert north-flowing rivers, keeping them from entering the Arctic Ocean, all with the express intent of reducing or eliminating the Arctic sea ice cover; strategically pump cold water on the ocean surface to steer hurricanes away from populated areas; artificially create a stratospheric dust layer; create artificial lakes in Africa; and cut down tropical forests.[96] The last one in particular seems startling given all the reasons why the loss of tropical rain forests in the intervening years has been so damaging. By now, many additional schemes have been proposed as well, including, in addition to those already discussed, sequestering excess CO_2 as ocean bicarbonate, sequestering excess CO_2 by increasing ocean alkalinity, sequestering excess CO_2 in land formations, capturing CO_2 from the air with plastic mesh sheets, capturing carbon in land-based ecosystems, and artificially creating reflective clouds.

Kellogg and Schneider warn against pursuing any of the schemes that they mention in their 1974 publication until their long-term effects can be adequately predicted. They further caution that while a scheme might have a desirable impact on a targeted regional climate, at the same time it would likely have undesirable impacts elsewhere.[97] These types of warnings have been standard fare in discussions of geoengineering, and reasoned analysis has generally prevailed.

However, as of the early twenty-first century, the tone has changed at least somewhat. More people are taking an interest in geoengineering, and more seem amenable to considering the actual implementation of one or more of these schemes within the next few years. Additionally, the former standard-fare warnings against geoengineering are now sometimes omitted, with statements instead explaining that the negative consequences might not be so bad. For a case in point, in an article advocating a strategy combining mitigation (reducing greenhouse gas emissions) and geoengineering, Tom Wigley of NCAR writes that injecting aerosols into the stratosphere at a level similar to the Mount Pinatubo emissions should "present minimal climate risks" because, after all, the Mount Pinatubo eruption "caused detectable short-term cooling . . . but did not seriously disrupt the climate system."[98] The analogy is not ideal, as the Mount Pinatubo eruption occurred in June 1991 and was not repeated in later years, whereas the proposed aerosol injections would occur repeatedly. Still, the key point here is the shift from explicit warnings against geoengineering to a statement saying that the potential negative side effects are perhaps not a serious concern.

The combination of an increased interest in geoengineering possibilities, the availability of technologies that now make some geoengineering schemes feasible, and the increased confidence of some scientists and policymakers in themselves and their predictions (see also chapter 11) makes it increasingly likely that large-scale geoengineering might be undertaken in the fairly near future. In fact, in modeling the possible impacts of aerosol injections into the stratosphere, at least two published studies assume hypothetically that these intentional injections begin in 2010.[99]

In working on this book, I found myself at moments becoming more rather than less inclined to consider that one or more of the geoengineering schemes might have some merit if indeed the situation becomes as dire as some people predict. I found this especially so in considering small-scale versions of the large-scale proposals. For instance, despite my horror at the thought of covering huge desert areas with white plastic, when I read the numbers for the impact of a small-scale version, I found myself much less adamant in my objections to the small-scale version than in my objections to the large-scale version. According to Gaskill, a square mile of covered desert would offset the emissions from 7,000 sport-utility vehicles (SUVs) over a 15-year period.[100] Somehow in that context (or rather its adjustment from SUVs to a far larger number of energy-efficient cars, as I do not feel that we should destroy deserts in order to allow for SUVs), it seemed that a square mile of destruction might become acceptable. However, I fear the slippery slope of accepting 1 square mile, then 10 square miles, then 100 square miles, and finally entire deserts. We should be able to do better than destroy other ecosystems in order to maintain our energy-consumptive lifestyles.

Much as I hope that we will be able to avoid large-scale geoengineering of the type discussed in this chapter, I recognize the value of analyzing geoengineering possibilities and their consequences. After all, whether caused by humans or not, changes could conceivably occur in the Earth system that would be unbearable or nearly so. In such an eventuality, having studies available that lay out potential schemes and their consequences could reduce the possibility of jumping too quickly to a proposed solution without adequately considering the consequences. Further, having and discussing such studies should increase the chances that better alternatives will be brought forward. For instance, with regard to the example in the previous paragraph, surely before we place reflective coverings over deserts or oceans, we should consider placing them where they will do less ecological damage and might even have additional benefits. A seemingly obvious choice would be to place the reflective coverings on the roofs of buildings in warm-weather climates (or at least on the new buildings being built in such climates, as retrofitting old buildings will require decisions by the individual owners, who will need to weigh costs as well as environmental concerns). Placing the reflective covers on the roofs of buildings would not displace ecosystems in the way that placing them on deserts or oceans would, and placing them on buildings would have the significant additional advantage of reducing the need for air-conditioning, thereby increasing comfort levels and/or reducing air-conditioning costs and emissions.

Just as I and other scientists have different and changing perceptions of geoengineering, discussion of geoengineering possibilities can generate vastly different responses within the general public as well. Concern about the very substantial potential dangers of geoengineering can lead some people to become considerably more serious about reducing greenhouse gas emissions and lessening the unintentional human impact on climate. Other people, however, are more likely to view geoengineering as a potential "cure-all," concluding that we can do whatever we want to the climate system, as our engineering prowess will allow us to undo any damage that might arise. This latter view would give us free rein to continue unabated our emissions to the atmosphere and oceans, unfairly leaving to future generations the task of fixing the resulting problems. This concern has been voiced before, at one level or another,[101] and I reiterate it here with the firm warning that geoengineering not only will not be able to undo whatever damage we might inflict but also has the potential of creating far greater problems.

CHAPTER 8

The Record on Smaller-Scale Attempted Modifications

The record so far of purposeful modifications to weather and/or climate is not encouraging. Despite numerous attempts and impressive large-scale efforts like Project Cirrus, the National Hail Research Experiment, and Project Stormfury, no one has yet developed any consistently reliable technique to enhance rainfall during droughts, suppress hail, tame hurricanes, or otherwise modify the weather in ways desirable for local communities or larger-scale regions.

IN CONSIDERING geoengineering, as in so much else, it is important not to ignore the past. Although no global or continental-scale geoengineering programs have yet been implemented for modifying weather and climate, many smaller-scale local and regional weather modification schemes have been attempted. Here as elsewhere, learning from history is far easier and wiser than repeating similar mistakes and learning from those.

Overall, the history of attempted weather modification is not impressive in terms of accomplishing desired goals. This adds yet another reason, beyond the danger of unintended consequences (chapter 6), to be cautious regarding geoengineering and its potential. Geoengineering the entire planet would be a gigantic step beyond small-scale weather modification. The failure to accomplish the intended goals could be enormously expensive, even if we were fortunate enough not to have any all-too-likely disastrous unintended consequences.

Besides the issue of scale, geoengineering and weather modification differ in that weather modification is generally aimed at modifying nature, while the proposed geoengineering schemes are more generally directed at modifying human impacts on climate. Still, both are aimed at intentional modification of weather and/or climate, and hence the success or failure of the weather modification attempts are relevant to considerations of future geoengineering for climate modification. Three key weather modification goals are considered here: rainmaking, hail suppression, and taming hurricanes.

RAINMAKING, OR ENHANCED PRECIPITATION

Among the most common goals of weather modification is rainmaking. Historically, rainmaking attempts have ranged from prayers and rain dances to sophisticated cloud-seeding experiments. Although the attempts extend back hundreds, likely thousands, of years, the emphasis here is on the scientifically based attempts in the twentieth century.

Two countries in which substantial scientifically based efforts at rainmaking took place during the twentieth century were the former Soviet Union and the United States. In the former Soviet Union, considerable activity was under way as early as the 1930s. In fact, an Institute of Rainmaking was established in Leningrad in 1932, and cloud-seeding experiments using calcium chloride were undertaken from 1934 to 1939. After World War II, additional experiments were undertaken using dry ice (solid carbon dioxide) in 1947 and silver iodide in 1949, and interest continued to be high during the 1950s and early 1960s.[1] In 1961, the Twenty-Second Congress of the Soviet Communist Party highlighted climate control as one of the most urgent problems for scientific attention. The Soviet interests encompassed a wide range of desired modifications of both climate and weather, including warming the Russian climate, most specifically through removing the Arctic sea ice cover, as discussed in the previous chapter, and developing the capability to produce rain at will.[2]

In the United States, scientific experiments on the effects of cloud seeding began in earnest at the General Electric Research Laboratory in Schenectady, New York, in 1946, prompted by the 1946 discovery by General Electric's Vincent Schaefer that ice crystals formed after he inserted dry ice into a small freezer unit containing supercooled cloud droplets (i.e., cloud droplets at a temperature below 0°C). Schaefer had been studying ice crystals in connection with such cold-weather problems as the icing of aircraft and had used a small refrigeration unit to undertake experiments on supercooled clouds. In the experiments, the unit was supercooled down to temperatures ranging from approximately –20°C at the bottom of the unit to –10°C at the top. Through more than 100 experiments, including the insertion of various types of particulate matter into the refrigeration unit, the supercooled clouds never formed ice crystals until a piece of dry ice was inserted, then, spectacularly, within 10 seconds the entire supercooled cloud consisted of ice crystals. Schaefer immediately recognized the potential for practical applications and concluded his initial scientific announce-ment of his discovery with mention of plans for testing the possibility of modifying atmospheric clouds by dropping dry ice into the cloud from an overflying aircraft.[3]

On November 13, 1946, Schaefer carried out tests in the atmosphere by drop-ping six pounds of dry ice from a small Fairchild airplane into a supercooled alto-stratus cloud having a temperature of −18.5°C.[4] The results of the cloud seeding were dramatic, as cogently summarized by Schaefer in the following comment in his laboratory notebook: "It seemed as though [the cloud] almost exploded, the effect was so widespread and rapid."[5] He continued his laboratory experiments and established that billions of ice nuclei could be generated with the insertion of a total of only 1 gram of dry ice.[6]

Schaefer's colleague at General Electric, Bernard Vonnegut, systematically searched for substances crystallographically similar to ice that might induce the conversion to ice crystals at higher temperatures, closer to the freezing point, than seemed possible with dry ice. Vonnegut identified both lead iodide and silver io-dide as promising substances. Subsequent experimentation successfully established that insertion of silver iodide could create ice crystals at temperatures as warm as −4°C. Furthermore, Vonnegut described the construction of a smoke generator that produced 10 trillion ice nuclei in 1 second from 1 milligram of silver iodide.[7] This and other work at General Electric suggested the possibility of artificially in-ducing precipitation from supercooled clouds by inserting either dry ice or silver iodide, which, if everything worked as planned, would initiate the formation of ice crystals that would then grow further by vapor deposition. Everything was limited by the amount of water vapor in the cloud, but getting whatever water was there to fall as rain could be a major benefit for water-strapped land beneath the cloud.

Besides Schaefer and Vonnegut, another key member of the General Electric team was Nobel Laureate Irving Langmuir, a visible spokesperson expounding the potential feasibility and benefits of large-scale weather control, including bringing rain to arid regions, elimi-nating damaging ice storms, and taming hurricanes. However, others at General Electric were fearful of lawsuits in the event of weather-related damages that could be attributed (rightly or wrongly) to the weather modification experiments, and Langmuir's team was instructed to shift from their publicized cloud-seeding experiments to work instead with the U.S. military's classified Project Cirrus.

> The nation which first learns to plot the paths of air masses accurately and learns to control the time and place of precipita-tion will dominate the globe.
> —General George C. Kennedy[8]

Under the military's Project Cirrus contract, General Electric continued to undertake research into cloud physics and chemistry and further investigated the possibility of cloud modification for military purposes, which include not only bringing rain to areas in need of it but also causing heavy rain in order to disrupt the activities of the enemy, causing rain to precipitate out early prior to reaching enemy agricultural fields, and, even more nefarious, using precipitation to deliver biological or radiological agents.[9] Taming the weather at the home base is of course also of both military and civilian importance, for instance, for keeping airfields free of fog and aircraft icing. General George C. Kennedy, commander of the Strategic Air Command, stated the military interest succinctly in 1947: "The nation which first learns to plot the paths of air masses accurately and learns to control the time and place of precipitation will dominate the globe."[10]

As part of Project Cirrus, cloud-seeding experiments were carried out at several locations, among the most important being experiments near Albuquerque, New Mexico, using a ground-based silver iodide generator. From December 1949 to mid-1951, the New Mexico experiments were performed on a weekly schedule of cloud seeding 8 hours a day on Tuesday, Wednesday, and Thursday and no cloud seeding on Friday through Monday. When several regions to the east of New Mexico, including the Ohio River basin, exhibited 7-day periodicities in their precipitation, Langmuir and his colleagues attributed this as well as rain within New Mexico to the New Mexico cloud seedings. Others were more skeptical, pointing out other 7-day periodicities in the meteorological records of the previous half century.[11] Here they encountered a common difficulty in Earth science studies, namely, how to establish cause and effect with no twin Earth to check out what would have happened in those weeks in the absence of the cloud-seeding experiments.

Approximately 180 field experiments on clouds were carried out between 1947 and 1952 as part of Project Cirrus. The mixed results included enough instances of apparent at least partial success in affecting the clouds and the rainfall from them that the U.S. military continued funding research into cloud and weather modification,[12] and people outside the military also became more interested in the possibilities. Among the U.S. military follow-ons to Project Cirrus in the 1950s were separate efforts by the Office of Naval Research, the Army Signal Corps Engineering Laboratories, an air force contract with the University of Chicago, an air force contract with Stanford Research Institute, and an army contract with Arthur D. Little, Inc. The results were inconclusive and failed to establish new military applications.[13]

Cloud seeding was used for military purposes by the United States during the Vietnam War in the 1960s and 1970s, with annual budgets of approximately $3.6 million in some years and a total of over 2,600 cloud-seeding airplane flights.[14] In particular,

Operation POPEYE performed aircraft-based cloud seeding with silver iodide and lead iodide in 1967–1972 in an attempt to increase rain along the Ho Chi Minh Trail and turn the trail into a muddy mess. As is typical, some claimed that the cloud seeding increased the rain amounts, while others (including General William Westmoreland, commander of the U.S. forces in Vietnam from 1964 to 1968 and U.S. Army Chief of Staff from 1968 to 1972) disagreed. In any event, the operation was carried out under extreme secrecy, and apparently no scientific data were collected.[15]

When the military-based weather modification efforts in Vietnam became public knowledge after columnist Jack Anderson broke the story in March 1971, a public outcry ensued. Although the argument that enhancing rainfall was more humane than dropping napalm had some merit, the stronger sense was that using weather modification techniques for military purposes was immoral.[16] With pressure from the public, the press, and Congress, President Nixon directed the National Security Council to undertake a study regarding whether there should be international restrictions on environmental warfare. The committee report in May 1974 provided three options, the two extremes being 1) no restrictions of any type and 2) quite comprehensive restrictions on all hostile uses. The Nixon administration opted instead for the middle position, and that one was essentially agreed on by President Nixon and Soviet General Secretary Leonid Brezhnev at their Moscow Summit on July 3, 1974, when they signed a joint agreement stating the desire to limit environmental modification for military purposes when the impacts would be "widespread, long-lasting and severe."[17] This wording permitted attempted rain-making, fog dispersal, and other tactics under many circumstances.

Recognizing the loopholes in the Moscow Summit agreement, the Soviet Union took the initiative and drafted a statement presented to the United Nations in September 1974 to outlaw all environmental modifications for military purposes. The United States insisted on modifying the statement to include the qualifiers "widespread," "long-lasting," and "severe," after which the resulting statement, named the UN Convention on the Prohibition of Military or Any Other Hostile Use of Environmental Modification Techniques, was finalized in May 1977 and opened for signatures. It went into effect on October 5, 1978, after ratification by 20 nations.[18] Although the various qualifiers limited its applicability, the convention did put the United Nations on record as opposing weather modification techniques for military purposes, at least when they would have severe long-lasting effects.

In the meantime, private entrepreneurs were setting up and operating cloud-seeding businesses, starting in the late 1940s and spurred on at that point by the early work at General Electric. In one such case, Wallace Howell was hired by New York City to increase the rainfall sufficiently to fill the city reservoirs. The city was initially pleased with the results but changed its mind when in 1951 it faced 169 claims from communities and individuals for flood and other damages. At that point, the city officials reversed their stance and attempted to establish that the cloud seeding had been ineffective. No damages were awarded for any of the claims, but there was a permanent injunction against New York City's engaging in further cloud seeding.[19]

Another private entrepreneur, Irving Krick, conducted cloud seeding for wheat farmers, ranchers, and others in the western United States during the regional drought of the early 1950s. The Bureau of Reclamation credited his efforts with increasing river flow by 83%, but the U.S. Weather Bureau discredited this and other claims. The many controversial results here and elsewhere led in 1953 to the formation by the U.S. Congress of an Advisory Committee on Weather Control. The committee issued a cautiously optimistic report in 1958, indicating that 10% to 15% increases in rainfall had been induced through cloud-seeding efforts in the western United States and recommending increased governmental funding of basic meteorological research. Following these recommendations, the U.S. National Science Foundation became the lead federal agency in the U.S. funding research into weather modification.[20]

The United States was hardly the only country testing cloud seeding and other weather modification tactics for military and civilian purposes. The extensive work in the Soviet Union—with considerably more people and funding involved than in the United States—has already been mentioned.[21] Although some of the Soviet attempts at cloud seeding with crushed carbon dioxide (dry ice) dispersed from aircraft in the 1960s appeared to increase the rainfall amounts, the increases (on the order of about 10%) were insufficient to make the seeding economically viable. In contrast, seeding with dry ice became routinely used in several Soviet airports for the alternative purpose of dispersing supercooled fog. More success was also found in hail suppression, discussed in the next section, than in rainfall enhancement.[22]

Commercial companies sprang up around the world intent on seeding clouds in order to affect rainfall, fog, and hail, and all of the following became important topics of scientific research: the physics of clouds, the factors affecting which clouds would be the best candidates for successful seeding, and the potentially best seeding substances for different cloud types (dry ice and silver iodide have become popular for cold clouds, while water and hygroscopic aerosol particles, such as salt, are popular for warm clouds). Much was learned scientifically, but this did not convert into unqualified success in controlled weather modification.

As with the Langmuir experiments in the 1940s and early 1950s, subsequent experiments also have elicited widely varying views of how successful the experiments were. Among the most promising results came from two extended experiments carried out in Israel in 1961–1967 and 1969–1975, with precipitation enhancements estimated at 15% to 22% from the first experiment and 13% to 18% from the second experiment.[23] However, a reanalysis of these results by Arthur Rangno and Peter Hobbs of the University of Washington in 1995 concluded that "neither of the Israeli experiments demonstrated statistically significant effects on rainfall due to seeding."[24] The detailed analyses by Rangno and Hobbs considerably tarnished the view of these experiments as successfully enhancing precipitation.[25] Similarly, encouraging results from cloud-seeding experiments undertaken near Climax, Colorado, in 1960–1965 (the Climax I experiment) and 1965–1970 (the Climax II experiment) from cold orographic clouds[26] were also reanalyzed by Rangno and Hobbs, again with far less encouraging conclusions, suggesting that the cloud seeding had little or no effect on the precipitation.[27]

In a review of a half century of cloud-seeding experiments carried out in Australia between 1947 and 1994, Brian Ryan and Warren King of Australia's Commonwealth Scientific and Industrial Research Organization found cloud seeding to have been ineffective over the plains of Australia but somewhat more successful in more mountainous regions, especially with clouds undergoing orographic uplift.[28] Discouragingly, the cases considered successful were limited to an extremely narrow range of conditions. In particular, the successes came with stratiform clouds undergoing orographic uplift, airflow from the southwest, and temperatures at the top of the cloud between −10°C and −12°C. Furthermore, even with all those limitations, the encouraging results occurred only in the mountainous regions of Tasmania, with considerably less convincing results in the mountainous regions elsewhere in the country.[29]

Moving to the global scale, in a 2001 article in the *Bulletin of the American Meteorological Society*, Bernard Silverman presents an assessment of four decades of efforts from locations around the world where convective clouds were seeded with dry ice and silver iodide in attempts at rainfall enhancement.[30] He applies the success criteria recommended in a 1998 American Meteorological Society (AMS) Policy Statement on Planned and Inadvertent Weather Modification, requiring both statistical evidence and physical rationales. Although some of the experiments Silverman summarizes had apparent positive results, these results were not confirmed in subsequent experiments and did not satisfy the 1998 AMS success criteria. Among the experiments with the most encouraging results were the previously mentioned 1961–1967 and 1969–1975 experiments in Israel, but even these failed to meet the success criteria, and attempts to reproduce at least their limited successes in a third Israeli randomized experiment in 1976–1991, in a World Meteorological Organization Precipitation Enhancement Project experiment in Spain in 1985, and in an experiment in Puglia, Italy, in 1988–1994 all failed.[31] Similarly, some apparent success in the Florida Area Cumulus Experiment 1 (FACE 1) in 1970–1976 was not confirmed by the much more disappointing results in FACE 2 in 1978–1980, and partial success in Texas in 1986–1990 and 1994 was not confirmed by experiments in Thailand in 1994–1998. Seeding in experiments in 1965 in the Caribbean and in 1968 and 1970 in South Florida appeared to cause the clouds to grow taller but, unfortunately, not to produce more rain. Some seeding experiments in South Africa, Thailand, and Mexico that employed hygroscopic seeds rather than dry ice or silver iodide registered statistically positive results but with puzzling aspects in which the sequence of events in the clouds did not match the theory. Considering the results as a whole, Silverman concludes that cloud seeding remains promising and worth pursuing yet still unproven despite decades of efforts. The many attempts have not resulted in either the necessary statistical evidence or the necessary physical evidence,[32] although limited encouraging results continue to come in for individual cases, such as analyses by William Woodley and Daniel Rosenfeld of cloud-seeding projects in Texas in 1999–2001.[33]

A significant part of the problem with devising a successful cloud-seeding strategy is the inherent variability among clouds and the fact that what works well in one cloud will not necessarily work well in another cloud. Whether cloud seeding will be effective in an individual instance and how much seeding will be favorable before an unfavorable overseeded state is reached are both highly dependent on the liquid water content of the cloud, the ice nuclei in the cloud, and other cloud-specific aspects. Hence, it is not particularly surprising that the numerous cloud-seeding experiments over the decades have yielded many cases with inconclusive results, some cases where rainfall seems to have been enhanced by the seeding, and some cases where rainfall actually seems to have been reduced by the seeding.[34] Still, it is possible that eventually researchers will sort through the critical factors determining what works and under which conditions so that eventually cloud seeding might become routinely effective, at least for specified cloud types and circumstances.

> Perhaps the most convincing evidence that cloud seeding has not been a resounding success is that it has not come into routine practice.

Perhaps the most convincing evidence that cloud seeding has not been a resounding success so far is that it has not come into routine practice even for major applications, like reducing severe drought conditions or steering clouds away from locations in imminent danger of major flooding.

HAIL SUPPRESSION

Just as humans have long sought to control rainfall, we have also long sought to control hail. Among the techniques have been bell ringing, prayers, firing guns and cannons into the atmosphere, animal and human sacrifices, and, as with rainfall, cloud seeding.[35] In view of the devastation that hail can cause to crops and other property, successful hail suppression could have major economic implications.

Shooting cannons directed at the atmosphere to reduce hail was practiced in Europe starting in the fourteenth century, part of the rationale being that the noise itself created a favorable change that discouraged hail formation. However, deaths and wounds resulting from the cannon firings in both Austria and Italy led to a reaction against this particular method of weather control. Because of the accidents and varied complaints that the hail and rainfall outside the targeted area could be negatively affected by the hail suppression techniques, Empress Maria Theresa in 1750 forbade the Austrian peasantry from using cannons for hail suppression. The ban was revoked later in the century, but reestablishing the efforts was not automatic, and so the ban effectively reduced hail suppression attempts for far longer than the duration of the ban itself.[36]

Renewed efforts at hail suppression were undertaken in Europe in the period 1895–1906. Many of these efforts tested a basic concept articulated by an Italian professor of mineralogy in 1880 that hailstone formation might be preventable by using cannons to fire smoke particles into thunderstorms, the intent being to have the smoke particles serve as condensation nuclei for the formation of raindrops and small ice particles. In the mid-1890s, winegrower and politician M. Albert Stiger experimented with this concept in his native Austria, being particularly desirous to reduce the enormous damage routinely caused by hail to the winegrowers and others in his province. Over several years of experimentation, Stiger developed a vertically pointing megaphone-like cannon that, when fired, sent a large smoke ring to a height of about 300 meters. This cannon had a spectacular initial test season, as Stiger had six of these cannons operating in his province during the 1896 hail season, and no hail at all fell during the entire season. The following year, the experiment expanded, with 30 cannons and again spectacular success, no hail falling in the cannon-protected region and severe hail losses in surrounding provinces. Word of the successes spread, and by the end of 1898, the megaphone-like cannons were selling in Italy and France as well as Austria. By 1899, 2,000 cannons were operating in Italy, and a major program using the cannons existed in Germany as well. By the end of 1900, over 10,000 cannons were distributed among Italian vineyards, and the cannon use had spread to Hungary and Spain. However, despite the rapid spread, the spectacular apparent initial successes in Austria were not matched in any of the other countries, and all this artillery had the quite unfavorable consequence of numerous accidental wounds and deaths. For instance, in Venice and Brescia, seven deaths and 78 injuries were reported in 1900 as a result of accidents from the firings of these hail-suppressing cannons. Still, in some arenas optimism ran high, and at the second International Congress on hail suppression, held in November 1900 in Padua, Italy (the first was in November 1899 in Casale, Italy), the Congress adopted a concluding statement indicating that the efficiency of the cannons in protecting against hail had been "proved beyond all question."[37]

The attitudes regarding the success of the cannons in suppressing hail were considerably more mixed the following year, in 1901, at the third International Congress, held in Lyons, France. With additional deaths reported in 1901, more instances of hail despite cannon firings, and an increase in the expression of scientific doubts regarding the impact of the cannons, the fourth International Congress, held in Gratz, Austria, in 1902, concluded that the success or failure of the cannons remained unclear and recommended controlled and coordinated field experiments to resolve the uncertainties. Government-coordinated experiments were then undertaken in both Austria and Italy in 1903–1904, with the resulting analyses in both countries concluding that the cannons failed, as both regions experienced damaging hail during periods when the cannons were being used. Other areas in Europe where the cannons were operating also experienced damaging hail, and as a result of these numerous failures, by 1905 the use of cannons to suppress hail had largely come to an end. The optimism early on had perhaps been too high, as the expectation from the initial Austrian successes was that the cannons would eliminate all hail damage. When they failed to perform to that impossible standard, they were rejected.[38]

After 1905, major hail suppression activities did not reemerge until the late 1940s and 1950s, when efforts arose in the United States using ground-based and aircraft-based cloud seeding, in the Soviet Union using cannons and rockets for cloud seeding, and in Italy and France using rockets to shatter hail that had already formed.[39] Because a large part of the damage from hail is due to the large size and hardness of the hailstones, limiting the growth of the hailstones would likely limit the hail damage, hence explaining the use of cloud seeding for hail suppression, the hope being to form multitudes of raindrops or ice particles and thereby prevent (or at least impede) the heavier, more damaging hailstones from forming. Near the end of a 1949 paper on his pioneering work on cloud seeding, Vincent Schaefer succinctly stated the central principle behind the potential application of cloud seeding to hail suppression, explaining that in the face of the tremendous quantities of ice crystals produced by dry-ice seeding, "the competition between particles for the available moisture would be so keen that no particles could grow large."[40]

Soviet experiments in the 1950s and 1960s were reported to have reduced hail damage by 50% to 100%.[41] These experiments were performed by several institutes in the Soviet Union, employing a variety of methods. Scientists at the Transcaucasian Hydrometeorological Research Institute used artillery shells to seed hailstorms with salt in the warm lower part of the cloud and with ice nuclei in the supercooled upper part of the cloud, the object being to wash out precipitation from the lower part of the cloud and to produce more and smaller hail particles in the upper part of the cloud. The Georgian Academy of Sciences' Institute of Geophysics seeded potentially hail-forming clouds in the Alazani Valley with lead iodide and silver iodide injected into the cloud through rocket launches. The perceived success of the cloud seeding was evidenced by an expansion in the area seeded from 15,000 hectares in 1961 to 460,000 hectares in 1969. Reports estimated a 70% reduction in hail frequency as a result of the cloud seeding. Even more impressive numbers were reported for the hail suppression efforts of the High Altitude Geophysical Institute in one region of the North Caucasus. There hail damage had spread over at least 20,000 hectares in each of the 20 years preceding cloud seeding but spread over only 500, 500, and 25 hectares, respectively, during 3 years of cloud-seeding efforts.[42]

Multiple attempts were made to reproduce the impressive Soviet results elsewhere, but all such attempts failed.* The cause of the failures could be in part because of local differences in the hail-producing storms and in part because the later efforts used different rocket-launching techniques for injecting the seeding material into the storms.[43] In the 1970s, Yugoslavian farmers did employ the Soviet Georgian Academy of Sciences' techniques, shooting silver iodide into suspect clouds from ground-based rockets, and many of the farmers of Yugoslavia seemed confident that the hail suppression efforts were successful. However, the lack of scientific controls kept skepticism alive and vocal.[44]

---

* See Federer et al. (1986) for a detailed analysis of an intensive 5-year effort by French, Italian, and Swiss research groups to reproduce in Switzerland the Soviet hail suppression successes.

In the meantime, reports in the early 1960s of the apparent major successes in the Soviet Union hail suppression attempts spurred renewed interest in the United States. This interest contributed to the initiation of both the Hailswath program in South Dakota and Colorado in 1966 and the U.S. National Hail Research Experiment (NHRE), with field campaigns in 1972–1976.[45]

In 1977, David Atlas of the U.S. National Center for Atmospheric Research and former director of the NHRE wrote an overview of hail suppression attempts, describing very mixed results. Many reports of successful hail suppression in the Soviet Union, South Africa, and elsewhere were countered by many reports of unsuccessful attempts, some where hail actually increased rather than decreased and others where the results were inconclusive and often controversial.[46] Among the records of unsuccessful efforts were a 7-year record of silver iodide seeding in Switzerland that showed a greater frequency of hail on the days when seeding took place than on those with no seeding and a 3-year record of silver iodide seeding in Colorado that similarly showed greater hail mass on the cloud-seeding days. Atlas mentions that by the mid-1970s, even the Soviets were voicing concerns, as they disclosed in particular that they had experienced increased hail during some failed individual hail suppression attempts and more generally that they had a lack of success in suppressing hail during severe storms.[47]

As Atlas explains, depending on the particular characteristics of an individual cloud, the insertion of silver iodide or other cloud-seeding particles could either successfully decrease damaging hail by creating the desired competition within the cloud for the cloud's finite water supply or unsuccessfully increase hail by providing nuclei around which the hailstones can grow. The successful hail suppression programs might have had a substantial element of luck in having clouds particularly conducive to hail suppression. Recognizing this, the challenge becomes to determine the critical cloud characteristics for distinguishing which clouds should and which should not be seeded. Atlas mentions two attempts along those lines, in both of which numerical models were used to examine the contrast between seeding clouds with cold cloud bases versus seeding clouds with warm cloud bases. Adding to the confusion, the two models produced opposing results.[48]

It is highly unlikely that any single characteristic—whether cloud base temperature or something else—will suffice for identifying appropriate clouds for hail suppression seeding, but it is certainly possible that eventually experts will obtain a handle on which clouds to seed based on a cluster of characteristics. They will also need to obtain a much better handle on how much seeding to do, as it is quite possible that mild seeding would increase hail in the same cloud that greater seeding would suppress hail or vice versa.

Overall, as in the case of attempted rainmaking, attempted hail suppression has had individual cases of apparently great success but has had no consistent scientifically validated wide-scale success. Correspondingly, as in Europe around 1905, in other regions also substantial optimism and activity has been followed by disappointment. In the United States, funding for research into hail suppression was up at $5 million per year in the early 1970s but was down to nothing by 1978.[49] Hail suppression remains a long-term constructive goal of applied science, but the enthusiasm is now tempered by a recognition of the unfavorable consequences that can occasionally result from hail suppression attempts. These include not only increased rather than decreased hail but also flooding from enhanced rainfall and in some regions even an increased chance of tornado formation;[50] without a doubt, hail suppression attempts gone wrong can have very unfortunate unintended consequences.

TAMING HURRICANES

In view of the destructiveness of hurricanes that reach landfall in the vicinity of cities and towns—some killing thousands of people and/or wreaking billions of dollars of damage—the desire to tame their destructive power is hardly surprising. To gain a sense of the magnitude of this power, consider the fact that the suggestion to explode a nuclear bomb into a hurricane in order to blow it apart has been rejected specifically because of the relatively small power generated by an artificial bomb compared to the power in the hurricane itself. In fact, the power in a hurricane is so great that it would take an estimated 400 20-megaton hydrogen bombs to release the amount of energy that a moderate-sized hurricane releases in a single day.[51]

Nuclear bombs being too feeble for hurricane suppression, an alternative suggestion has been to cut off the hurricane's energy supply from the underlying ocean. A hurricane is powered by the energy that water absorbs during evaporation, when the water converts from a liquid to a vapor state. This energy remains latent in the vapor molecules until being released within the hurricane when the water converts back to a liquid or converts to a solid form, becoming water droplets or ice particles, respectively. If the water in the regions prone to hurricanes could be prevented from evaporating and lifting into the atmosphere, this could well solve the hurricane problem although perhaps in the process creating even greater problems. Cutting off the water supply with a solid sheet placed over the ocean would be impractical because of the area involved and extremely unwise anyway because of the consequences to marine life and to the ocean/atmosphere system. A simpler approach would be to cover the sea with a chemical-retardant film, but this would be difficult to maintain in the face of wind and wave action[52] and likely also would be damaging to marine life and an environmental blight.

Instead of blocking the water supply from below, William Gray of Colorado State University and others have considered modifying the atmosphere surrounding the hurricane. More specifically, Gray has proposed injecting very small black carbon particles into the atmospheric boundary layer around the outside of the hurricane. The objective would be to have the particles absorb solar radiation, thereby enhancing the activity at the edge of the hurricane and reducing the activity in the main, more destructive portion of the hurricane. Gray is well aware of the drawbacks of this scheme, including the pollution from the carbon particles, and so would not recommend it except in the event of a hurricane immediately threatening a high-population area.[53]

The one scheme for hurricane modification actually attempted in the twentieth century is cloud seeding of storm clouds near the hurricane center.[54] As early as 1947, General Electric's Irving Langmuir and his colleagues considered the possibility of cloud seeding to alter the course or reduce the intensity of the storms, and on October 10, 1947, the theory was tested by seeding a hurricane off the East Coast of the United States with about 102 kilograms of dry ice. Soon thereafter, the hurricane changed direction. No one knows whether this shift was or was not influenced by the cloud seeding, but in any event the changed direction resulted in the hurricane's slamming into the coast of the state of Georgia, with the quite undesired further effect of a flurry of lawsuits against General Electric,[55] highlighting one of the standard risks of intentional weather modification.

Despite the disappointing results of the October 1947 test case, by the early 1960s further hurricane research led to the hypothesis that considerable supercooled liquid water existed in clouds within a hurricane's eye wall and just outside it and that injection of silver iodide crystals into these clouds could start a sequence of events that would send some of the warm moist air upward or outward rather than spiraling inward. Theoretically at least, this should have the effect of decreasing maximum wind speeds and thus decreasing hurricane-caused damage.[56] Project Stormfury, a joint effort of the U.S. Weather Bureau's National Hurricane Research Project and the U.S. Navy, tested these concepts by seeding selected Atlantic hurricanes in the period 1961–1980, starting with preliminary experiments on Hurricanes Esther in 1961 and Beulah in 1963, prior to a more comprehensive experiment on Hurricane Debbie in 1969.[57] Seeding of Hurricanes Esther and Beulah on September 16, 1961, and August 24, 1963, respectively, was done in each case with silver iodide inserted into the cumulonimbus clouds surrounding the eye of the storm to cause an outward movement of the clouds and a lessening of the storm's intensity. Both hurricanes appeared to respond to the seeding, with similar results in the two cases. In Beulah, the eye wall dissipated soon after the seeding and then reformed 4 to 10 miles outward from the center of the hurricane, and wind speeds on average were reduced by 14%.[58]

Results from the seeding of Hurricane Debbie in August 1969 seemed particularly promising, with reductions in wind speed from 50 meters per second to 35 meters per second on August 18 and from 50 meters per second to 42 meters per second on August 20, both attributed at least in part to cloud seeding. Such success generated calls for the project to go operational, with seeding of any major hurricane threatening landfall. Project Stormfury's director, Robert Cecil Gentry, however, cautioned against inordinate optimism, reiterating the uncertainties regarding cause and effect and stating that it was still unclear whether any technology existed that could reliably weaken a hurricane. Gentry's cautions were strongly supported by the U.S. National Academy of Sciences' Stormfury Advisory Panel, which recommended in 1970 that more data be collected from both seeded and unseeded hurricanes. The U.S. Congress consequently appropriated $30 million for the needed aircraft and instrumentation. Numerous flights were made in the 1977–1980 period with the new airborne capabilities. However, the timing was quite unfortunate, as no appropriate Atlantic hurricanes formed during that period, and so no seeding took place. New questions arose regarding whether the changes that occurred in Hurricane Debbie in 1969 were caused by the seeding or by natural forcings, and this, along with the expenses involved and the lack of appropriate hurricanes for seeding over the previous several years, led to the discontinuation of Project Stormfury* in 1980. The project resulted in considerably

---

* Project Stormfury undertook seeding of non–hurricane clouds as well, with some notable cases of cloud growth. In August 1963, after being seeded with silver iodide, several supercooled cumulus clouds grew explosively, for instance expanding vertically an additional 10,000 to 20,000 feet within 10 minutes of the seeding and more than doubling in horizontal diameter within 40 minutes (Simpson and Malkus 1964). Two years later, in July and August 1965, a carefully constructed experiment randomly seeded 14 out of 23 selected tropical oceanic cumulus clouds, focused on determining whether seeding tropical cumulus clouds with cloud-top temperatures between $-5°C$ and $-20°C$ would increase the cloud height. Seeding was done with silver iodide, and the results showed that the seeded clouds on average grew vertically by 1.6 kilometers more than the control clouds (Simpson et al. 1967). Further, the observed results matched quite well the earlier predictions from a numerical model. Only one of the nine unseeded clouds had a vertical growth in line with the seeded clouds, and that one cloud may well have grown through natural glaciation within the cloud. The careful randomized experiment structure allowed the researchers to conclude with high statistical confidence that the silver iodide seeding increased the vertical growth of the seeded clouds (Simpson et al. 1967).

improved knowledge about hurricanes but did not lead directly to an operational hurricane seeding effort.[59]

In the United States, federal funding for weather modification of all types, including taming hurricanes and suppressing hail, peaked at nearly $19 million per year in the mid-1970s but fell to about $12 million per year by 1985 and to about $2 million per year by 2007.[60] Overly optimistic expectations, lack of success, and fear of lawsuits all contributed to the reduced efforts.

One further point to note regarding intentional hurricane manipulation is that despite the destruction wrought by hurricanes, not everyone wants them tamed. In the 1970s, the Japanese in particular were insistent that no Stormfury operations be done on typhoons (which is the term typically used for hurricanes in the northwestern Pacific) with the potential of reaching land, as these storms bring hugely beneficial rainfall to Japan's agricultural fields.[61] In April 2009, I discussed this topic with Stephen Leatherman, the director of the International Hurricane Research Center at Florida International University, and Leatherman indicated that the Japanese are not unique in appreciating the value of these massive storms, stating that "people living on low-lying islands in the Caribbean pray for tropical storms and category 1 and 2 hurricanes for the rain." Typhoons/hurricanes additionally play an important role in heat and moisture transfers within the Earth's atmosphere, so that taming them on a routine basis would have other, uncertain consequences to the global weather system.

LEGAL COMPLICATIONS

All the schemes for weather modification and other local or regional environmental adjustments are fraught with potential legal as well as technological and environmental problems. Failures can result from too much as well as too little of the desired change (such as increased rainfall), and even with "success," the schemes by and large will entail losses for some people in addition to the intended gains for others. For instance, when cloud seeding succeeds in producing rain for one community, it lessens the chance of rain for the communities directly downwind, as the water content rained on the first community will no longer be available for raining on the downwind communities. All the people and communities who regard themselves as damaged by a weather modification attempt become, in some circumstances, potential litigants against the weather modifiers.

Already mentioned were 169 claims against New York City in 1951 for flood and other damages following cloud seeding, claims that were unsuccessful in terms of any damages being awarded but successful in terms of stopping the city from further cloud-seeding attempts. Another relevant case occurred on June 9, 1972, in Rapid City, South Dakota, which experienced a disastrous flash flood a few hours after a cloud-seeding experiment was conducted near Rapid City by the Institute of Atmospheric Sciences at the South Dakota School of Mines and Technology under contract with the U.S. Department of the Interior. As reported on the Rapid City Library's website, 1 billion tons of rain fell, 238 people died, 3,057 people were injured, and financial damages in the affected region totaled $165 million.[62] Whether the flood was caused by the cloud seeding is not known, and even knowledgeable meteorologists have argued at length about it.[63] Based on interviews done later in the year in South Dakota, where operational weather modification efforts were being undertaken in 22 counties, generally with considerable public support, only 4% of the respondents felt that the cloud seeding was the sole or primary cause of the flood, and only 21% felt that it was a contributing cause.[64] Still, although the litigants were in the minority, a class-action suit was brought against the federal government for hundreds of millions of dollars.[65] In light of the uncertainties, the U.S. Bureau of Reclamation ruled that the flood was not connected to the cloud-seeding experiments, and the case was dropped. Still, it remained a source of emotional opposition to attempts at weather modification[66] and was cited, for instance, by opponents of weather modification in the Pacific Northwest.[67]

The lack of success of lawsuits brought against weather modification attempts does not eliminate concern about potential lawsuits, as even an unsuccessful lawsuit generally costs the defendants considerable time and expense. Further, in cases when lawsuits might not be feasible, for instance, when different countries are involved or when the damaged party might not trust or be willing to go through the legal system, some people might resort to violence against the weather modifiers or those supporting them.[68]

Recognizing the potential legal and other actions against weather modification attempts, in 1974 Will Kellogg and Steve Schneider, both at the time employed at the U.S. National Center for Atmospheric Research, an organization some of whose scientists were participating in weather modification experiments, proposed a concept of "no-fault climate disaster insurance." They illustrated specifically with the case of the June 1972 Rapid City flash flood, saying that in cases like this, where weather modification operations are being carried out to help the people of the state, it might be wise to issue beforehand a statewide no-fault weather modification insurance policy. The policy could compensate those whose climate deteriorated after the weather modification schemes were implemented, without needing proof that the fault lay exclusively with the weather modifiers. This would circumvent the otherwise high likelihood of inconclusive debates over cause and effect. Kellogg and Schneider do realistically warn, however, that the premiums for such insurance might end up unacceptably high.[69]

If and when massive geoengineering schemes begin to be implemented, the concept of no-fault insurance policies might be equally relevant to those efforts as to the smaller-scale efforts discussed by Kellogg and Schneider[70] (equally relevant, but even more complicated and expensive).

ENGINEERING THE NETHERLANDS

Before bringing to a close this chapter on past attempts at weather and climate modification, it seems appropriate to point out that some engineering efforts to modify the environment have by and large been successful. Probably the most well known of such efforts on a countrywide scale is the success (incomplete though it might be) of the people of the Netherlands in expanding their coastline seaward and reclaiming land as their country has been encroached on by the sea. Approximately a quarter of the land area of the Netherlands is below sea level, and the country borders the often stormy North Sea, placing it at major risk of flooding; so maintaining the land area has indeed required a major engineering effort. (Another location with a similar problem although on a smaller spatial scale is the American city of New Orleans.)

Starting as early as the twelfth century, the Dutch would surround a section of swampland with dikes and then, aided by windmills, would pump the land dry, creating fertile farmlands near and sometimes below sea level.[71] This has in large part been successful, as the country has survived and thrived despite its low altitude. Still, it has not been easy, cheap, or without major setbacks.

Among the notable historic floods in the Netherlands, one on December 14, 1287, in the northern part of the country killed more than 50,000 people and significantly impacted not just the outer coastline but also what had been a substantial freshwater lake. The lake was enlarged and converted into a saltwater gulf when the storm-driven flood broke through the strip of land forming the lake's northern boundary, dividing that boundary into a string of islands and linking the former lake to the North Sea to its north. Being a small and well-populated country, the Netherlands could ill afford to lose the land, and so, as they have now done for centuries, the people of the Netherlands proceeded to drain the water from many of the flooded areas and to strengthen the protective system of dikes and embankments.[72]

More recently, high seas and a raging storm on February 1, 1953, sent a wall of water crashing through the dikes and cascading across at least 500,000 acres, killing over 1,800 people and forcing the evacuation of an additional 72,000. Following the 1953 incident, the Dutch instituted a decades-long engineering program called the Delta Works Project, creating a more secure network of protective dams.[73] After an expenditure of $5.5 billion, the Delta Works project was completed in 1997. The Dutch have built over 9,870 miles of dikes, 8,000 of them along canals and the rest along rivers, lakes, and the sea, and for the most part, the dikes have held.[74] No one expects that these are forever indestructible, but the experience of the Dutch does show that with persistence and some major setbacks, there can be successful engineering efforts explicitly aimed at modifying the environment.

PART IV

Further Cautionary Considerations

CHAPTER 9

The Possible Fallibility of Even a Strong Consensus

As history amply reveals, experts can be wrong, and scientific consensus, however valuable, is not necessarily correct.

WHEN a strong consensus exists among experts, the overwhelming likelihood is that the consensus viewpoint is correct or at least is as correct as any understanding of the particular topic at the time. However, no matter how strong a consensus might be, its correctness is not guaranteed. Indeed, history is replete with examples of incorrect strong consensuses. The examples range over all fields, including science, technology, economics, politics, medicine, and many others. For instance, before Copernicus and for many years afterward, most experts thought that the Sun orbited the Earth rather than the other way around. Experts declared the impossibility of flying in a heavier-than-air machine before Alberto Santos-Dumont and the Wright brothers proved them wrong. Experts declared the *Titanic* unsinkable shortly before it sank calamitously on its maiden voyage when it struck an iceberg in the North Atlantic on April 15, 1912. Experts aggressively recommended purchasing stocks shortly before the stock market crash of 1929. Experts confidently predicted that Thomas E. Dewey would win the U.S. presidential election of 1948, repeating that prediction days and even just hours before the victory of Harry S. Truman; and 60 years later, with all the advantages of modern technology, experts predicted, again just hours before the voting, an overwhelming victory by Barack Obama in the New Hampshire primary of the 2008 U.S. presidential campaign, only to be foiled by a narrow but clear-cut victory by Hillary Clinton (prior to her eventual loss to Obama later in the primary season). Much more devastatingly, experts prescribed the drug thalidomide to pregnant women in the late 1950s and early 1960s with the result that thousands of children were needlessly born with severe malformities. The list could go on and on with examples from practically every field of human endeavor. The central point here is that experts should never be assumed to be all-knowing, even in the immediate field of their expertise.

> We would be blind not to realize that today's theories also will be overridden, as future progress will reveal them to be inadequate as well.

Despite the care given to obtaining correct results, science is by no means spared the possibility of a flawed consensus. In fact, the fact that scientific consensus is sometimes wrong has become a core tenet in the understanding of how major scientific advances emerge. Notably, one of the most cited of all works in the history of science is Thomas Kuhn's *The Structure of Scientific Revolutions*.[1] Kuhn examines the major turning points associated with Nicolaus Copernicus, Isaac Newton, Antoine Lavoisier, and Albert Einstein and explains that in each case the scientific revolution "necessitated the community's rejection of one time-honored scientific theory in favor of another incompatible with it."[2] In each case, all the experts holding to the earlier theory were found to be wrong. This included the experts prior to Copernicus who believed and stated unequivocally that the Sun revolves around the Earth, the experts prior to Lavoisier who believed that a burning object gives off a negative-weight substance called phlogiston, the experts prior to Einstein who believed in the absolute nature of space and time, and all the many other scientific experts emphatically expounding theories that have subsequently been abandoned. We would be blind not to realize that today's theories also will be overridden, as future progress will reveal them to be inadequate as well. Our state-of-the-art theories are the best that we have and should be relied on to help guide policy, but they should never be accepted as infallible. And, yes, this does include the theories of how Earth's climate might change in the decades ahead. This becomes particularly important when considering massive geoengineering schemes that could cause irreversible damage to the entire planet. Just like all other experts, experts advocating a particular geoengineering scheme could be right, but they could also be wrong; in view of the magnitude of the potential unintended consequences, it is important to be extra cautious before undertaking any massive geoengineering effort.

Accepted theories being overturned is such a fundamental element of the progress of modern science that an admonition to "beware the consensus" should be at the heart and soul of an innovative scientist's career. All scientists should have some level of skepticism despite the negative connotations that the term "skeptic" has acquired in recent years when it comes to climate change issues. As history has demonstrated repeatedly, the advance of science is an ongoing process that includes articulating a theory (sometimes by one person, sometimes by many), developing that theory (generally by many people) and its many consequences (definitely by many people), and then recognizing flaws, generally because new observations or newly interpreted old observations are not matching the theory as well as they should, and either adjusting the theory or coming up with a new theory to replace the old one. Newton's theory of gravitation, as laid out in detail in 1687 in his masterful *Principia Mathematica Philosophiae Naturalis* (*Mathematical Principles of Natural Philosophy*), was magnificent and spurred over two centuries of phenomenal progress, but it was not the final say, as became clear when aspects of it were overturned by Einstein. Similarly, no matter how masterful any of today's major scientific theories might be, it is extremely unlikely that any of them (not Einstein's or anyone else's) will not eventually get overturned at least in part.

Even in cases where nothing so grand as a "theory" exists, often scientific concepts widely accepted by the experts get overturned in the normal course of scientific discovery. For instance, in the late 1970s when stunning new ecosystems were found at the bottom of the ocean associated with the newly discovered hydrothermal vents, it became clear that large creatures—including 10-foot-tall tubeworms—were living at temperatures of hundreds of degrees Celsius and that plant life was surviving and thriving without any sunlight available for photosynthesis. These life forms totally overturned earlier consensus beliefs regarding where and how life can exist.[3]

Yet despite the history of accepted theories and concepts being overturned again and again, science textbooks, the general public, the media, and even many scientists frequently treat current scientific theories and concepts as though they are final and infallible, totally ignoring the reality that they almost certainly are not.

Of particular relevance to the topic of this book are examples regarding global climate. Two such examples are highlighted in this chapter: the changing concepts regarding the cause of the ice ages and the changing concepts regarding the importance of CO_2 to global temperatures and whether human impacts on climate are favorable or unfavorable. Among the other examples that could have been selected is the concept of continental drift, which was initially laid out in detail in Alfred Wegener's 1915 *Die Entstehung der Kontinente und Ozeane* (*The Origin of Continents and Oceans*). This concept sparked vigorous discussion in the 1920s, was rejected by most scientists until the 1960s, and then was widely although not universally accepted thereafter.

ICE AGE THEORIES

As detailed in chapter 2, the Earth has experienced periods with substantially more ice than exists today. Termed "ice ages" or "glaciations" or "glacial periods," these periods have been relatively rare in the Earth system's evolution, but they have been a major element of the past 2 million years, as the Earth has cycled repeatedly through glacial and interglacial episodes during that time frame.

Not long after the concept of the existence of full-scale past ice ages was first advocated in the 1830s, most forcefully at the time by the Swiss natural scientist Louis Agassiz starting with his "Discourse at Neuchâtel" presentation to the Swiss Society of Natural Sciences in 1837,* people began speculating on possible ice age causes. In 1842, well before many scientists even accepted that ice ages had ever occurred, the French mathematician Joseph Alphonse Adhémar proposed that the cause of the ice ages could be tied to changes in the Earth's orbit around the Sun. Many other theories have come and gone, but the tie to changes in the Earth's orbit is now widely accepted, and hence the long and distinguished but checkered history of that concept is particularly relevant.

Adhémar's 1842 speculation was that ice ages have a 22,000-year cycle in line with the 22,000-year cycle in the precession of the Earth's axis (or "precession of the equinoxes," in reference to the resulting shift in timing of the spring equinox and autumn equinox), earlier calculated, in the 1750s, by the French mathematician and philosopher Jean Le Rond d'Alembert. According to Adhémar's theory, the ice ages alternate every 11,000 years between the Northern and Southern hemispheres, and their timing is determined largely by the length and severity of the winter season.[4] The theory generated some interest and acceptance but lost favor in the 1850s when German naturalist Alexander von Humboldt and others expressed criticisms. Von Humboldt specifically countered the basic concept that the cold periods alternate between the two hemispheres.[5]

The self-educated Scottish geologist and climatologist James Croll improved on Adhémar's orbit-based explanation of the ice ages in the 1860s and 1870s by incorporating consideration of the 100,000-year cycle in the eccentricity of the Earth's orbit. Discovered earlier by Frenchman Urbain Jean Joseph LeVerrier, this cycle involves the slow transition from an orbit that is nearly circular, with the Sun at the center (a low-eccentricity orbit), to an orbit that is slightly flattened, with the

---

* For a fascinating history of how the existence and details of past ice ages were determined, the reader is referred to the highly readable book *Ice Ages: Solving the Mystery* by John Imbrie and Katherine Palmer Imbrie (Imbrie and Imbrie 1979).

Sun slightly off center (a high-eccentricity orbit). Like Adhémar, Croll considered winter conditions to be the critical determinant of when the ice ages occur, but now the flatness of the orbit comes into play, as the flatter the orbit, the more severe the winters would be for the hemisphere experiencing winter when the Earth is farthest from the Sun. Croll further argued that the orbital changes did not act alone but served as triggering mechanisms to set off a variety of changes within the Earth system, including changes in ocean currents and in snowfall, the latter leading to a positive feedback as more snowfall causes more reflection of solar radiation back to space and hence further cooling. In fact, Croll was a prime originator of the concept of positive feedbacks in the Earth system. Croll's orbital theory of the ice ages was highly acclaimed at the time, earning Croll major awards and recognition, and gained widespread acceptance. However, by the end of the nineteenth century, orbital forcings (constituting the core of the "astronomical theory" of the ice ages) had fallen out of favor once again because none of the proposed schemes matched the glacial/interglacial dates suggested by the growing abundance of geologic evidence.[6]

Other people proposed other possible causes of the ice ages, but the astronomical theory eventually reemerged. In the early twentieth century, Serbian mathematician Milutin Milankovitch returned to and expanded the theory, incorporating cycles in three parameters of the Earth's orbit (including the two identified as important by Adhémar and Croll): the obliquity, or tilt, of the Earth's axis, which varies with a period of about 41,000 years; the eccentricity of the Earth's orbital ellipse, which varies with a period of about 100,000 years; and the precession of the equinoxes, with periods of about 19,000 and 23,000 years. Importantly, Milankovitch argued that the timing of the ice ages is based not on when in these very long cycles there is particularly low winter insolation (i.e., cold winters), as Adhémar, Croll, and others suspected, but instead on when there is particularly low summer insolation, specifically at 65°N.[7] According to Milankovitch, the key to ice ages was having summers sufficiently cool that the winter snowfall would not completely melt away.

Many scientists favored the Milankovitch theory in the 1930s and 1940s. However, in the early 1950s, newly developed radiocarbon dating techniques were applied to dating geologic strata, and, similarly to what happened near the end of the nineteenth century, the new dates for the glacial periods did not match the dates determined in the theoretical calculations based on the orbit variations, this time the calculations of Milankovitch. As a result, by 1955 most of the scientists involved in paleoclimate research rejected the Milankovitch and all other astronomically based theories of the ice ages. The twists and turns in this story, however, were far from over.

In 1955, Cesare Emiliani from the University of Chicago published a highly relevant analysis of eight deep-sea cores that encompassed a 300,000-year record and showed a good correspondence with the Milankovitch curves.[8] In 1956, Hans Panofsky of Pennsylvania State University concluded a review of ice age theories by favoring essentially the Milankovitch theory with the addition of a mountain-building component.[9] The Emiliani work generated much additional research and debate, with some scientists finding that by tweaking the Milankovitch theory in one way or another, they could establish a better match with the observations. In one such case, Wally Broecker and colleagues found that by calculating the Milankovitch curves for 45°N instead of the commonly used latitude of 65°N, they obtained an improved match with paleoclimate data from Barbados, where distinct terraces reflect the history of sea level and hence the history in part of the changes in the amount of water locked in massive ice sheets on the North American and Eurasian continents. In fact, the match was sufficiently close for the authors to conclude that "the often-discredited hypothesis of Milankovitch . . . must be recognized as the number-one contender" in the effort to explain the ice ages.[10] Still, most scientists continued through the 1950s and 1960s to reject the Milankovitch theory.

New data continued to be uncovered, analyzed, interpreted, and reinterpreted in relationship to the Milankovitch theory. In the continuing interplay of observation and theory, just as Broecker and colleagues had adjusted the ice age calculations to be based on insolation at 45°N instead of 65°N,[11] additional adjustments were made to Milankovitch's astronomical theory, in particular to account for the dominance of the 100,000-year cycle in the data and the indication from the data that the interglacials are short relative to the glacial periods.[12] In 1972, the theory was mentioned favorably although not exclusively by Mikhail Budyko, director of the Voeikov Main Geophysical Observatory in Leningrad, in an overview of factors influencing long-term climate.[13]

In a key paper published in 1976, James Hays from Columbia University, John Imbrie from Brown University, and Nicholas Shackleton from Cambridge University concluded from analysis of two deep-sea cores from the Indian Ocean—even farther away from the Northern Hemisphere ice sheets than the Barbados data— that major changes in climate over the past 500,000 years correspond well with cyclings similar to those in the Milankovitch theory. More specifically, Hays, Imbrie, and Shackleton find that approximately 50% of the climate variance in the Indian Ocean deep-sea records can be explained by a cycle of approximately 100,000 years, approximately 25% of the variance can be explained by a cycle of approximately 42,000 years, and approximately 10% of the variance can be explained by a cycle of approximately 23,000 years.[14] These three periods correspond closely with the Milankovitch cycles for the eccentricity of the Earth's orbit, the tilt of the Earth's axis, and the precession of the equinoxes, respectively. The Emiliani, Broecker-led, and other pre-1976 studies had also provided evidence for the importance of the orbital forcings,[15] but the Hays et al. (1976) paper essentially clinched the argument in the eyes of many scientists, through its convincing presentation using a mathematical technique known as "bandpass filter analysis" that allowed them to separate out and highlight the different cycles. Since the 1976 Hays, Imbrie, and Shackleton paper, the astronomical theory, explaining the glacial/interglacial cycling on the basis of orbital forcings, has been the favored theory of the ice ages, in fact so much so that it is often presented as though it is completely established, with no lingering debate. In recognition of the major advances made by Milankovitch, it is commonly referred to as the Milankovitch theory.

Both the tweaking of the Milankovitch theory prior to the 1976 Hays et al. publication and the widespread acceptance of the theory in the late twentieth century were centered on evidence from within the past 1 million years, a period for which the data suggest the dominance of an approximately 100,000-year ice age cycle. Records now suggest that the glacial/interglacial cycles had periods of approximately 41,000 years in the late Pliocene and early Pleistocene, between 3 million and 1 million years ago, before switching to the approximately 100,000-year cycles of the past million years.[16] Here is another instance in which additional data necessitated an adjustment or further elaboration of the previous theoretical explanation.

Maureen Raymo and Lorraine Lisiecki from Boston University and Kerim Nisancioglu from Norway's Bjerknes Center for Climate Research published a study in 2006 offering a hypothesized explanation for the shift from about 41,000-year cycles to 100,000-year cycles, both of which individually are deemed astronomically forced (Milankovitch theory). Specifically, Raymo and her colleagues suggest that it was approximately 1 million years ago that the growing Antarctic ice sheet became sufficiently large that the edge no longer terminated on land but instead extended outward over the surrounding oceans, as today, with ice shelves around much of the continent.[17] (Others place the shift somewhat earlier, between 3 million years ago and 2.6 million years ago, a timing that would not fit well with the Raymo et al. explanation. The growth of the ice sheet itself had begun much earlier, as discussed in chapter 2.) With a fully land-based Antarctic ice sheet 3 million years ago to 1 million years ago, according to the Raymo-led study the ice volume changes in the Northern and Southern hemispheres were each controlled by summer insolation and were therefore out of phase with each other, with a globally integrated record dominated by the 41,000-year cycle in the tilt of the Earth's axis. Once the Antarctic ice sheet expanded outward to terminate in ice shelves, its fluctuations became, in the Raymo et al. theory, heavily impacted by the major fluctuations in the Northern Hemisphere ice and in phase with them, as sea level rise from massive ice decay in the Northern Hemisphere would cause the retreat of the grounding lines of the marine-based ice around Antarctica, resulting in rapid ice shelf disintegration in the Southern Hemisphere. Conversely, growth of ice in the Northern Hemisphere would produce sea level fall, which would allow the Antarctic ice sheet to expand out further onto the continental shelf.[18] This proposed link between the Northern Hemisphere and Southern Hemisphere land-based ice growth and decay once the Antarctic ice sheet grew large enough to extend over the ocean provides a mechanism to explain why a shift in periodicity would have occurred at about 1 million years ago, as long as that is indeed the correct timing for when the Antarctic ice sheet formed ice shelves.

Raymo and colleagues further explain that the full development of the 100,000-year glacial/interglacial cycle may also have needed enhanced atmospheric CO_2 and albedo feedbacks.[19] Their evaluation thereby encompasses the Milankovitch cycles, ice sheet dynamics, atmospheric CO_2, and albedo feedbacks all as important factors. Others have offered alternative hypotheses. For instance, in 2006, Peter Huybers offered a new parameter, still involving summer insolation, that might better explain the 41,000- or 40,000-year glacial/interglacial cycling that appears from the paleoclimate record to have been in effect for the million years or so before 1 million years ago.[20] In 2009, Bärbel Hönisch from the Lamont-Doherty Earth Observatory and several colleagues examined and rejected a hypothesis that the transition from an approximately 40,000-year cycle to a 100,000-year cycle was caused predominantly by changes in atmospheric CO_2.[21] In 2005, William Ruddiman offered an explanation that is independent of the Antarctic ice sheet, Huybers's new parameter, or CO_2, suggesting that the shift from the 41,000- to 100,000-year cycles arose approximately 900,000 years ago because the cooling trend reached a new threshold at that time, such that the Earth system had become cool enough that the ice sheets no longer retreated back on the basis of the increased availability of summer radiation every 41,000 years, instead requiring the larger solar radiation amounts that occur every 100,000 years.[22] A full understanding will likely require further modifications to the Milankovitch and post-Milankovitch theories.[23]

Although as of the early twenty-first century the Milankovitch theory remains the most widely supported overarching explanation of the ice age cycles, it is not universally accepted, even in modified form. For instance, Ross Johnson of the

University of Minnesota suggests that atmospheric circulation changes were likely more important than orbital forcings for the initiation of the last ice age, postulating that the ice sheet growth was probably triggered by increased precipitation and summer cloudiness in Baffin Bay and surrounding regions.[24] A Milankovitch component does enter Johnson's full sequence leading to the ice age but not in the single-focus way that it enters in the more widely accepted Milankovitch theory. In Johnson's sequence, Milankovitch-based insolation decreases at 25°N latitude lead to a weakening of the African monsoons, a reduction of freshwater flow from the Nile River to the Mediterranean, and hence a rise in the salinity of Mediterranean waters. This increased salinity increases the outflow at the Strait of Gibraltar, which increases upwelling off Scotland, directs warm surface water to the Labrador Sea, increases moisture advection into Canada, and initiates Canadian ice sheet growth. It is the same basic sequence, starting at the point of the reduced Nile discharges, that led Johnson to recommend damming the Strait of Gibraltar, as described in chapter 7. In the distant past, the sequence was affected only by natural forces, whereas in the modern era, human activities are viewed as having an impact.[25]

These various theories of the ice ages demonstrate many relevant factors. First, an intricate interweaving plays out between observations and theory, with new observations sometimes necessitating either the rejection or the adjustment of a stated theory but also with the theory often steering the collection and interpretation of the observations (or data). Second, theories come and go in acceptance and popularity, with orbital forcings as a central cause of the glacial/interglacial cycles being one example of a theory that has gone in and out of favor repeatedly since the 1840s. Third, the Earth system has many intertwining elements, greatly complicating the interpretations. From the broad-scale overview of Earth's 4.6-billion-year history provided in chapter 2, it is clear that no single cause can fully explain the ice age cycling. Importantly, although orbital changes have presumably been occurring throughout Earth's history, the ice age cycling definitely has not. Before the Milankovitch cycles could affect ice coverage, the configuration of land and ocean around the globe had to be amenable to ice accumulation (chapter 2).

Many of the elements contributing to the glacial/interglacial changes were recognized by the end of the nineteenth century, although often the different possible causes were seen as contending theories rather than important complementary elements. In 1895, Italian scientist Luigi De Marchi critiqued the major extant theories of ice ages and interglacials in his book *Le cause dell' era glaciale*. Among the proposed causal factors were changes in the Sun's radiation reaching the Earth, the obliquity of the Earth's axis, the eccentricity of the Earth's orbit, the geography of continents and oceans, the vegetative cover of the Earth, ocean currents, the timing of the equinoxes, the positioning of the North Pole and the South Pole, and the transparency of the atmosphere.[26] Other theories centered on volcanic eruptions and vertical movements of the Earth's crust.[27] In contrast to Adhémar, Croll, and later Milankovitch, De Marchi favored changes in the atmosphere's transparency as the central cause,[28] particularly the importance of the greenhouse effect.

By the start of the twenty-first century, the greenhouse effect was no longer considered a prime contender as a central cause of the ice ages, especially in view of ice core data suggesting that changes in greenhouse gases have followed rather than preceded changes in temperature. Still, the greenhouse/climate connection or, more specifically, the CO_2/climate connection remains very much a topic of interest and in fact one of the foremost research topics in Earth sciences.

CO_2/CLIMATE CONNECTION AND WHETHER HUMAN IMPACTS ARE FAVORABLE OR UNFAVORABLE

Another scientist besides De Marchi who viewed changes in the atmosphere's transparency as a central causal element in the coming and going of the ice ages was the Swedish chemist Svante Arrhenius, who has become well known in climate circles because his consideration of the impact of CO_2 on the ice ages led also to the earliest calculation of projected temperature changes in the event of a future doubling of atmospheric CO_2.

Calculating the temperature response to a CO_2 doubling now has a substantial history, going back to Arrhenius's calculations in the 1890s. Recognizing the potential climate implications of the laboratory experiments of John Tyndall, an Irish scientist who had demonstrated that both water vapor and CO_2 (as well as other chemical compounds) are strong absorbers of infrared radiation,[29] Arrhenius considered the temperature consequences of altered atmospheric CO_2 amounts. Of most interest to scientists a century later, Arrhenius calculated that a doubling of atmospheric CO_2 would raise annual mean surface temperatures for most of the globe (55°S–65°N) by 5°C to 6°C, depending on latitude. This was part of a larger set of calculations in which he also showed the effect of tripling atmospheric CO_2, of raising it to 150% and 250% of its baseline value, and of decreasing it to 67% of its baseline. Arrhenius does not present the temperature increases in the polar regions for a CO_2 doubling but indicates that Arctic temperatures would rise 8°C to 9°C for a CO_2 increase of 2.5 to 3.0 times the baseline value. The 8°C to 9°C range is close to Arrhenius's tabulated values for 55° and 65° latitude for the same level of CO_2 increases.[30] Assuming that polar temperature increases with a CO_2 doubling would also be comparable in Arrhenius's scheme to the increases at 55° and 65° latitude, Arrhenius's polar temperature increases for a CO_2 doubling would be approximately 6°C, and his global temperature increases would be 5°C to 6°C. In a later publication, Arrhenius placed the surface warming in the event of a doubling of CO_2 at 4°C,[31] which is impressively well within the range of 2.0°C to 4.5°C given in the 2007 assessments of the Intergovernmental Panel on Climate Change (IPCC), based on a suite of estimates from major state-of-the-art global climate models in the early twenty-first century.[32]

In great contrast to the concerns in the early twenty-first century, Arrhenius and colleagues such as fellow Swedish scientist Nils Ekholm—not realizing how fast atmospheric CO_2 would rise—viewed the projected temperature increases due to rising CO_2 from human activities as favorable, producing a milder, more pleasant climate and delaying a return to ice age conditions. These expected climate benefits complemented the expected benefits that increased CO_2 would have on plant growth (encouraging it through increased CO_2 availability for photosynthesis), the latter being particularly important in terms of helping to provide adequate food for the Earth's growing population.[33] Increased atmospheric CO_2, at the (underestimated) anticipated levels, was considered a very good thing.

> Arrhenius viewed the projected temperature increases due to rising CO_2 as favorable, producing a milder, more pleasant climate and delaying a return to the ice ages.

Arrhenius was hardly the first person to suggest that human activities might be changing climate or that these changes might be favorable. In about 300 b.c., the Greek Theophrastus suggested that clearing land for agriculture resulted in a warmer climate, and in the 1660s, John Evelyn from London's then recently established Royal Society expressed the view that major deforestation improves both the climate and the healthfulness of a region. Many in both Europe and America explicitly stated in the 1700s that the American colonists were lessening the severity of the climate of the New World through their settlements. Among these, David Hume, around 1750, felt that the climate of Europe had been ameliorated over the centuries through the gradual increase in cultivation and that the same amelioration was now occurring more rapidly in the Americas with the widespread clearing of forests. By the 1790s, such ideas had become common among Enlightenment thinkers. In the Americas themselves, Thomas Jefferson was one of several prominent people also expressing the belief that the harsh American climate was being improved as a result of human settlements. In his *Notes on the State of Virginia*, published in 1785, Jefferson specifically mentions the much-reduced frequency of river freeze-ups and snowfalls and the fact that sea breezes were extending farther inland. However, widespread as such views might have been, there were always skeptics, and in 1850 Alexander von Humboldt mentioned in his widely read *Views of Nature: Or Contemplations on the Sublime Phenomena of Creation* that the earlier views that deforestation along the U.S. East Coast had moderated the climate were by then largely discredited.[34]

Arrhenius's favorable view toward the warming expected to be brought about by continued CO_2 emissions from human activities is fully understandable in the context of his and others' concerns in the 1890s and early 1900s about the possibility of an upcoming ice age. Among the relevant newspaper headlines conveniently compiled by Warren Anderson and Dan Gainor in 2006 are the following from October 7, 1912: "Fifth Ice Age Is on the Way: Human Race Will Have to Fight for Its Existence against Cold" in the *Los Angeles Times* and "Prof. Schmidt Warns Us of an Encroaching Ice Age" in the *New York Times*. Both stories discuss the work of Cornell University Professor Nathaniel Schmidt. A decade later, an August 9, 1923, *Chicago Tribune* first-page headline warned that a "Scientist Says Arctic Ice Will Wipe Out Canada," the scientist this time being Yale University Professor Gregory. Picking up on the *Chicago Tribune* story, the *Washington Post* followed the next day with the headline "Ice Age Coming Here."[35] How effective CO_2 emissions might be in warming the climate and potentially saving us from the coming ice remained contentious; in 1929, Arrhenius's view of the warming potential of CO_2 was declared by British meteorologist George Clark Simpson to have been largely discarded.[36]

Concerns about a coming ice age or, later, about global warming quite understandably rise and fall in large part depending on the most recent long-term (decadal or longer-scale) temperature trends. In the twentieth century, global temperatures were generally falling in the first decade, then rising from about 1910 to about 1940, then decreasing until the late 1960s/early 1970s (with major fluctuations), and by and large increasing since the early 1970s.[37] Both scientific and media attention have reflected these trends, although often with a lag, as it takes time for a trend to become apparent.

By the late 1930s, among those aware of the ongoing warming was British engineer Guy Stewart Callendar. In 1938, Callendar reexamined the issue of the potential impact of CO_2, finding like Arrhenius (and in opposition apparently to the majority view at the time) that emissions of CO_2 from human activities could cause significant temperature increases. Callendar estimated that 150 billion tons of CO_2 had been injected into the atmosphere in the previous 50 years through fossil fuel combustion, that approximately 75% of this injected CO_2 had remained in the atmosphere, and that this was a causal factor in the temperature increases during those 50 years, accounting for approximately 0.003°C warming per year of the estimated average actual warming of 0.005°C per year. Callendar also plotted the expected temperatures in the face of various atmospheric CO_2 levels ranging from below 50 parts per million (of the atmosphere) up to 600 parts per million. From that plot, a doubling of CO_2 either from preindustrial values or from the estimated then-current value of approximately 300 parts per million produces a warming of approximately 2°C. Like Arrhenius, Callendar viewed warming favorably, considering it likely to be beneficial to mankind, especially in expanding the range of productive agriculture poleward and in delaying the return of ice age conditions. However, Callendar underestimated the rate of rise of CO_2 and hence the rise in temperature, projecting temperature increases of only 0.39°C from the mean of the nineteenth century to the mean of the twenty-first century and only 0.57°C from the nineteenth-century mean to the twenty-second-century mean.[38] If he had anticipated greater changes, his attitude toward their beneficial versus detrimental nature conceivably would have been different.

By 1950, global warming was a topic of public interest, along with related phenomena such as rising sea levels, the retreat of the Greenland ice sheet and mountain glaciers, and changes in ocean fisheries.[39] These are all concerns that once again are hot topics in the early twenty-first century.

However, by 1950, temperatures had begun to cool, and during the 1950s the concern in some arenas shifted anew to the possibility of an upcoming ice age. This was not only because of the cooling, which was not widely recognized at the time, but also because of a new theory of the ice ages proposed by Maurice Ewing and William Donn of Columbia University. Ewing and Donn include as part of their hypothesized ice age cycle a sequence in which a warming Arctic leads to an ice-free Arctic Ocean, which provides a source of moisture for evaporation and thereby leads to increased snowfall and glaciation.[40] This was publicized in the popular press in an extended article in *Harper's Magazine* in 1958 by Betty Friedan, who at the time was an active although still relatively unknown journalist. Friedan warned that, according to the Ewing and Donn theory, a new ice age could be on its way, as the Arctic Ocean warming of the previous 50 years and the resulting thinning of the Arctic sea ice could lead to a removal of the sea ice, with resulting increased snowfall and glacier advances.[41]

On the scientific front, reminiscent of Simpson in 1929, noted American meteorologist Hans Panofsky in 1956 declared that further changes in the atmospheric CO_2 levels would not have much effect on atmospheric temperatures[42] so that we could not depend on the CO_2 emissions to save us from further cooling. The following year, Roger Revelle and Hans Suess indicated that Callendar in his 1938 study had wrongly assumed that nearly all the CO_2 from fossil fuel usage would remain in the atmosphere and had thereby greatly overestimated the atmospheric CO_2 increases due to fossil fuel usage. Reviving and expanding on a possibility mentioned by Arrhenius in 1903, Revelle and Suess explained that the vast majority of the emitted CO_2 would instead be absorbed into the ocean.[43] They also recognized the incompleteness of our understandings and made their much-referred-to statement that by burning fossil fuels and thereby returning carbon to the atmosphere and oceans, "human beings are now carrying out a large scale geophysical experiment" with unknown consequences.[44]

Even though overall temperatures had been decreasing since the early 1940s,[45] there were fluctuations from year to year, and occasional headlines still emphasized warming, among them a February 15, 1959, New York Times headline "A Warmer Earth Evident at Poles" and a February 20, 1969, New York Times headline "Expert Says Arctic Ocean Will Soon Be Open Sea."[46] The February 20, 1969, article quotes polar explorer Bernt Bachen as indicating that the Arctic could be free of ice within 10 to 20 years.[47] This projection was not realized, but four decades later, in the early twenty-first century, similar projections have a greater chance of being realized because of having less sea ice to begin with.

By the early and mid-1970s, cooling seemed to be a greater concern than warming. At least on January 11, 1970, when the Washington Post ran the headline "Colder Winters Held Dawn of New Ice Age," global temperatures had not yet reversed from their overall downward trend. This was no longer the case 5 years later, when on March 1, 1975, Science News warned in its cover story of a possible coming ice age, indicating that the cooling since 1940 "will not soon be reversed,"[48]* or when two months later, on May 21, 1975, the New York Times contained the following headline: "Scientists Ponder Why World's Climate Is Changing; A Major Cooling

---

* In addition to indicating the likelihood of continued cooling and the possibility of an approaching ice age, both of which are concerns that were also raised in a then-recent report of the U.S. National Academy of Sciences, the March 1, 1975, Science News article also summarizes the complexity of the Earth system and mentions the possibility that warming from CO_2 increases could conceivably eventually override the various cooling tendencies (Douglas 1975).

Widely Considered to Be Inevitable."[49] Journalist George Will wrote in a January 24, 1975, article in the *Washington Post* that some climatologists are predicting the possibility of a 2° to 3° cooling of the Northern Hemisphere by 2000. "If that climate change occurs," wrote Will, "there will be megadeaths and social upheaval" because of its impact on grain production.[50] No wonder George Will has not been leading the charge in warning about global warming in the late twentieth and early twenty-first centuries; he probably learned well from the 1975 experience that climate predictions (even from the experts) are not necessarily reliable, no matter how scary and seemingly media-worthy they might be. Journalist Lowell Ponte went beyond Will, arguing in his 1976 book *The Cooling* that the decreasing temperatures in the Northern Hemisphere could lead to a collapse of the Soviet grain harvest, followed by a Soviet invasion of western Europe and perhaps the start of World War III,[51] illustrating that scary scenarios based on climate projections are not new with the twenty-first century.

One of the scientists most vocal in expressing concern about the cooling Earth was Reid Bryson of the University of Wisconsin. Bryson recognized that CO_2 emissions contributed a tendency toward warming, but he also recognized that this tendency, until the 1970s, was being outweighed by other forces, as the global average temperatures were decreasing rather than increasing. Bryson attributed the cooling from 1940 to the early 1970s to a combination of the particulate matter injected into the atmosphere during volcanic eruptions and the particulate matter poured in through human industrial activities and agriculture.[52] This particulate matter produced a cooling effect by increasing the reflectivity (albedo) of the Earth/atmosphere system. Bryson calculated that in 20 years the Earth had gone one-sixth of the way toward a new ice age.[53]

Bryson was hardly the only scientist in the early 1970s sounding alarms regarding an upcoming ice age. S. Ichtiaque Rasool and Steve Schneider, both at the time at NASA's Goddard Institute for Space Studies, published a study in which they examined the potential impacts on temperature of both CO_2 emissions and increases in particulate matter (or aerosols). Regarding CO_2, they concluded that any temperature increases from it would not be a concern, in fact finding that "even an increase by a factor of 8 in the amount of CO_2, which is highly unlikely in the next several thousand years, will produce an increase in the surface temperature of less than 2°K [= 2°C]."[54] Regarding particulate matter, however, their conclusions were quite different, finding that dust increases in the atmosphere could decrease average surface temperatures by 3.5°C as early as sometime within the twenty-first century and that just a few years of such temperatures "could be sufficient to trigger an ice age!"[55] Scientists George Kukla and Robley Matthews were concerned enough to write a letter to U.S. President Richard M. Nixon in December 1972 warning that the "present rate of the cooling seems fast enough to bring glacial temperatures in about a century, if continuing at the present rate."[56]

John Gribbin, a climate researcher at Cambridge University in England, summarized many of the ice age concerns in his book *Forecasts, Famines, and Freezes: The World's Climate and the Future of Mankind*, originally published in 1976 with a paperback edition the following year. Knowledgeable about the many factors influencing climate and the reality of climate changes fluctuating back and forth, Gribbin was not as fearful of an imminent ice age as some of the other writers of the time. Still, he does declare, "We are indeed entering a period of at least a mini ice age."[57] Gribbin suggests that the cooling since the 1940s might have been caused in part by the effects on the atmosphere of the volcanic eruptions of Alaska's Mount Spurr in 1953 and Kamchatka's Mount Bezymiannyi in 1956 and that the 11-year solar cycle has a marked impact on global climate as well, while he rejects the concept that dust emissions from human activities had a significant impact. Gribbin also addresses the issue of CO_2 emissions, describing their warming potential through the greenhouse effect, but concludes, "According to the latest calculations, we have nothing to worry about from the greenhouse effect."[58] Gribbin labels himself an optimist, clarifying that he does not share the pessimists' angst that the coming full-scale ice age is imminent, instead feeling that it may arrive within the next few hundred years and that our global society may well be able to adapt to it.[59]

> Warnings in the early and mid-1970s about the coming ice ages were replaced in the mid- and late 1970s by warnings about global warming.

Warnings in the early and mid-1970s about the coming ice ages were replaced (once again) in the mid- and late 1970s and beyond by warnings about global warming. The shift started slowly and certainly had precursors in the 1960s and early 1970s.[60] Several scientists in this time frame were recognizing the likelihood of anthropogenically induced warming, although not necessarily concerned about it. For instance, William Sellers of the University of Arizona performed noteworthy calculations with a primitive (by later standards) global climate model and stated that "man's increasing industrial activities may eventually lead to the elimination of the ice caps and to a climate about 14C warmer than today."[61] However, it is only in hindsight that this becomes a "warning," as Sellers nowhere indicates a concern about the likely temperature increases. Quite the contrary, he calculates that an ice age could be created by a reduction of about 2% in the solar constant and indicates that a 3% decrease in a particular coefficient, "m" (specifically, the "atmospheric attenuation coefficient"), could bring us to the "brink of an ice age," a statement followed immediately by: "Fortunately, because of the increasing carbon dioxide content of the atmosphere, m is more likely to increase than decrease. Hence, the global mean temperature should slowly rise."[62] Note in particular the word "fortunately." The renowned Soviet climatologist Mikhail Budyko, in a 1972 article, "The Future Climate," matter-of-factly states that the ongoing increases in atmospheric CO_2 will increase near-surface temperatures and reduce polar ice coverage and that the changes could have significant practical consequences, hence warranting careful study. Like Sellers, he does not explicitly indicate great concern about the expected warming, instead pointing out the likely descent into a major glaciation without it, even mentioning the possibility of a complete global glaciation in several hundred thousand years.[63]

The shift toward consensus and forceful alarm about global warming picked up speed in the 1980s and has not yet reversed. Among the prominent U.S. scientists voicing concerns in the 1970s were Wally Broecker of Columbia University and Will Kellogg and Steve Schneider of the National Center for Atmospheric Research in Colorado, Schneider having moved to Colorado from his position at the Goddard Institute for Space Studies in 1972. Broecker in 1975 projected that the cooling since about 1940 would "bottom out" and be succeeded by a "pronounced warming induced by carbon dioxide."[64] He was certainly correct about the cooling being succeeded by a warming and was quite likely also correct about the warming being induced by CO_2. By the time his paper was published, the cooling had in fact already "bottomed out." Another of the projections in Broecker's 1975 paper is that the upcoming warming would be sufficient by 2000 to raise the average global temperature to a level higher than it had been at any time in the past 1,000 years, and although there is not universal agreement, many scientists in the early twenty-first century do feel that temperatures have risen enough to validate Broecker's 1975 projection. Similarly, cautiously given estimates by Kellogg and Schneider in 1974 that atmospheric carbon dioxide might rise to about 380 by 2000 and that global temperatures might rise in response by about 0.5°C[65] are not considered far off the mark.

Schneider published a book that came out during the latest major transition from concerns centered on cooling to concerns centered on warming. Wonderfully titled *The Genesis Strategy*, the book encourages maintaining large margins of safety, in line with the advice given by the Hebrew slave Joseph to the Egyptian Pharaoh in the Old Testament book of Genesis. The Pharaoh wisely heeded

Joseph's warning that 7 years of feast might be followed by 7 years of famine and stored food to prepare for that eventuality. Schneider lays out the cooling/warming issues, including the following: the fact of cooling since 1940, perhaps due to volcanic and other aerosols; the uncertainty regarding whether the net effect of aerosols is one of cooling or one of warming; and the possibility that the cooling since 1940 could be replaced by a warming due to the continuing increases in CO_2. He presents concerns and the rationales behind them while carefully steering clear of predicting that we are actually headed either toward severe cooling or toward severe warming.[66] After it became clear in subsequent years that the overall global cooling from the 1940s to the early 1970s had reversed to a warming, Schneider published a new book that was unequivocal in its concerns about the current and projected continued warming and the importance to it of greenhouse gas increases due to human activities.[67]

In the 1980s, another scientist, Jim Hansen, emerged in the limelight regarding global warming issues. By the early twenty-first century Hansen, a NASA scientist and director of NASA's Goddard Institute for Space Studies, has become probably the scientist most widely known in the United States for expressing concerns about warming and the likely damaging impacts. Hansen has published prolifically in the scientific literature, but it was his June 1988 testimony at a U.S. Senate hearing on climate change that more than any other single event catapulted him to the forefront as a scientist associated with global warming issues. At the hearing, Hansen voiced "99% confidence" that the Earth is warming and high confidence that the warming is connected to the greenhouse effect. At the time, quite a few scientists objected to Hansen's testimony, especially his statement of 99% confidence,[68] but over the course of the succeeding two decades, the scientific mainstream has increasingly praised Hansen's courage and foresight rather than denouncing effective sound bites that might lack scientific rigor.

Hansen and many others have used major global climate models to calculate what can be expected to happen to global temperatures in the event of a doubling of atmospheric CO_2. Interestingly, Arrhenius's 1896 calculation of a 5°C to 6°C temperature rise for such a doubling is remarkably close to the modern estimates, even though his was done without the benefit of computers or modern measurements. Callendar's 1938 estimate of a 2°C warming, also done without the benefit of computers or modern measurements, is also close to the early twenty-first-century results. In the modern era of computer calculations, Syukuro Manabe and Richard Wetherald from the Geophysical Fluid Dynamics Laboratory at Princeton University performed doubled-CO_2 calculations in 1967 with a one-dimensional computer model, obtaining a 2.3°C global warming,[69] then repeated the calculations in 1975 with a more sophisticated three-dimensional model, obtaining a 2.9°C global warming.[70] Many individuals and groups have done doubled-CO_2 calculations since that time, with different models and different baseline assumptions but with most of the results obtaining global temperature increases somewhere between 1°C and 6°C in the event of a doubling of atmospheric CO_2. Hansen's group in 1984 calculated a range of 2.5°C to 5.0°C,[71] and the IPCC provides a range of 2.0°C to 4.5°C in its 2007 assessments, based on results from a suite of major global climate models.[72] The projected warming is not uniform geographically, instead tending to be markedly larger in the polar regions than at lower latitudes, due in large part although not entirely to the positive feedbacks associated with the snow and ice covers at polar latitudes.[73]

I am pleased to have participated with American climatologist Will Kellogg in one of the earliest computer modeling studies on the projected impact of doubled CO_2 on a variable other than temperature. Using 5°C as a conservative estimate for the warming in the Arctic in the event of doubled CO_2, Kellogg and I found that with such a temperature rise, the model-simulated Arctic sea ice disappears completely in August and September but re-forms in mid-autumn.[74] Little did I know at the time the flood of doubled-CO_2 studies about to inundate the scientific literature over the following decades, with increasingly sophisticated models and results, or the amount of attention that would be given in the early twenty-first century to the possibility of a late summer ice-free Arctic. A few years after the 1979 study, I used the same sea ice model in a similar study for the Antarctic ice, done in collaboration with NASA colleague Robert Bindschadler. Bindschadler and I found that with a 5°C temperature increase, the model simulates a midsummer ice-free Antarctic ocean (or Southern Ocean) in all except one grid square plus a wintertime Antarctic sea ice cover reduced by half from the values simulated without the temperature increase.[75] In the subsequent decades, doubled-CO_2 results have become available for many variables, from a variety of climate models.

By the early twenty-first century, a strong scientific consensus exists that global temperatures are rising and that part of the cause is the anthropogenic emission of CO_2 and other trace gases into the atmosphere. The evidence seems compelling that the Earth has warmed, overall, over the past 150 years and that human activities have contributed to the warming. Although warming has favorable as well as unfavorable consequences,[76] in view of the projections of extremely unfavorable consequences (see chapter 5) if human activities continue in a business-as-usual mode, it would certainly be prudent to adjust our activities to become less potentially damaging.

However, we should also be duly cautious about our predictions, recognizing, as the cases in this chapter illustrate, that people have been quite certain in the past also and have been mistaken. The fact of a large consensus among scientists is relevant but not conclusive, as even a position with a large consensus can be wrong. Furthermore, as detailed in the next two chapters, the models on which the climate predictions are based are far from perfect (chapter 10), and there are many pressures contributing to making the consensus seem to be larger than perhaps it is (chapter 11).

QUALIFIER

The examples provided in this chapter are not meant to denounce all experts or all consensus positions. Experts and scientific consensus by and large remain our best chances for getting correct information. However, geoengineering is so fraught with the potential for worldwide calamity that people need to understand not only that the geoengineering schemes, if implemented, could lead to serious unintended consequences (chapters 6 and 7) but also that experts advocating geoengineering, like all experts, could be wrong even about a subject at the core of their expertise. As a child, I naively saw science as more consistently cumulative than I now recognize it to be, thinking then that each new piece of information added further to the grand edifice and not understanding that at times major chunks of the edifice crumble, to be replaced by a new, sparser skeleton that then, bit by bit, gets fleshed in and filled out until eventually it too crumbles. This wobbly route toward scientific advance is not entirely satisfying, but it is the way that science works, and it is important to understand this before accepting too dogmatically any of today's favored theories. On the other hand, it is equally important not to become paralyzed by uncertainties. We need to move forward based on the best information we have (generally coming from consensus positions of the experts), but we also need to be cautious, recognizing that even our best information might be flawed. Before implementing any large-scale geoengineering scheme, the imaginable potential consequences should be examined very carefully.

CHAPTER 10

The Unknown Future: Model Limitations

The computer models used to simulate future climates are the best tools available for climate predictions but are far from perfect.

THE CENTRAL tool used today for predicting future climate is computer modeling. Global climate models, abbreviated as GCMs,* come in a variety of forms and different levels of sophistication but typically include such key components as the atmosphere, the oceans, sea ice, and land vegetation. Some of the models perform calculations for 20 or more levels in the atmosphere and 20 or more levels in the oceans, doing so at horizontal resolutions around the globe of 100 kilometers or better. The GCM calculations are based on fundamental laws of physics, simplifications of those laws, numerical techniques for solving the equations, and parameterizations (i.e., simplified representations) of many of the processes that for one reason or another are not able to be incorporated precisely.[1]

GCMs have proven to be wonderful tools. The best of them successfully simulate, at least approximately, many of the basics of global climate, including the fundamental geographic patterns and seasonal cycles of atmospheric and oceanic temperatures, the jet stream and trade winds in the atmosphere, and the Gulf Stream and other major currents in the oceans. They can simulate at least some key aspects of modern large-scale atmospheric and oceanic phenomena and at least some key aspects of past climate states around the globe.[2]

---

* "GCM" is alternatively sometimes spelled out as "general circulation model," especially in the case of models that center exclusively on the atmosphere and/or oceans.

Because of their success in simulating current and past climates, GCMs have been widely used to simulate possible future climates as well. In particular, they have been used in numerous simulations that have been incorporated into the assessment results of the Intergovernmental Panel on Climate Change (IPCC) (see chapter 5). The IPCC efforts have involved examining projected future climates under a variety of different scenarios, including different possible levels of greenhouse gases and aerosols, and many of those projections and analyses are widely referred to in both the scientific literature and the popular press. Results from several of the projections are included in chapter 5.

However, no computer model fully mimics the entire climate system, and all modelers are aware that the models have limitations and that the projections from the models should not be assumed to predict precisely what the future will be. In fact, some scientists are extremely doubtful about the credibility of the predictions, as epitomized by William Cotton and Roger Pielke Sr. of Colorado State University and the University of Colorado, respectively, in their statement that the problems "are so complex that it may take many decades, or even centuries (if ever!), before we have matured enough as a scientific community to make credible predictions of long-term climate trends and their corresponding regional impacts."[3]

This chapter describes some of the major limitations of the models and why their predictions should be viewed with at least some caution. Because of the importance of the international IPCC efforts, involving many hundreds of scientists (including myself as a minor contributor), I first summarize both the strengths and the limitations of climate models as identified in the 2007 IPCC report.

SELECTED IPCC-HIGHLIGHTED STRENGTHS AND LIMITATIONS OF CLIMATE MODELS

The 2007 IPCC 74-page chapter "Climate Models and Their Evaluation" summarizes notable progress in model development in the early twenty-first century and some of the remaining limitations.[4] Early twenty-first-century progress has come through enhanced scrutiny and more comprehensive testing of the models as well as changes in the model formulations and improvements in model resolution (providing greater spatial detail). A key element of the resulting progress is enhanced understanding of the differences in the results from different models, with cloud feedbacks being confirmed as a primary reason for the differences in simulated climate sensitivity to a doubling of atmospheric CO_2 concentrations. Other areas of progress include improved simulation of the influential El Niño/Southern Oscillation (ENSO) events by some of the models and improved simulation of the transport of heat in the oceans.[5] The increased capacity and speed of computers has greatly facilitated the incorporation of finer horizontal and vertical resolutions, the generation of multiple simulations under a variety of different conditions, and the incorporation of additional processes into the models. Quite understandably, far more advanced, more complete models can now be used to perform long-term simulations in far less time and with far less effort than in the early days of computer climate modeling in the 1960s and 1970s.

The IPCC process itself has substantially contributed to the progress being made in understanding differences among model results. Specifically, the IPCC has encouraged extensive analysis and intercomparisons of results from a coordinated set of simulations done by many modeling groups. Important model intercomparisons had been done earlier, with two primary examples being 1) an examination of the responses of 14 atmospheric general circulation models to changes in prescribed sea surface temperatures, finding that the 14 models produced comparable results in locations without clouds but very different results under cloudy conditions and concluding that a primary need in modeling efforts is improved treatment of clouds,[6] and 2) the extensive international Atmospheric Model Intercomparison Project, initiated in 1989 to test and intercompare atmospheric model performances.[7] Nonetheless, the IPCC intercomparisons have made a gigantic additional contribution through the widespread participation in the IPCC process by scientists around the world and the widespread distribution of the results.

The 2007 IPCC report makes especially strong use of 23 major atmosphere/ocean GCMs from Australia, Canada, China, France, Germany, Japan, Korea, Norway, Russia, the United Kingdom, and the United States, with some countries housing two or more of the models. For those 23 models, improvements in the first several years of the twenty-first century included the following: improved resolutions; incorporation of more processes regarding aerosols, the land surface (including carbon cycling through vegetation and soil, feedbacks between soil moisture and precipitation, and multilayer snow models), and sea ice dynamics; and, for many of the models, the elimination of "flux adjustments"* that had artificially reduced the amount of undesired climate drift in the simulations.[8] (Undesired climate drift

---

* A flux adjustment is an artificial change made in the model to the calculated amount (or "flux") of some quantity, such as heat or water, transferred between the ocean and the atmosphere, the purpose being to produce more realistic results and in particular to reduce undesired climate drift.

occurs, for instance, when the simulated climate slowly drifts toward warmer or cooler conditions because of something internal to the calculations. Reducing such drift is always desired, but doing so by artificially adjusting fluxes to prevent the model results from drifting frequently makes the results look much better than the model itself is capable of simulating. Hence, it is preferable to avoid flux adjustments and, if possible, instead reduce the undesired climate drift by correcting the flaws within the model that cause the drift in the results. Determining and correcting these flaws is an important aspect of model improvement.)

Among the model limitations mentioned in the 2007 IPCC report are the following: many important processes are included in the models only in approximate form, partly because of limitations of the computers and partly because of limitations in our scientific understanding or in the observational database; major uncertainties exist regarding the representation of clouds; most of the GCMs used for the IPCC climate predictions do not include ice sheet calculations; no major GCM includes calculations for smaller glaciers and ice caps; inadequate horizontal resolution hinders the proper simulation of upwelling in the oceans; "serious systematic errors in both the simulated mean climate and the natural variability" limit the success of ENSO predictions despite the considerable progress made; systematic errors exist in the simulation of the Southern Ocean, and these propagate to lower latitudes because of the importance of the Southern Ocean in global ocean heat exchanges; and the simulation of the Madden-Julian Oscillation, an important oscillation in the lower atmosphere of the tropics, remains "unsatisfactory."[9]

The IPCC report presents many simulation results and indicates where they match observations well and where they fail to do so. One notable weakness in the simulations is the underestimation of precipitation in "intense" rainfall events, defined conservatively as events with greater than 10 millimeters of precipitation in a day and hence including many rainfall events that would be fairly routine in some locations.[10] (World rainfall records include a 1-minute rainfall of 31.2 millimeters in Unionville, Maryland, on January 4, 1956, and a 24-hour rainfall of 1,825 millimeters [1.825 meters] in Foc-Foc, La Reunion Island, during the passage of Tropical Cyclone Hyacinthe on January 18, 1966.[11])

Despite the acknowledged limitations, the IPCC authors express strong confidence in the models and in their ability to make meaningful projections: "There is considerable confidence that climate models provide credible quantitative estimates of future climate change, particularly at continental scales and above."[12] They specify three major sources for their confidence: the fact that the models are based on established physical laws, the fact that the models correctly simulate many aspects of the current climate, and the fact that the models correctly simulate some of the key features of past climates.

CONCERNS

I agree with the majority of the IPCC points mentioned in the previous section, including the specific limitations of the models and the specific major improvements made in the models during the first few years of the twenty-first century. I also recognize that the model improvements have led to measurable improvements in model performance, as detailed, for instance, in the case of the simulated average modern climate state by Thomas Reichler and Junsu Kim in 2008.[13] Still, I am less confident that today's models "provide credible quantitative estimates of future climate change," as indicated in the 2007 IPCC report.[14] They provide our best guesses and hence are extremely important, but even our best guesses could be wrong, and they could be wrong in the broad overview as well as in the details.

I applaud the successes (while recognizing their imperfect nature) in the simulations of past and present climates. However, success in simulating the past and present does not necessarily translate to success in simulating the future, one fundamental reason being that for simulations of the past and present, we have the possibility of tuning the model to improve the match between the simulated results and a variety of observational data records, but until the future arrives, there are no observational data records to allow a similar tuning for the simulated future. As a result, simulations of the future cannot reasonably be expected to be as accurate as simulations of the past or present. Further, although the models are based on established physical laws, they are necessarily based on only a subset of the physical, biological, and chemical laws, processes, and constituents acting within the Earth system, as the Earth system has far more intricacies than any model can fully handle either now or in the foreseeable future. As our understanding improves, we might find that some of the laws, processes, and constituents left out of the models are critically important.

> Success in simulating the past and present does not necessarily translate to success in simulating the future. For simulations of the past and present, we have the possibility of tuning the model to improve the match with observations, but there are no observations to allow a tuning for the simulated future.

The next three sections detail more fully some of the issues related to tuning and to the incompleteness of the models. These sections are followed by a section specifically addressing the issue of the uncertainty in the many widely referenced predictions of the amount of global warming that can be anticipated in conjunction with a doubling of atmospheric CO_2 from preindustrial levels.

TUNING

The concept of tuning and the issues involved can perhaps best be explained through an example, so I present here an example with which I am intimately familiar, specifically one that arose during my PhD dissertation research. This research was performed in the mid-1970s and centered on the development of a computer model of the sea ice covers of the Arctic and Antarctic. One of the parameters needed for the model was a value for the flux of heat upward from the underlying liquid ocean to the underside of the floating ice. This heat contributes to melting the ice and hence is important for calculating the sea ice thickness and its progression through time. As I developed the model, I sought a reasonable value for this and other parameters that were needed as model inputs. I found a value for the ocean heat flux, at 2 watts per square meter (W/m^2), from a study[15] done by two noted Arctic sea ice researchers, and I inserted that value in the model. After performing the simulations, however, and comparing them against sea ice observations, I found that the simulated ice cover was fairly realistic, overall, for the Arctic but showed far too expansive an ice cover in the Antarctic. I was doing this work at the National Center for Atmospheric Research (NCAR) in Boulder, Colorado, and as I puzzled over this problem, I was fortunate that a renowned oceanographer, Arnold Gordon from Columbia University's Lamont-Doherty Geological Observatory (now the Lamont-Doherty Earth Observatory), was visiting NCAR. So I approached him about the problem of the overly expansive simulated Antarctic ice cover. Dr. Gordon contemplated the problem and asked what value I was using for the ocean heat flux. When I replied, he explained that I should try the simulation with a much higher value, perhaps as high as 25 W/m^2, for the Antarctic because of the significantly different ocean heat availability directly under the sea ice in the two polar regions. So I tested three values, including 25 W/m^2, and, lo and behold, the results were so much better with the 25 W/m^2 value in the Antarctic that I proceeded to complete the dissertation using an ocean heat flux of 2 W/m^2 in the Arctic and an ocean heat flux of 25 W/m^2 in the Antarctic. This is an example of "tuning" a model to get a desired result.

Although in reality the ocean heat flux to the overlying sea ice varies considerably throughout each polar region and varies considerably through the course of a year, using a single value in the Arctic and a single value in the Antarctic, as I did in my dissertation, is the type of simplification that is standard procedure in the early development of attempting to model a geophysical feature such as sea ice. This instance, however, was the first time I personally saw the extent to which I could adjust the model results to what I wanted by tuning the model, in this case adjusting only one variable, the ocean heat flux. Note that I could do this only because I knew what the result should be, namely, the current distribution of sea ice, which I was trying to match in the simulation. This is the type of tuning that can be quite effective in getting realistic results when the desired results are known. It is an entirely different situation when a person is trying to simulate the unknown future and hence does not already have the answer to which he or she wants to tune the results. It reminds me of physics labs in high school, when the correct answers are often known beforehand and students occasionally work quite diligently to make whatever adjustments are necessary to obtain the already known correct answers.

A few years after the dissertation work and the ocean heat flux adjustments, my colleagues and I at NASA Goddard Space Flight Center were contemplating a large open-water region, termed a "polynya," that appeared in the midst of the sea ice cover of the Weddell Sea off the coast of Antarctica in the winters of 1974, 1975, and 1976. This wintertime Weddell polynya became known through the satellite data that we had recently compiled and were analyzing. At the time, in the early 1980s, 1974–1976 were three of the four years (1973–1976) for which we had analyzed satellite data for the Antarctic sea ice, and it was thought that perhaps this wintertime Weddell polynya was a common occurrence.* It occurred to me that I could probably simulate such a polynya with my sea ice model simply by adjusting the wind fields used in the simulation. So I made the attempt, and sure enough, I was able to simulate a Weddell polynya under some sets of wind conditions and no polynya under other sets of wind conditions.

I liked the Weddell polynya study and the resulting paper,[16] as it was a straightforward and clean-cut experiment, with the explicit result that both a Weddell polynya and its absence could be successfully simulated with this sea ice model through a simple manipulation of the wind fields, with no other differences in the model inputs or the model formulations. However, while liking the cleanness of the result, I also realized that the model simulations, with the polynya appearing

---

* The wintertime Weddell polynya proved not to be common at all, as became clear through additional years of satellite data. By 2009, the large Weddell polynya had still never reemerged in full force after its last appearance in 1976.

and disappearing based exclusively on the wind fields, might be totally disconnected from the causes of the actual Weddell polynya. Other scientists, led by Doug Martinson, concluded from other considerations that the polynya was caused by oceanographic conditions,[17] and the *Globe* tabloid speculated that it was generated by the heat coming from a UFO base "housing a city of alien creatures" located underneath this "mysterious warm-water lake" in the midst of the Antarctic ice pack.[18] The *Globe*'s UFO speculation is an imaginative hypothesis although extremely unlikely, whereas the Martinson-led conclusion is far more solidly based. Most important, the fact that the Weddell polynya simulated by me with a sea ice model was directly caused by manipulation of the wind fields in the model by no means established that the actual Weddell polynya was caused by winds. I recognized that and made it clear in the published paper.[19] In huge contrast, it seems that in many cases, other modeling results are being interpreted by the researchers or the media as "proving" something about the actual physical world when they too are only establishing something about the modeled or simulated world, which is, at best, only an approximation to the physical world in which we live and breathe.

The ocean heat flux study, the Weddell polynya study, and similar cases strengthened my recognition that a skilled computer modeler can, with fairly minor manipulations, simulate quite a range of results, whether they concern sea ice or oceans or future atmospheric temperatures or any of numerous other variables. I doubt that many (if any) scientists simulate something with the intention of being misleading in their resulting publications, but it is probably fairly common for a scientist hearing of an interesting possible phenomenon or a possible cause for an observed phenomenon to manipulate the model to simulate that phenomenon or that causal relationship. Such efforts can be constructive, valid, creative scientific research. The problem arises when people believe that because the model has simulated a phenomenon or proposed relationship, the simulated version must be an accurate representation of what happens in the real world, as this is simply not the case. The fact that I was able to adjust the wind fields to simulate either the presence or the absence of a Weddell polynya established a cause-and-effect relationship within the model simulation, but it by no means established that the winds caused the actual Weddell polynya. Computer models

> The fact that I was able to adjust the wind fields to simulate either the presence or the absence of a Weddell polynya established a cause-and-effect relationship within the model simulation, but it by no means established that the winds caused the actual Weddell polynya.

are tremendously powerful and helpful tools for helping us understand the world, but they are not replicas of the world, and it is extremely important that people interpreting and/or using the results understand that. On an intellectual level, probably every scientist does fully understand that, and yet, sometimes in relaying the results, the scientists do not project that particular understanding, instead giving a false impression that the simulated results mean more than they do.

INCOMPLETENESS OF THE MODELS

Each GCM incorporates only the components and processes that the model developers insert into the model. The Earth system is very large and very complex, and it would be impossible to incorporate everything into the calculations. Even if the computer capabilities could handle every molecule in the entire system, no observational program could establish where every molecule is and its critical characteristics. Hence, when modeling the Earth system, some simplification is inherently essential. Among the most important tasks that a modeler faces is the determination of which simplifications are appropriate. This varies considerably, depending on the computer capabilities and the purposes intended for the model. For instance, if the purpose is to simulate weather patterns a few hours in advance, the need for accurate simulation of ocean-to-atmosphere heat fluxes is far less than if the purpose is to simulate long-term global climate.

Early GCMs had huge simplifications even in the basic land/sea geography of the Earth. For instance, in a renowned early study of the impact of doubling atmospheric CO_2, Syukuro Manabe and Richard Wetherald used a model geography that extends only from the equator to 81.7°N and across 120° of longitude, with a single block of land and a single slightly smaller block of ocean.[20] This looks nothing at all like a map of the actual Earth. In contrast, by the early twenty-first century, most GCMs have quite realistic depictions of the land/sea boundaries of each continent and of many large islands. The land surface topography and ocean bottom topography tend to be less realistic, with the mountains and ocean trenches generally considerably more blocky in the models than in the physical world.

Many early computer models centered exclusively on atmosphere or ocean calculations, not both, and excluded calculations for ice and for vegetation. Bit by bit, progress has been made in incorporating more elements into the models, and this has been vitally important to opening the possibility of simulating more of the connections within the Earth system and hence obtaining more complete and eventually more realistic results. However, this evolution has not been without its difficulties. The inevitability of at least partial setbacks can be illustrated by considering a smoothly running atmosphere model with, after years of effort, quite realistic results for the atmosphere, based in part on care taken to incorporate realistic boundary conditions at the bottom of the atmosphere. When such a model is enhanced by incorporating into it calculations also for the ocean (e.g., by "coupling" the atmosphere model with a preexistent ocean model), suddenly the boundary conditions at the bottom of the atmosphere, such as sea surface temperatures, are no longer conveniently set at realistic values but are instead calculated as part of the model calculations. Any errors in the ocean calculations now get propagated into the atmosphere calculations, and, correspondingly, errors in the atmosphere calculations get propagated to the oceans. Hence, the atmosphere results in the coupled atmosphere/ocean simulations understandably initially are less realistic than those calculated in the original finely tuned atmosphere-only simulations. Still, getting a model appropriate for long-term climate simulations requires having both atmosphere and ocean calculations, so modelers deal with the initial drawbacks of the coupling. This is sometimes done through temporary means, such as the flux adjustments mentioned earlier in this chapter, and it is at other times done through accepting the less realistic results while working to improve the model formulations.

By the 1980s, many GCMs included both atmosphere and ocean calculations but not much in the way of ice or land/vegetation calculations. By the early twenty-first century, many GCMs include atmosphere, ocean, sea ice, and land/vegetation calculations but no land ice calculations and varying levels of completeness in the atmosphere, ocean, sea ice, and vegetation calculations. For instance, in the case of sea ice, some of the GCMs include calculations for thinning and thickening of the ice (ice thermodynamics) but not for movement of the ice (ice dynamics), while others include ice dynamics but at differing levels of detail. Furthermore, in addition to not including land ice, most early twenty-first-century GCMs do not include thermokarst lakes (chapter 2) and the potentially important methane emissions to the atmosphere from them,[21] land ecosystem details that can affect the fluxes between the land and the atmosphere,[22] and other features of the climate recognized as important at least in some locales and time periods. By not including these elements, the models cannot simulate accurately changes that depend heavily on them.

Supplementing the problem of the omission altogether of some elements from the models is the problem of the incomplete treatment of other elements. Here I highlight clouds and dust while referring the interested reader elsewhere[23] for additional examples.

The treatment of clouds in particular, including such details as the effect of cumulus convection, is widely recognized as a major source of uncertainty in predictive simulations.[24] Clouds are well known to have multiple and varied impacts in the Earth system. They both reflect radiation from the Sun, which has a cooling effect, and absorb and emit radiation, which has a warming effect. With some clouds, the warming effect outweighs the cooling effect, but with other clouds, the cooling effect outweighs the warming, and indeed whether a particular cloud has a warming or cooling effect at any moment often depends on the time of day. For instance, a large bulky cloud on a hot summer day in the midlatitudes tends to keep the near-surface air cooler than it otherwise would be during daytime hours (because of reflecting sunlight back to space) but warmer than it otherwise would be during the night (because of a combination of blocking Earth radiation from getting out and emitting radiation back toward the surface). Probably everyone in the GCM community agrees that clouds are important and that a complete model must include them, but exactly how to incorporate (or "parameterize") them is a topic of considerable ongoing research and uncertainty. Their constantly changing sizes and shapes and ephemeral nature make clouds particularly difficult to nail down, and indeed different models parameterize them and their radiative interactions with the rest of the Earth system quite differently, with different resulting simulated cloud feedbacks.[25] With so much uncertainty about how cloud processes should be parameterized in models and with the contrasting results depending on those parameterizations, the models are quite unable at this point to predict convincingly how clouds will change as the climate continues to evolve and what the full consequences of those changes will be.

> With so much uncertainty about how cloud processes should be parameterized, the models are quite unable at this point to predict convincingly how clouds will change as the climate continues to evolve and what the full consequences of those changes will be.

Dust is another important element in the Earth system—affecting, for instance, air quality and health, hurricane formation, and the transmission through the atmosphere of solar and Earth radiation—that climate modelers are actively working on incorporating into their models. Although improvements have been made in the early years of the twenty-first century, the simulated spatial distributions of dust tend to be notably flawed, often with the dust spread too uniformly over large regions.[26]

Regarding simulating past climates, Gisela Winckler and colleagues mention cases in which models simulated that the dust amounts during the peak of the last ice age were 12 to 105 times greater than now versus the estimate of 2.5 times greater based on available observational data. Even more surprising, they mention a newer simulation that actually simulates less dust during the peak of the last ice age than today, in stark contrast with the observational data.[27] Clearly, work remains to be done.

Complementary to incorporating all the relevant important features in a model and to handling them adequately is the issue of incorporating the necessary spatial detail. Model calculations are often done on a three-dimensional grid, with the globe's surface divided into a two-dimensional horizontal grid and the third dimension extending that grid upward into the atmosphere and downward into the ocean. If the horizontal grid elements have, say, 200 kilometers on a side, then an individual grid square represents 40,000 square kilometers of the real world, with one value accepted for the grid square's height above sea level, one value calculated for its surface temperature, one value calculated for its near-surface wind speed, and so forth, despite the fact that the 40,000-square-kilometer area in the real world likely contains a large variety of surface types, surface topography, temperatures, wind speeds, cloud characteristics, and so forth. As the horizontal spacing, or "resolution," improves to lower values, for instance from 200 kilometers on a side to 50 kilometers on a side, then considerably more spatial detail is possible in the simulation, but at the same time the computer calculations need to be done at many more points, increasing the number of calculations, the time that it takes to complete the calculations, and the space needed in the computer to store all the results. The situation is similar with the model's vertical resolution as well, with improved resolution having the advantages of greater detail and more realism but the disadvantages of increased computer expense and storage requirements and increased time needed to run the model. Fortunately, the disadvantages have been tremendously lessened over the years by the vastly increased speed of computer calculations and increased computer storage capacities. As a result, improved resolution has been not only a major goal for modeling groups over the years but also a strong trend.

As of the early twenty-first century, one of the world's highest-resolution GCMs is Japan's Earth Simulator, which began operations in March 2002 and can handle 96 layers in the atmosphere, 54 layers in the ocean, and horizontal resolutions of approximately 10 kilometers in the atmosphere component and 0.1° of latitude and of longitude in the ocean component. This reflects a gigantic improvement over almost all previous models, and the high resolution allows the Earth Simulator to simulate many fine-scale details in the oceans and atmosphere that could not have been simulated with models with considerably coarser resolution. Still, there was a problem as the Earth Simulator began major Earth simulations. The desired high-resolution simulations could be run in a reasonable amount of time, but the output data sets were too voluminous for convenient storage and handling, resulting in the need to "thin" the stored output data (retaining only a subset of the results) in some cases and to limit the length of the simulations in other cases. In other words, when the Earth Simulator began operating, although the huge number of calculations involved could be performed well, the computer storage requirements for the full retention of the results proved to be a major limitation.[28] In addition, significant as the improvement down to a 10-kilometer horizontal resolution was, even that fine a resolution requires considerable simplification, as almost all 10-by-10-kilometer areas on the Earth contain considerable variety (to personalize it, consider your own neighborhood and the variety there).

Despite the quite substantial progress that has been made in computer modeling over the past several decades and the huge advances incorporated in the Earth Simulator and other major GCMs, even the most sophisticated and advanced climate models cannot come close to simulating the full complexity and interconnectedness of the global climate system. They cannot incorporate all the chemical elements and their reactions or the full details of the surface topography or any of numerous other details within the system. Essentially, they incorporate, when feasible, the elements and processes that the modelers view to be the most important for the purposes they have in mind. Furthermore, the sophistication of the incorporation strongly depends on how easy it is to model the particular element or process. After the simulated results are obtained, comparisons with observational data can reveal where the results are noticeably flawed and which aspects of the model warrant the most attention for further improvement, at least regarding current observations. These improvements can come in many ways, including through tuning of different parameters, addition of new elements or processes, adjusting how various elements are calculated, and improving the model resolution, allowing smaller-scale features to be simulated. This is all very much an evolving process, with all major models expected to evolve further and improve over time.

IMPLICATIONS OF THE
INCOMPLETENESS OF THE MODELS

When evaluating the climate projections provided by models, it is important to recognize how strongly those projections can depend both on which components and processes are incorporated in the model calculations, knowing that not all can be incorporated, and on exactly how the incorporation is done, as in almost all cases there are many levels of sophistication possible. For a sobering reality check on how readily simulated results can change, recall the case described in chapter 7 when researchers modeling the impact of ethane and propane injections on stratospheric ozone depletion added one additional photochemical reaction to the 130 or so they already had included and found that their model turned from projecting a decrease in ozone depletions to projecting an increase.[29] This is quite a turnaround. Although few additions into major GCMs would be expected to have such a dramatic effect, surprises continue to arise regarding one phenomenon or process or another that becomes recognized as having a previously unrecognized importance, necessitating adjustments to the models in order to have even a chance of simulating climate change accurately. This section presents a selection of examples regarding the importance to the simulations of including various elements and processes.

Polar Amplification of Climate Change

There is a widely recognized concept called "polar amplification" that asserts that climate changes, such as warming from increased atmospheric CO_2, tend to be more pronounced and apparent earlier in the polar regions than at lower latitudes. The concept might or might not be valid, but the basis behind it centers on the high reflectivity (i.e., high albedo) of ice and snow and the fact that the ice and snow covers vary in conjunction with other climate changes. Basically, with warming, ice coverage overall can be expected to decrease, and this means less of the highly reflective ice surface and hence more retention/absorption of solar radiation within the Earth system and a further warming, a classic instance of a positive feedback (chapter 1). As long as a model incorporates the high albedos of ice and snow, the lower albedos of the underlying surfaces, and the ability for ice and snow coverages to increase and decrease, it incorporates these ice-albedo and snow-albedo feedback mechanisms and their tendency to enhance climate changes in polar regions and other regions with ice and snow covers.

Because the ice-albedo and snow-albedo feedbacks have been recognized for many decades (in fact, extending back to James Croll's work in the nineteenth century[30]), they are incorporated in major GCMs, and indeed many major GCMs have simulated for both hemispheres a polar amplification of the warming at the Earth's surface and in the lower atmosphere.[31] As demonstrated in a study with the GCM of the Goddard Institute for Space Studies, if the ice-albedo feedback is removed from the model calculations, the polar amplification in the results disappears as well.[32]

However, as model development has progressed, additional processes and feedbacks have been incorporated, and some of these have reduced or even eliminated the polar amplification simulated for the Southern Hemisphere (while retaining polar amplification in the Northern Hemisphere), even though the ice-albedo feedback remains in the model.[33] The lack of polar amplification simulated for the Southern Hemisphere derives in part from more sophisticated ocean calculations that changed the amount of warm water advected into the Southern Ocean from lower latitudes and allowed a deeper mixing of the waters of the Southern Ocean. With deeper mixing, the Southern Ocean warming is spread over a greater volume, thereby reducing the amount of warming at the surface. Additionally, NCAR scientists David Pollard and Starley Thompson found that by the single change of incorporating sea ice dynamics into their model, they changed their model results from having strong polar amplification of surface warming in both hemispheres to having strong polar amplification in the Northern Hemisphere but considerably less amplification in the Southern Hemisphere.[34] More generally, by the start of the twenty-first century, major GCMs tended to be simulating less polar amplification in the Southern than the Northern Hemisphere. This had the encouraging feature of matching better with the observations for the previous several decades, which showed a more pronounced warming in the Arctic than in most of the Antarctic. In 2001, however, a further twist was added when a new study tested an improved parameterization for ocean mixing and found that adding the new parameterization greatly increased the Southern Hemisphere's polar amplification.[35] The twists and turns are likely far from over, and it is uncertain whether polar amplification will end up being increased or decreased by further model developments.

Another factor that can affect polar amplification and how well it is simulated is the role of clouds. In fact, under certain circumstances, clouds can reduce and potentially even eliminate the normally strong positive ice-albedo and snow-albedo feedbacks.[36] Specifically, if the lessened ice and snow covers are accompanied by an increase in cloud cover—for instance, because of greater evaporation from the newly ice-free water areas—the high albedo of the clouds can result in as much or even more solar radiation getting reflected than before the ice and snow retreats. An examination of the results from 17 different GCMs indeed did find that in some cases (although not the majority), the presence of clouds reversed the snow feedback from a positive feedback to a weakly negative feedback.[37] Hence, models that incorporate the ice-albedo and snow-albedo feedbacks but not the potentially countering further feedback from the cloud changes could be giving unrealistic weight to the ice-albedo and snow-albedo feedbacks and the enhancement they provide to an initial warming (or to an initial cooling in the case of a cooling climate).

The studies cited in the previous three paragraphs and additional studies in the scientific literature provide an ample supply of experiments to give a skilled modeler a sense of how to increase or decrease the amount of simulated polar amplification to any level desired. In general, the decisions on which processes to include in a model are made quite independently of any considerations regarding polar amplification, but if a skilled modeler were intent specifically on simulating high, medium, or no polar amplification, perhaps differing in the two hemispheres, he or she would be able to adjust the model to do that. Whether it is done intentionally or not (and in general it is not intentional), a key point is that the amount of polar amplification simulated by a model depends quite directly on which processes the modelers decide to include in the calculations. Polar amplification is hardly unique in this regard; whether the topic is polar amplification or another characteristic, the models do not independently unveil the future to us, as the simulations tell us only the simulated consequences of the assumptions, equations, and other inputs programmed into the model.

> The models do not independently unveil the future to us, as the simulations tell us only the simulated consequences of the assumptions, equations, and other inputs programmed into the model.

Polar Stratospheric Clouds and Ozone Depletion

The second illustrative example of the importance to the simulations of including specific elements and processes also has a polar focus, although now well above the Earth's surface, in the upper atmosphere. Polar stratospheric clouds (i.e., clouds high in the atmosphere—in the stratosphere—over the polar regions) have been recognized at least since 1911, when they were described over Antarctica by the polar explorer Robert Scott. However, prior to the 1980s, their importance for ozone depletion (see chapter 4) was not recognized,[38] and GCMs had not included these clouds or the processes associated with them. Consequently, the models had not predicted the magnitude of the stratospheric ozone losses determined by British scientists[39] when they discovered what came to be known as the Antarctic ozone hole. Modelers now realize that to simulate future stratospheric ozone levels realistically, the models need to incorporate, at least in some manner, the role of polar stratospheric clouds. (Stratospheric ozone depletion is a quite separate issue from

global warming but is still an important part of the changing and heavily intertwined global Earth/atmosphere system and, moreover, is known to be caused at
least in substantial part by human activities.)

Propagating Impacts of Changes in the Stratosphere

What happens in the stratosphere affects the rest of the Earth system in more ways
than just through the impacts of ozone and ozone depletion. In the late twentieth
and early twenty-first centuries, an increased awareness has arisen of the wide-scale
importance of events in the stratosphere to the climate and weather conditions
underneath. In particular, Drew Shindell of NASA's Goddard Institute for Space
Studies and Columbia University and his colleagues highlight in one paper the importance of the stratosphere to the Arctic Oscillation and North Atlantic Oscillation, two closely associated large-scale atmospheric oscillations with many impacts
on happenings at the Earth's surface,[40] and in another paper the possibility that
changes in the upper stratosphere play a key role in amplifying the effects of the
11-year solar cycle on the Earth's climate.[41] These studies imply that for a model to
simulate properly the influential Arctic and North Atlantic oscillations and impacts
of the solar cycle, it needs to include at least a somewhat realistic stratosphere,
something that many models do not even attempt to include. Simulations from
models without calculations for the stratosphere could be generating results that
are meaningless for several long-term climate change issues.

Moulins and Sea Level Rise

As discussed in chapter 5, prior to the early twenty-first century, no one realized
that surface ice melt draining downward through an ice sheet can quickly reach the
bottom, lubricate it, and speed the outward flow of the ice. Scientists had neither
seen nor expected such a phenomenon on a major ice sheet, and hence it was of
course not incorporated into any GCMs or even into any stand-alone ice sheet
models, thereby omitting a potential mechanism for accelerating climate change
and sea level rise. Now, with the evidence from Greenland that surface melt can
rapidly reach the bottom through moulins, if modelers want to simulate accurately
the outflow of the ice sheets and the resulting sea level rise, they eventually need to
incorporate this additional mechanism.

Incorporation of moulins into models, however, is not straightforward, as the
moulins are only a few meters across at most, making them much smaller than the
horizontal resolution of the models, and although there is considerable scientific
literature linking the ice velocity at the base of an ice sheet or glacier to the amount

of lubrication, there is no widely accepted flow law to quantify the relationship.[42] Byron Parizek and Richard Alley of Pennsylvania State University have tackled the problem with a model of the Greenland ice sheet in which they have incorporated the basic mechanism although in a highly parameterized fashion and with many uncertainties and assumptions.[43] Further model development should eventually lead to a resolution of some of the uncertainties and incorporation of the moulin-based enhanced outflow into one or more full-scale GCMs. Tuning the parameterizations could either increase or decrease the amount of outward ice flow, so that the modeler will have more control than most nonmodelers realize over how much ice is simulated to flow into the ocean and thereby over the simulated amount of sea level rise. In this regard, the modelers who incorporate the lubrication mechanism encouraging outward ice flow should also be careful to incorporate the constraints on that flow, such as those that Tad Pfeffer and others argue will limit the sea level rise in the twenty-first century to no more than 2 meters[44] (chapter 5).

Sources of the Greenhouse Gas Methane

Models generally do not incorporate many of the known or suspected sources of atmospheric methane (CH_4), such as the bubbling of methane upward from northern lakes (thermokarst lakes, glacial lakes, peatland lakes, and alluvial floodplain lakes)[45] and of course do not incorporate sources that remain undiscovered. New sources of atmospheric methane keep being identified, and additional sources are likely to be identified in the future.[46] Furthermore, methane bubbling up through thermokarst (thaw) lakes in northern Siberia could prove to be a far more important source of atmospheric methane and hence a greater contributor to enhanced greenhouse warming than anyone recognized prior to the early twenty-first century, with these lakes potentially capable of releasing tens of thousands of teragrams of methane to the atmosphere.[47] Also important among the recent discoveries is the finding reported by Frank Keppler and colleagues in 2006 that significant amounts of methane can be emitted from living and dead plants, even under aerobic conditions, that is, in the presence of oxygen. They provide rough estimates that this previously unidentified source of methane could account for 62 to 236 teragrams per year from living plants and 1 to 7 teragrams per year from plant litter, adding to the approximately 600 teragrams per year of methane from previously recognized sources.[48] Leaving these additional sources of methane out of the models leaves out potentially important sources of greenhouse warming.

Underestimating versus Overestimating Climate Change

In both the moulin case and the methane case, eventual incorporation into the models of the phenomenon would lead to an enhancement in the simulation results of an expected global change. In the first case, it would be moulins leading to greater simulated sea level rise; in the second, it would be additional methane sources leading to greater simulated greenhouse warming. Certain other additions inserted into the models would instead have a moderating effect; this would include the incorporation of any negative (dampening) feedback that is not already included.

Roy Spencer in particular argues that the climate system most likely contains important stabilizing processes currently left out of the model formulations. He specifically suggests that a more complete incorporation of precipitation processes could result in a considerable dampening of the simulated climate changes.[49] If he is right, then the failure of the models to incorporate these processes adequately could indeed lead to quite flawed predictions. Precipitation, after all, controls how much water vapor, Earth's primary greenhouse gas, remains in the atmosphere. By removing water vapor, precipitation increases could contribute toward reducing the increased greenhouse forcing.

At least one observation-based study suggests that precipitation is likely to increase at as high a rate as atmospheric water amounts, revising earlier estimates that predicted a much lesser rate of precipitation increase.[50] If this is the case, then the increased precipitation could serve as a partial stabilizing factor from the point of view of greenhouse gas forcing. (However, at least in the Arctic, it appears from a study comparing Arctic results from 21 models that GCMs may already be simulating too much precipitation, not too little,[51] so that in the Arctic simulations there may not be an underestimation of this particular partial stabilizing factor.)

Spencer's suggestion that, by and large, GCMs are overly sensitive to many perturbations and consequently overestimate future change[52] is by no means a settled issue. In fact, there are documented cases in which GCMs have decidedly underestimated past changes. For example, GCMs did not simulate as large a decrease in Arctic sea ice as satellite data reveal occurred during the period 1979–2006 or as large a sea ice decrease as ship, aircraft, and satellite data suggest occurred during the longer period 1953–2006.[53] Models have also been found to underestimate precipitation trends in the tropics,[54] precipitation changes over land areas,[55] increases in evaporation over the ice-free oceans,[56] the increase in extreme rainfall events with warming,[57] and rates of sea level rise.[58] The underestimation of precipitation trends is in line with Spencer's suggestion that the models do not capture the full potential effect of precipitation changes on greenhouse forcing, but the other underestimates do not support the contention that GCMs overestimate climate changes.

Recognizing 1) how much difference one phenomenon or mechanism or tuning can make to model results, 2) the fact that many phenomena and mechanisms known to be important in the climate system are not yet included in the models or not yet included adequately, and 3) the fact that there are likely other important phenomena and mechanisms that are thus far not even recognized and so surely are left out of the models, I am personally quite uncertain whether our current models are underestimating or overestimating future climate change. I fear that they are underestimating, although I realize that it could be either way, or that the models are actually doing quite well. If they are doing well or are underestimating future changes, then we have reason to be concerned about future climate.

UNCERTAINTIES IN MODELED CLIMATE SENSITIVITY

The "CO_2/Climate Connection" section of chapter 9 introduced, from a historical standpoint, the topic of the calculation of the response of global temperatures to a doubling of atmospheric CO_2 concentrations and pointed out that in 1896 Svante Arrhenius calculated a 5°C to 6°C warming, in 1908 Arrhenius estimated a 4°C warming, and in 1938 Guy Stewart Callendar estimated a 2°C warming. Such calculations have become standard fare for many major GCMs, and, in fact, the simulated global surface temperature response to doubled CO_2, termed the "climate sensitivity," is now one of the most frequently featured quantities derived from modern GCM simulations. (All of these Arrhenius and Callendar results, done without computers or the benefit of modern observations, are impressively in the same general vicinity as the more recent results.) Because of how frequently the climate sensitivity is calculated and referenced, it becomes relevant to consider how well bounded the number might be.

Climate sensitivity calculations from modern GCMs include a broad range of results, although the more narrow range of 2.0°C to 4.5°C given in the 2007 IPCC assessments[59] is probably quite reasonable based on current understandings. Still, the uncertainties far exceed this range. In addition to the many limitations mentioned earlier in this chapter, all the models used to obtain the sensitivity estimates contain many parameters whose values are not yet known with certainty (as well as other parameters with well-known values) or whose values vary widely but are simplified in the model to the point where they are held to one or a few prescribed numbers. Running the same GCM with a selection of different feasible values for a particular parameter helps to generate a sense of how far off the standard results of the model might be just because of the uncertainties in that particular parameter. Such studies have been undertaken by James Murphy of the United Kingdom's Hadley Centre for Climate Prediction and Research, David Stainforth of the University of Oxford, and several of their colleagues.

In 2004, James Murphy and several colleagues from the Hadley Centre (plus one from the University of Oxford) presented the results of simulations that examine the impacts of changes in not just one uncertain parameter but 29 of them, doing so with the Hadley Centre's atmospheric model HadAM3 coupled to a relatively simple ocean model. (The ocean calculations involve only the upper portion of the ocean, with temperature, salinity, and other properties of the ocean all assumed to be well mixed vertically in this upper portion, called the "mixed layer.") Murphy and colleagues ran a sequence of simulations, changing the 29 parameters one at a time and running the model twice for each selected parameter suite, once to simulate the present climate and once to simulate a climate with doubled CO_2. Weighting the different cases according to the estimated likelihood of accurate predictions, they obtain a median climate sensitivity of 3.5°C, solidly within the IPCC's range of 2.0°C to 4.5°C, with 90% of their simulated climate sensitivity values lying between 2.4°C and 5.4°C and some outlier values exceeding 8°C.[60]

In a further effort involving several of the same people as the Murphy-led 2004 study, David Stainforth and colleagues published a new study in 2005 using a different GCM, the United Kingdom's Met Office Unified Model, to perform 2,578 simulations, all with various combinations of perturbations in just six parameters related to clouds and precipitation and all calculating the climate sensitivity.* They obtain climate sensitivities ranging from 1.9°C to 11.5°C for the simulations not excluded for one reason or another, with most of the sensitivities clustering around 3.4°C. Although the range of sensitivities is large, at least all the simulations, as expected, produced warming rather than cooling in the event of doubled CO_2.[61] Expanding from six to more parameters would likely produce a wider range of calculated sensitivities, as would expanding from one to numerous GCMs.

Many people have high hopes that the models and our understanding will advance to the point of significantly narrowing the uncertainties in the calculations of climate sensitivity and other quantities. However, some scientists, including Gerard Roe and Marcia Baker of the University of Washington, have suggested that this hope is unrealistic, arguing in particular that even a reduction in the uncertainties in the underlying climate processes will not significantly reduce the uncertainty in climate sensitivity estimates.[62]

---

* Interestingly, the 2,578 Stainforth et al. (2005) simulations were performed on hundreds of computers from around the world. In fact, many participants without any science expertise volunteered their personal time and their computer time to be part of this activity, knowing that they would thereby be participating in a state-of-the-art scientific study.

> With all the imperfections, it is hard to be confident that the models will be capable of giving fully convincing predictions any time soon.

With all the imperfections remaining in the models, including the many mentioned in this chapter, it is hard to be confident that the models will be capable of giving fully convincing predictions of climate sensitivity or other aspects of climate change any time soon. Still, despite their many limitations, GCMs remain the best tools we have for projecting the Earth's climate into the future, and we would be unwise to ignore what they predict. We need to be reasonable, recognizing that the simulations provide important but not flawless information.

DATA LIMITATIONS AND CLIMATE COMPLEXITY

This book includes a chapter on model limitations for two reasons: first, the central role the model predictions play in our expectations of future climate change; and second, the lack of understanding of the model limitations outside of the modeling community. However, this inclusion of a chapter on model limitations is in no way meant to suggest that the model limitations are more severe than the limitations of the data or the limitations brought about because of the complexity of the climate system. After all, if we had more complete data sets, there would be far less confusion about the climate system; and if the climate system were a good deal less complex than it is and we had the data to understand it, then the modelers would have a far easier task of simulating it.

Unfortunately, the data sets on which we base our understandings of past and present climates are, just like the models, very far from perfect. Most of our data sets are too short, too incomplete spatially, and too lacking in consistency both from place to place and from time to time. If a data set succeeds in not having all three of these problems, then it has at least one or two of them. Data about the distant past are particularly lacking, and uncertainties about the distant past appear throughout chapter 2. However, uncertainties by no means stop at the start of recorded history, or the period of instrumental records, or even the period of the most sophisticated state-of-the-art records.

Some of the major limitations of observational data sets include: satellite records are tremendously limited by the short length of the record, only going back a few decades at most; ice-core and deep-sea-core records are tremendously limited spatially, only having records in the few places where cores are drilled, which in the case of ice cores is tremendously confined by being limited to the high-latitude and high-altitude locations where sizeable ice amounts exist; tree ring records are confined to land areas; and anecdotal records are confined to locations and times with people who had the interest and know-how to make the records. With satellite and other instrumental data and with anecdotal records, a precise time line is typically available; but with tree rings and even more so ice cores and deep sea cores, accurate dating of the records is a major and often highly uncertain part of the data interpretation, so that, for instance, something thought to be 100,000 years old might actually be considerably older or younger. Moreover, the satellite instruments generally directly collect information only about radiation, whereas greater interest lies in information about temperature, vegetation, ice, and other Earth-system variables. The conversions from radiation information to other Earth-system information are never perfect and sometimes turn out to be seriously flawed.* In the latter case, the flawed conversion schemes are generally iteratively improved, although could also instead be replaced by different conversion methods altogether.

Prior to artificial satellites, which first appeared in the second half of the twentieth century, there were no feasible means of getting spatially comprehensive global records of any climate variable. Gallant efforts have been made to create global temperature records extending back to the 1800s based on instrumental measurements,[63] but the data incorporated in presatellite global temperature estimates are well known to be biased toward land areas and especially toward Europe and North America. The completeness issue is only one of several serious problems with the temperature records. A second involves the lack of precise matching of the records as new instruments replace old ones, and a third involves the siting of stations and the fact that a trend determined from one station's data might reflect the changing local landscape much more than it reflects any broader temperature change. For instance, a station placed in a rural region that subsequently underwent major population growth would certainly have a data record that reflected the urbanization. Although some partial adjustments can be made in light of the changing landscape, it is unlikely that errors arising from (in hindsight) unfortunately sited stations can be corrected completely.[64]

When considering data for periods extending back before the instrumental records, the difficulties are even greater, as illustrated by the major controversy summarized in chapter 4 regarding the "hockey-stick" plot[65] of Northern Hemisphere

---

* For an introduction to satellite observing of the Earth and some of its limitations, see Parkinson (1997).

temperatures over the past 1,000 years. Chapter 11, although focused on the social aspects of science, contains additional examples of problems with the data and our interpretations of it.

I turn now from the issue of data limitations to the issue of climate complexity. Many portions of the climate system are now known to be much more variable than they had been thought to be prior to having the database to examine the variability. This is true for many variables, from the surface to the top of the atmosphere, and I certainly encountered it while doing research on sea ice with satellite data. When satellite technology first allowed the full annual cycle of the polar sea ice covers to be viewed in detail, the observations obtained refuted earlier simplistic concepts of a smooth annual advance and retreat of the ice cover, instead showing considerable unevenness in the annual cycles in both the Arctic and the Antarctic as well as considerable variability from year to year.[66] Explorers had seen and mentioned variability before, but the full scope of the variability was not recognized. Similarly, although more recently, the satellite data reveal a variability in the radiative energy budget (involving the radiation entering and leaving the Earth/atmosphere system) at the top of the atmosphere in the tropics that is much greater than had been thought, with this variability perhaps caused by changes in cloudiness.[67] Even the "solar constant," which is the solar energy reaching the Earth's outer atmosphere per unit area per unit of time and was initially thought to be constant, has for decades now been well known not to be constant at all. There are also all sorts of variabilities on medium scales, as reflected in the famed Indian monsoons and ENSO and the lesser-known monsoons and Arctic Oscillation, North Atlantic Oscillation, Pacific Oscillation, and multiple others, and there are all sorts of variabilities on long-term scales, as reflected in glacial/interglacial cycles and all the changes in ocean and atmospheric circulation necessitated by continental drift. The climate system is full of variabilities, those mentioned here being only a sampling. It is also full of incompletely understood interactions—between atmosphere and oceans,[68] vegetation and atmospheric trace gases,[69] upper atmosphere and lower atmosphere,[70] high latitudes and low latitudes,[71] and on and on—that are areas of active research. Further, disturbances like fires, earthquakes, and volcanoes can dramatically change the fluxes between the land and the atmosphere and the reflectivity of the surface.[72] All the variabilities and interactions and jolts to the system add considerably to the complications of understanding the global climate and hence of modeling it. If the system were a good deal less complex, the models would almost certainly be less flawed.

CHAPTER 11

Compounding Social Pressures

Science is conducted by scientists, all of whom, being human, are influenced by pressures from within and outside the scientific community, affecting what science they do and at times even what results are obtained.

Being an endeavor carried out by humans, science is necessarily subject to the full range of human flaws. Very few, if any, scientists can proceed throughout their careers on a totally objective course guided solely by the scientific issues. Instead, a vast array of additional factors affects job, research, and communication choices. Different scientists are affected to a greater or lesser extent by different factors, but among the important factors for many scientists, in addition to their personal scientific interests, are what jobs are available, where the jobs are located, what research is likely to get funded, what applications the research might have, the likely level of media and public interest, acceptance by colleagues, and esteem in the broader scientific community. Having collaborated with dozens of other scientists and having been attuned to issues raised by a far greater number and broader range of scientists over the past several decades, I have encountered much that impresses me but also much that concerns me regarding the state and validity of the science being done today.

Science is a tremendously exciting field of endeavor and one that I energetically encourage young people to consider for a career path. However, there are problems, and some of those problems have direct relevance to a central issue of this book, namely, the issue of whether we know enough about the Earth's climate to attempt to modify it significantly in one way or another. In this chapter, I highlight several of those problems, including ones that are entirely internal to the scientific community, such as those concerning peer review, and ones that involve external factors, especially the media.

Issues involving the media can be particularly difficult because, by and large, scientists have not been brought up with media savvy. Most scientists in the climate change research community who became scientists before the mid-1980s had no inkling that their work would eventually receive media attention. We were trained as scientists, not as publicists, public figures, or policymakers. For many of us, it has been difficult to make the transition to a more public role, and there has been a wide variety of responses, some scientists avoiding the media altogether, others cautiously responding to questions, and others much more aggressively pursuing media attention and working hard to express points in ways that the media will utilize, often involving the intentional creation and enunciation of memorable sound bites. Naturally, there are occasional blunders, sometimes with misstatements that the scientist involved would like to retract but also sometimes with statements that do not stand up to scientific scrutiny but that individual scientists willingly state and repeat because of their effectiveness in gaining attention and/or getting the base message out. These latter statements arouse the admiration of some and the ire of many others.

To lay out some specifics, I begin with the discipline in which I am most experienced, namely, polar sea ice.

THE CASE OF POLAR SEA ICE

I am probably not unusual among climate researchers in terms of preferring to obtain results that would place my work in the forefront of mainstream climate change research. I pursue and report whatever the results show, but I find it far more comfortable and enjoyable when the results fit in the mainstream than when they risk placing me as a potential "skeptic" regarding some of the mainstream climate change concepts, especially in view of how negatively the skeptics are treated (more about that later). Fortunately for me, many of the results that my colleagues and I have obtained during decades of research on sea ice from satellite data do provide important pieces of evidence regarding serious changes in the Earth's climate system in line with the mainstream scenario of global warming. However, by no means do all of our results do that. Without going into the many details, the satellite-derived sea ice results for the Arctic region by and large show a prominent trend toward reduced Arctic sea ice coverage from the late 1970s at least through

2009. These results are solidly in line with warming of the Arctic and have gained particular interest both because the Arctic has frequently been highlighted as one of the regions where global warming has become most prominent and because the Arctic sea ice decreases have wider-scale climate and ecosystem impacts, including impacts on polar bears, which use the ice as a platform from which they hunt for seals and other marine-based life. In fact, the dependence of polar bears on the ice is so great that there has been speculation regarding whether polar bears can survive as a species if the Arctic ice cover continues to retreat. Climatologically, the sea ice decreases are notable also because as the ice retreats, the retreat has the strong amplifying effect of reducing the amount of solar radiation reflected back to space (as the white ice is a strong reflector), hence increasing the amount of radiation absorbed into the polar climate system, causing further warming in a classic positive feedback (chapter 1). If my research focused exclusively on Arctic sea ice, I would be well positioned to provide in a powerful way strong evidence supporting the majority viewpoint that global warming, produced at least in part by human activities, is generating serious changes within the climate system, with propagating further impacts on climate and on life.

The problem with the scenario of placing my (and others') Arctic sea ice results at the forefront of the global warming mainstream is that I also examine Antarctic sea ice, and the Antarctic sea ice does not show the strong global warming signal that the Arctic sea ice does. The Antarctic sea ice cover experienced major reductions in the early and mid-1970s, in line with warming, but since the late 1970s, at least through 2009, it has, overall, instead undergone some expansion rather than retreat. The ice expansion in the Antarctic is not nearly as much as the ice retreat in the Arctic—the Arctic retreat averages about 45,000 square kilometers a year (losing each year ice of an area considerably exceeding the area of Switzerland), whereas the Antarctic expansion averages about 13,000 square kilometers a year—and it is compounded by some limited regions of the Antarctic sea ice (especially in the vicinity of the Antarctic Peninsula) experiencing a marked retreat. Still, it is, overall, an expansion of Antarctic sea ice cover since the late 1970s, which, on the face of it, is hardly strong additional evidence for global warming.

In view of how intertwined the Earth system is, the changes in the Antarctic sea ice will eventually be explainable in the context of the changes in the rest of the climate system, but, for the moment, the increases in the Antarctic ice cover are not nearly as readily explainable in a global warming context as are the decreases in the Arctic ice. I have made the conscious decision to try to present a balanced picture both in what I write and in what I say, while recognizing that those of my colleagues who have opted instead to place a heavy emphasis on the Arctic sea ice decreases have in some cases significantly enhanced their careers and media interest by doing so. (So far, the Arctic results have garnered considerably more media coverage than the Antarctic results, but the discussion of the Antarctic results will likely increase in the event of either of the following: the Antarctic sea ice cover begins to lose ice, or the explanations of why the Antarctic sea ice coverage has increased become stronger and tied more closely to other changes in the Earth system, especially changes having a direct human cause. Explanations fitting this latter characterization could be on the horizon, as the spatial pattern of changes in the Antarctic sea ice are consistent with changes in atmospheric flow patterns tied at least in part to the human-caused Antarctic ozone hole, and researchers are beginning to examine the possible connections.[1])

Part of my reluctance to speak out more forcefully on the topic of the Arctic sea ice decreases, despite my concerns about them, derives from an instructive sequence that occurred in the early 1980s regarding the Antarctic ice. Specifically, George Kukla and Joyce Gavin of Columbia University's Lamont-Doherty Geological Observatory published a paper in 1981 highlighting marked decreases in Antarctic sea ice coverage from 1973 through 1980. The decreases were so large that in the eight-year period 1973–1980, the Antarctic lost more areal coverage of ice[2] than the Arctic lost in the entire period 1979–2006.[3] When Kukla and Gavin's paper was published, media interest was aroused, and the Antarctic sea ice losses were mentioned as conceivably the first real geophysical evidence of the global warming that was beginning to come to the fore as a topic of concern. However, by the time the Kukla and Gavin paper appeared, it was already clear that the major sea ice decreases, which had occurred mostly in the 1973–1977 period, were not continuing. Jay Zwally, myself, and Josefino Comiso from NASA Goddard Space Flight Center proceeded to publish a paper updating the results and showing that the decreases had not been maintained after 1977 and mentioning also that the ice decreases of the mid-1970s had been preceded by sea ice increases between 1966 and 1972.[4] Witnessing how the decreases in the 1973–1977 period generated media interest and strong statements about how they conceivably provided definitive evidence for global warming, then watching over the years as the Antarctic ice slowly rebounded, contributed toward keeping me more cautious than others in making forceful statements regarding the Arctic sea ice decreases (which, after all, by and large have been less than the decreases in the Antarctic sea ice during the 1970s) and declaring them to be irreversible.

Other sea ice experts have taken very different approaches, ranging from focusing exclusively on the research and rarely if ever speaking to the media to focusing predominantly on the results that obtain media attention. I am most concerned when the scientists comfortable with or even pursuing media attention lose their scientific objectivity and attention to accuracy. Some statements made or at least quoted as having been made by scientists are outright absurd. To illustrate, I use an example from an engaging Earth-system scientist who on many occasions is duly cautious and quite careful in not jumping to attention-grabbing conclusions but, like most of us, is more cautious on some occasions than others. This scientist was quoted by Gabrielle Walker in *Nature* as saying, regarding the Arctic sea ice, "Once you start melting and receding, you can't go back."[5] The quote is close enough to statements I have heard this scientist make in person that I suspect that Walker at a minimum correctly relayed the sense of what the scientist said and perhaps correctly relayed the exact words. However, effective as it is as a sound bite, the statement is an instance of an absurd statement that badly misinforms the listener who is not an expert in the subject matter. Every year in the spring and summer the sea ice significantly melts and recedes, and yet, come autumn and winter, it consistently reverses direction and expands outward again, demonstrating the absurdity of the statement as quoted. Moreover, the paleoclimate record shows the statement's absurdity on the longer time scales that are likely more germane to the speaker's original meaning, as there have been numerous instances of sea ice coverage greatly lessening over the course of years, decades, or centuries, then later reversing and again expanding outward. The quote is memorable, but it makes a statement that is simply not accurate.

The following somewhat similar statement appeared in a publication in a major scientific journal in 2006: "Extreme sea-ice losses in recent years seem to be sending a message: the ice-albedo feedback is starting. With greenhouse gas concentrations on the rise, there may be no counteracting mechanism in the climate system powerful enough to stop it."[6] The second sentence is appropriately qualified, but the statement still shares with the quote in the previous paragraph the sense that the ice retreats and/or ice-albedo feedback produce an unstoppable progression, which is clearly not the case, as the climate system has gone through many glacial/interglacial cycles (chapter 2) versus getting to one extreme and stopping. The quoted 2006 statement has the added complication of including that "the ice-albedo feedback is starting." The ice-albedo feedback (chapter 1) has been acting all along, as have several other enhancing (positive) and dampening (negative) feedbacks. Saying that the feedback is "starting" is attention grabbing, whereas saying that it continues is certainly not, but the desire for attention-grabbing statements should not override the desire to relay correct information.

Trained as scientists, not as politicians or media personnel, many Earth scientists are now struggling with how best to get key messages out. Although I can see major problems with the sound bites mentioned above and many others, I most certainly can see major problems with my responses as well, which tend to include too many qualifiers and attempts at balance for effective mass communication. Furthermore, I recognize the unfortunate but very real possibility that any of us could make major mistakes for a variety of reasons, including slips of the tongue, like saying "per year" when we meant "per day" or "per decade." None of us is perfect, and there are no perfect solutions to the problem of how best to get a message across. Perhaps those in subsequent generations, having grown up with the Internet, pervasive video clips, and a recognition of the media attention given to climate topics, will be more used to having their statements video and audio recorded and thereby will have an easier time traversing the media terrain.

THE BROADER CASE OF GLOBAL CHANGE

The sea ice example is only one of many that could illustrate the complicated picture revealed by the evidence and the differing responses of scientists to it. Whether examining sea ice or examining other portions of the climate system, scientists have differing results, opinions, and senses of urgency regarding the global warming issue.

The fact that some regions are cooling while others are warming and that all regions continue to experience both unusually cold spells and unusually warm spells provides plenty of data-based ammunition for the skeptics as well as for those most ardent about the urgency of addressing global warming and related climate changes. The Earth is a very large place, and the climate system is hugely complex, with many changes occurring at any time. As a result, selective data culling and articulate advocacy could lead to markedly different compelling arguments. Further, the facts that warming has some benefits, like lengthened growing seasons, reduced need for artificial heating, and fewer deaths from cold, and that temperatures are known to have been higher during extended periods in the distant past add to the difficulty of impressing on the public the concern that human activities are leading to dangerously unacceptable warming levels. Yet this could certainly be the case, as many scientists now believe.

In view of how dangerous some scientists and nonscientists believe the imminent future climate changes to be, how strongly others feel that the concerns are essentially baseless, and how expensive it would be to make the changes that some people call for in order to stave off the predicted upcoming disasters, it is not surprising that discussion of climate change has at times become heated. Unfortunately, in this heated atmosphere, objectivity has often been lost by scientists and nonscientists on both sides of the issues. On a number of occasions, I have listened with a mixture of horror and awe as scientists I know skillfully and successfully use carefully chosen superlatives to generate media attention, even when the superlatives are not warranted and the scientists know that they are not valid but feel that some exaggeration is acceptable. Both alarmists (those most concerned and vocal about the dangers of global warming) and skeptics have on occasion presented very biased pictures and go to extremes of showing only one side of the story. This is sometimes for the admirable reason that the individual's deep concern about the future of the planet or of civilization overrides considerations such as balance and objectivity. However, even when the reason is admirable, giving a demonstrably biased story is troublesome and can greatly confuse the issues, lowering rather than raising the level of constructive concern among the public and policymakers.

The demonstrably biased story opens a floodgate for each piece of counter-evidence. For instance, a statement that all the world's glaciers are retreating can readily be quashed by evidence that one individual glacier is advancing. This can seriously reduce confidence not only in the speaker but also in the main thrust of the argument that he or she is attempting to make. It can also lessen the confidence of the public in scientists and science in general, an outcome that could conceivably be favorable in the short term for those in the minority but would most definitely not be favorable for those in the mainstream majority.

I feel strongly that everyone would be better served if all the experts would place a high priority on presenting a balanced picture. When a nonexpert hears a speaker presenting only one side of an issue, knowing that the other side also exists, often the main conclusion he leaves with is simply which side the speaker is on. Importantly, a "balanced picture" here does not mean equal time for each side. A balanced picture means, for instance, that if 90% of glaciers are retreating and 10% are advancing, the statement would not be "all glaciers are retreating" but rather "90% of glaciers are retreating" or "by far the majority of Earth's glaciers are retreating" or something roughly similar.

It is my strong impression that both the alarmists and the skeptics deeply believe that they are on the correct side of the global change issue and that it is important that the global community respond accordingly (e.g., by severely reducing greenhouse gas emissions or by not doing so). If each side would treat the other with respect, would forgo making statements that exaggerate the respective positions, and would carefully listen to each other, with the object not of countering but of understanding, then there should be a reasonable chance of working through and at least lessening the differences.

TREATMENT OF THE SKEPTICS

Sadly, the controversies between the mainstream and the skeptics over global change issues have become so contentious that both sides have made unfair statements against the other, further hindering the needed wide-based civil discussion. Being in the minority, the skeptics have received the brunt of this exchange, and they are often quite severely criticized by some of the scientists comfortably situated in the majority.

Some of the criticism of the skeptics seems quite unfair, even attacking their integrity, for instance implying that they express the minority position for financial gain. The presumption is that they have received money from the oil and/or other industries to make statements opposing the mainstream view. I have no direct knowledge that this criticism is valid for any of the skeptics, although conceivably it is. It is not, however, valid for all of them, and, from my perspective, the totality of these attacks against the integrity of the skeptics is a major blot against the science community for not being above bullying a minority group.

Some of the skeptics have indeed received funding from segments of the corporate world with a stake in maintaining the status quo. Although that does not imply that their scientific positions were affected by the funding, scientists would be prudent to avoid such funding, if feasible, as, at a minimum, it has the appearance of a possible impropriety. However, it is important to note that some of the scientists most concerned about global warming, on the other side of the fence from the skeptics, also receive funding from the corporate world, and this happens without a comparable level of criticism directed at them. The fiscal reality is that scientists generally need money to carry out scientific research. Much of the research funding in the United States now comes from government agencies, such as the National Science Foundation, NASA, the National Oceanic and Atmospheric Administration (NOAA), the Environmental Protection Agency, the Department of Energy, and the Department of Defense. Those sources generally (although not always) make their selections of whom to fund based on peer review of proposals, and I am not aware of the selected individuals ever being criticized for taking the money. In the past, before government funding for science was widespread, it was more common for scientists (and artists) to be funded by private individuals or corporate or other groups. Some of that continues, but today scientists run a risk when they accept private or corporate funding, especially if the funder is perceived—rightly or wrongly—as seeking results in only one direction.

In the field of Earth sciences, the funding-related criticisms leveled against scientists and private and corporate funders seem extremely biased against the skeptics and the oil industry in particular. By now, I have so frequently heard statements that one skeptic or another is funded by the oil industry that it seems that some people automatically suspect that oil industry funding is behind all skeptics. This is certainly not the case, and even those skeptics who have received oil industry money might not have received much. For a specific example, I have repeatedly heard criticisms of skeptic Fred Singer of the Science and Environmental Policy Project (and professor emeritus at the University of Virginia) for accepting oil industry funding, yet apparently the only such funding he ever received was one $10,000 check from the Exxon Foundation in the 1990s. This was confirmed by Singer on October 30, 2008, by e-mail, with the further comment that the $10,000 Exxon check is the only money he ever received from any industry, not just the oil industry. (Perhaps there are important indirect sources; if so, I am not aware of them.) Interestingly, Wally Broecker—solidly in the mainstream on the global warming issue—also received money from Exxon,[7] and, as far as I know, no one criticized that.

I might not be sufficiently informed, and I recognize that there are major concerns regarding actions that elements of the oil and tobacco industries have taken with regard to environmental science,[8] making those funding sources legitimately suspect. However, some of the criticisms of the skeptics regarding funding sources are unfair, either because of being based on incorrect information or because of applying more difficult standards to the skeptics than to mainstream scientists. The criticizers of the oil funding seem to be claiming that the oil industry wants only one line of results and that the funded scientists are no longer objective in their science, instead producing only the

> As long as the funding does not go to the skeptics, corporate funding is generally not criticized. The criticism of corporate funding seems to be extremely unevenly applied.

desired results. I am neither certain that that is a fair charge against the oil-funded scientists nor certain that environmentally oriented organizations and individuals funding mainstream scientists are any more indifferent than the oil industry regarding what type of results they hope the funded researchers will obtain.

For the record, I have never received private or corporate funding, as I have been extremely fortunate in that I have always had fully adequate funding from NASA or, before my employment at NASA, from the National Science Foundation. However, some of my friends in the climate change mainstream have received corporate funding, and I am not aware that any of them has ever been criticized for that. As long as the funding does not go to the skeptics, corporate funding is generally not criticized, at least as far as I know.

Funding is hardly the only arena in which the skeptics are criticized. As Roy Spencer puts it, "Global warming skeptics like me are being increasingly demonized."[9] From what I have witnessed, I feel that Spencer's comment is valid, although Spencer himself is not above the fray in terms of making inappropriate attacks, as shown by quite a few snide and sneering comments mixed in with many excellent points in his book *Climate Confusion: How Global Warming Hysteria Leads to Bad Science, Pandering Politicians and Misguided Policies That Hurt the Poor*,[10] many key points of which are summarized earlier in this book in the concluding section of chapter 5.

Others of the skeptics also express dismay at how they and the other skeptics are treated. Among these is Massachusetts Institute of Technology meteorologist Richard Lindzen, who is perhaps the most highly regarded scientifically of all the well-known skeptics. Lindzen agrees with the mainstream regarding the reality of global warming but disagrees regarding the extent of human responsibility for it or the level of concern that is warranted. Although the strength of his scientific credentials has generally kept the criticisms of him relatively muted compared to the criticisms of the other skeptics, he too has certainly endured the wrath of the mainstream and has repeatedly witnessed the unfair treatment of his fellow skeptics. Lindzen has summarized as follows: "Scientists who dissent from the alarmism have seen their grant funds disappear, their work derided, and themselves libeled as industry stooges, scientific hacks or worse."[11] He mentions not only attempts to discredit himself but also the dismissal of one skeptic from his position as the director of the Royal Dutch Meteorological Society, the labeling of a former director of the World Meteorological Organization as a tool of the coal industry because he questioned the rising climate alarmism, and the loss of funding for two respected Italian professors after they also raised questions. Loss of funding is difficult to attribute definitively to one cause or another, as, for mainstream as well as nonmainstream scientists, funding tends to be uncertain to say the least. Nonetheless, obtaining funding generally becomes noticeably more difficult when one's work and positions are outside the mainstream. For one thing, not being in the mainstream makes it less likely that the proposal reviewers will be in agreement with the proposer regarding the science issues. Again quoting Lindzen, "Alarm rather than genuine scientific curiosity, it appears, is essential to maintaining funding. And only the most senior scientists today can stand up against this alarmist gale, and defy the iron triangle of climate scientists, advocates and policymakers."[12]

I share Lindzen's concern about the "climate alarmism" that has become all too common, although we have different perspectives on its cause. When discussing why scientists are participating in generating the hype and alarmism, Lindzen emphasizes that by doing so, they are hoping to maintain or increase federal funding for climate research.[13] From what I have witnessed, I sense that it is not so much a desire for funding as a desire to get an important message out mixed with a desire for career advancement (promotions, awards, and so on) and for recognition (getting on television and radio and mentioned in the print and online media). Either way, Lindzen and I agree that it is not always for the noblest of scientific reasons.

Lindzen has labeled some of the mainstream work "junk science,"[14] which is part of the "tit for tat" between skeptics and alarmists that I dislike. Still, we agree on many aspects not only regarding how the skeptics are treated but also regarding why many scientists who have doubts are generally keeping quiet. As stated by Lindzen, "It's my belief that many scientists have been cowed not merely by money but by fear."[15] Perhaps "cowed" and "fear" are too strong, but the point that some scientists with doubts have stayed quiet is valid, and it quite likely has led to making the mainstream consensus appear stronger than it really is.

> The fact that some scientists with doubts have stayed quiet (at times in fear of repercussions) quite likely has led to making the mainstream consensus appear stronger than it really is.

The unfair treatment of skeptics extends beyond those skeptical of global warming, human contributions to it, or the fact that it represents a serious problem. Even people like Bjorn Lomborg who agree that global warming is real and serious and that humans are part of the cause have been subjected to considerable criticism and derision focused on their disagreement with the proposed actions to be taken in light of global warming. I had heard enough by word of mouth from the global warming mainstream about Lomborg's book *Cool It: The Skeptical Environmentalist's Guide to Global Warming* by late 2007 that when I read the book at that time, I was expecting at least a few strong and unreasonable statements questioning the reality of global warming. Quite the contrary: the book repeatedly states unequivocally that global warming is occurring, that it is partly the fault of human activities adding CO_2 to the atmosphere, and that, yes, we should do something about it.[16] Lomborg further states that our best information about future climate comes from the mainstream, consensus reports of the Intergovernmental

Panel on Climate Change (IPCC). Where he falls out of favor with the mainstream is when he examines the costs involved in the mainstream suggested solutions and resoundingly concludes in case after case that the reality of the economics makes the mainstream solutions unwise options, offering what he sees as economically more reasonable ways (see chapter 5) for achieving the central goals. This is the type of thoughtful consideration that everyone should embrace as part of the discussion that should be taking place. It should not be outright rejected because it fails to fit neatly into today's mainstream position. In line with the mainstream, I do not agree with Lomborg's conclusion that only very moderate CO_2 reductions are warranted at this point, but I strongly feel that his argument is important and should be part of the mainstream discussion.*

Despite the reasonableness of much of what he recommends, Lomborg is viewed with disdain by many in the mainstream. In his words, "At present, anyone who does not support the most radical solutions to global warming is deemed an outcast and is called irresponsible and is seen as possibly an evil puppet of the oil lobby."[17] The statements quoted from Spencer, Lindzen, and Lomborg all illustrate quite clearly that scientists who have some doubts but are not anxious to incur the wrath of colleagues and the scientific community (and sometimes the media and politicians as well) have reason to pause before articulating their concerns. Certainly, that has been heavy on my mind as I have written this book. I have no wish to be denounced and demonized. In fact, I am quite horrified by the thought, but I am even more averse to standing by and saying nothing in the face of serious issues that need to be raised.

PEER REVIEW AND ADVANTAGES OF BEING IN THE MAJORITY

The entire science research system makes it easier and more comfortable to be in the majority than the minority, especially when the consensus grows to as large a majority as it is with the global warming issue. Most scientists, like most people in general, prefer to be liked and spoken well of than to be disliked and spoken ill of.

---

* Importantly, I have heard from experts in specific disciplines that Lomborg has made mistakes regarding those disciplines, and that also needs to be considered, as do the mistakes made by the mainstream scientists and authors. So much contradictory material exists that mistakes can understandably be made even by the most conscientious of researchers. I do not recall ever reading a book on a topic that I am knowledgeable about without finding at least some mistakes.

Being in the majority provides a strong advantage in terms of collegial relationships and in terms of lessening the likelihood of being denounced both to one's face and behind one's back. Peer pressure can be a powerful influence for scientists as well as for all other groups of people.

Being in the majority also provides a decided and very practical advantage in terms of getting one's work funded and published. In general, both for funding and for publication, scientists go through a "peer-review" process. For those un-familiar with the process, the following sentences provide a brief summary for the case of a paper submitted to a journal for publication. To start off, when scientists submit manuscripts for publication in journals that have peer review (which in-cludes most major science journals), the journal editor assigned to the manuscript generally sends it to two or more scientists familiar with the general topic, asking them to read it and send in written comments concerning its accuracy, quality, and appropriateness for publication in the particular journal. The reviewers proceed to do this, and then the editor decides on the basis of the various reviews (which sometimes differ markedly in their recommendations) and his or her own opinion whether 1) the paper should be rejected, 2) the paper should be accepted, or, as is often the case, 3) the authors should be encouraged to revise the paper in line with the reviewers' comments, after which the editor will likely accept the revised paper, and it will proceed to a copy editor and then to publication.

The peer-review process for proposals written to obtain research funding is similar to that for publications, although with some important contrasts. For pro-posals, there are often more reviewers assigned and sometimes additionally a re-view panel that evaluates several proposals and groups or ranks them according to their perceived worthiness for funding. Instead of an editor, the decision maker for funding proposals is generally a program manager or other agency official. Most important for the submitters, the decision made on a proposal is generally either outright acceptance or outright rejection, without the third alternative of condi-tional acceptance or rejection along with a list of suggested revisions.

The review process is a very hit-and-miss affair, all quite dependent on who does the reviews. A proposal that one reviewer finds creative and exciting another reviewer might reject as too vague or impractical, and another proposal that one reviewer might support as necessary and important work another might reject as too routine and lacking in creativity. Similarly, with submissions to journals, the same manuscript might have one reviewer recommend outright rejection, another recommend acceptance with almost no revisions, and yet a third recommend that the editor delay a decision until quite substantial revisions are made. Even in the case of a conditionally accepted manuscript, the particular reviewers and their attitudes regarding their role as reviewers (and occasionally their moods while doing the review) can make a difference ranging from several hours to a year or more in the amount of time it takes to revise the paper prior to its acceptance for publication. Sometimes this results in a paper that has been greatly compromised and is quite different from the paper the authors originally intended. In the best instances, the review process improves the paper, but this is by no means always the case. The peer-review process is well intentioned and is probably the best method we currently have for weeding out questionable science, but it is very imperfect and is uneven in its implementation.

Of most relevance here, when a strong consensus exists on a topic, such as global warming, there are features of the review process that can, by and large, make it easier on the proposals and papers that are in line with that consensus view. This comes down to a numbers game: the larger the consensus, the more likely the reviews will be written by people adhering to the majority viewpoint. A few editors, program managers, and reviewers bend over backward to make sure the minority viewpoint can be heard, if argued compellingly, but still, on a practical level, it is often easier to get funding and published through the peer-review process when your research is in line with the majority view. That is my very strong impression, after years of working within the system, although obtaining statistics to confirm this (as requested by one of the reviewers of this book) is not feasible because of the restrictions on the information revealed by funding agencies; for example, NASA routinely reveals the list of successful proposals for any given proposal solicitation but does not publicly list unsuccessful proposals, a policy that is kind to the proposers in terms of not publicly embarrassing those who are rejected.

OTHER PROBLEMS INTERNAL
TO THE SCIENCE COMMUNITY

A variety of factors currently inherent in the research community can, in practice, be detrimental to optimizing the quality of research publications. In addition to the flawed review process described in the previous section, other factors include the following: pressure placed on scientists by their institutions to increase the quantity of their publications (which tends to greatly exceed any pressure to increase the quality of the publications, admittedly much harder to measure), as the number of publications is a metric commonly used during deliberations regarding tenure, promotions, awards, and job offers; stringent page limitations in several prestigious journals, frequently preventing the researchers from presenting their methods in sufficient detail for others either to reproduce the results or to fully understand the results; a frequent preference in the review process for writing that comes across as sophisticated and obscure rather than writing that emphasizes simplicity and clarity; and the importance placed on being the first to report a finding, thereby encouraging rapid publication, sometimes at the expense of a careful and complete analysis. Each of these problems has subtleties that might not at first be apparent.

One of the subtleties concerns when to publish results from incomplete data records that continue to be updated. In the Earth sciences, in some cases there exists a finite, well-bounded data source, such as an ice core or the data from a completed field season. However, in other cases, the data collection is open ended, with no predetermined end point, such as the data collection from a still-operating ground-based or satellite-based instrument. In the latter case, a decision must be made regarding when enough data have been collected to warrant publication. With the pressure to publish, papers are sometimes published from very limited data despite the ongoing data collection. In other cases, the researchers do wait until they have a substantial chunk of data, with important results to report, but even then, the initial papers sometimes present results that later need substantial revision in light of intervening data enhancements. In the meantime, erroneous conclusions from the initial papers can propagate through the scientific literature.

For an illustrative example of a case in which likely no one is guilty of even a minor indiscretion, consider the sea ice–thinning results published in 1999 by a group from the University of Washington led by Drew Rothrock. In the late 1980s and early 1990s, people recognized that submarine data collected by the U.S. Navy since the 1950s could potentially provide a valuable record of changes in Arctic sea ice thicknesses, if only the data were made available. Then-Senator Al Gore became aware of the issue, and I was pleased to brief him in 1990 about the importance of sea ice to the climate system. Throughout the hour and a half private briefing, I was impressed with Gore's keen focus, his insightful questions, and how quickly he assimilated all the key points (while focusing exclusively on the data suggesting ice reductions and warming). Gore proceeded to work with the navy and encourage declassification, bit by bit, of the submarine data that no longer warranted security classification. By the late 1990s, a sizable amount of data had been declassified, and the University of Washington team proceeded to analyze them, obtaining and publishing the startling result that, from the limited data available thus far, it seemed that the Arctic sea ice might have thinned by about 40%, from a mean ice draft* of about 3.1 meters in the period 1958–1976 to a mean ice draft of about 1.8 meters in the 1990s.[18] The submarine data are extremely spotty, being limited to when and

---

* The ice "draft" is the thickness of the portion of the ice (sea ice or iceberg) that is underwater. On average, the ice draft is about 90% of the total ice thickness.

where the submarines were cruising under the ice, so the University of Washington team correctly realized that the results would likely need to be adjusted after more data became available but would still probably indicate a sizable thinning. The navy continued releasing data after the 1999 publication, and the University of Washington group continued analyzing the enhanced record. In 2004, they published a new paper, centered on changes in the distribution of the amount of ice in various thickness categories between the two periods 1958–1970 and 1993–1997. Although the emphases were different in the two papers, the 2004 paper did include an important update, namely, that the "overall volume loss was about 32%, which is 8% less than the value reported by *Rothrock et al.* [1999]. The main cause of this difference is that more submarine tracks are included in this analysis."[19] Although the 2004 paper clearly included updated numbers, incorporating the more complete data set, the 1999 number, with the more severe 40% thinning, continued to be highlighted for years afterward as the amount of thinning, often with no mention of the 32% update.[20]

As is often the case, it is quite likely that no one purposely misrepresented anything regarding the ice thickness results. The authors of the 1999 and 2004 ice thickness publications[21] undertook the extremely difficult task of analyzing an important but very incomplete data set, published their 1999 results with full indication of the incompleteness of the data, and several years later published a new paper with an enhanced (although still very incomplete) record. They had substantial interesting new results in both cases to warrant publishing when they did. Furthermore, with the huge ongoing outflow of science publications, there is no possibility of anyone's being aware of everything that has been published, so it is quite likely that many of the people reiterating the 40% thinning subsequent to the 2004 publication were not aware that the 40% value had been revised to 32% and therefore are not guilty of intentionally reiterating the higher value because of its being more dramatic. Nonetheless, the result of this sequence is the frequent reiteration of a result (40% ice thinning) that has been shown to be too high by the very researchers who originally obtained it. My strong suspicion is that if the revised value had been substantially higher rather than substantially lower than the original value, then word of the revisions would have spread much more quickly, and the updated value would have been the one widely reported. The scientific enterprise is far less objective than most people assume, and that will remain the case for the foreseeable future. However, there are piecemeal improvements that could be made, and a relevant one for the example of the ice thickness results would be to take advantage of electronic publication and have authors who update their results notify the original journal and have the journal add a comment to the electronic version of the original paper. This comment could simply indicate that an update is available and give the appropriate reference to it.

As the interest in climate change has intensified in recent years, the rush to publish results has also intensified, adding to the overflow of publications and the difficulty of identifying and reading through the really important ones, even just in one's own specialty. The system is almost forcing this behavior of publishing quick results. Among the key causes are the wide interest in climate change issues and the desire to publish before other researchers obtain and publish similar results or members of the media release the results first. My colleague Josefino Comiso and I felt this pressure in the late summer of 2007, when the Arctic sea ice cover underwent an astounding decline—far exceeding a routine continuation of the Arctic sea ice decreases since the late 1970s—and we were anxious to report an analysis of it in the scientific literature. (The ice decreases were in part due to the wind fields in the summer of 2007.) In huge contrast to every other paper I have been an author on so far in my scientific research career, this paper[22] took less than 3 weeks from conception to submission, and revisions to accommodate the reviewers' comments took only a few days. I enjoyed working on the paper, although it did highlight for me how the system has become so much more conducive to quick results than to long-considered and thoroughly analyzed results. This is right in line with our modern, speed-oriented society, and I could see myself adapting and going with the flow here, as it is more exciting and a whole lot less work to get a paper out in such a short time frame. However, I do not confuse papers written in a rush to get out some result about a current event with the years-long effort to complete a major research publication.

The fact that the system is now favoring quick, less thorough results further heightens my concern that people are becoming more attuned to acting quickly and might rush into an ill-advised geoengineering scheme that could have far-reaching and perhaps horrifying ramifications. Furthermore, the rush to publish could additionally add to the risk that even carefully considered geoengineering decisions could be based on flawed published work.

Another important human factor is that once people, including scientists, speak out in favor of a particular position (e.g., that the Earth is currently experiencing a dangerous global warming), they can become additionally prejudiced toward that position, not wanting to counter what they are on record as having said. This can take many forms, including being blind to countering evidence and including also being inclined to fit results to the personally accepted position regardless of what the data show.

In the late 1980s/early 1990s, researchers from NASA and NOAA separately analyzed data from the radar altimeter on board the U.S. Navy's Geosat satellite to determine changes in the elevation of the Greenland ice sheet south of 72°N over an 18-month period in 1985–1986. Although using the same data set for the same geographic region, over the same time period, the two groups obtained diametrically opposite results: the NASA group determined a thickening of the ice sheet of 28 centimeters per year,[23] whereas the NOAA group, which inserted a 50-centimeter offset that the NASA group did not feel was appropriate, determined a thinning of 22 centimeters per year.[24] Such contrasts are of significant interest in and of themselves, as they highlight that even with sophisticated data sets, the results depend significantly on which corrections or other adjustments are or are not applied, and those choices can vary greatly among researchers and research groups. The fact that two groups using the same data set came to opposite conclusions regarding whether the Greenland ice sheet had thinned or thickened should provide a very strong caution against indiscriminately accepting each data-based "fact" one might hear.

> The fact that two groups using exactly the same data set came to opposite conclusions regarding whether the Greenland ice sheet had thinned or thickened should provide a very strong caution against indiscriminately accepting each data-based "fact" one might hear.

Also of immediate relevance here, each of the two groups obtaining opposite results regarding the Greenland ice sheet (or at least the lead authors of the two studies) placed their results within a global warming context. The NASA group explained that warming could lead to more snowfall and hence to thickening of the ice sheet,[25] whereas the NOAA group explained that warming leads to melting and hence thinning of the ice sheet. Both groups (or at least their lead authors) accepted global warming, and both therefore quite logically placed their results in the context of global warming. This fitting of new results into a larger picture is neither uncommon nor unexpected. When new results are obtained, they tend automatically to be placed in the larger philosophical, scientific, or personal context of the individual's or relevant community's beliefs. In the history of science,[26] this kind of occurrence, when done in line (as here) with the mainstream viewpoint, is known as accepting and further building the normal-science paradigm of the time. Frankly, if you know that you want your result to fit in a particular context, it is often fairly easy to devise an argument explaining that indeed your results do fit.

Relatedly, if results are obtained that fail to fit the desired expectations, it is far more likely that a researcher will double- and triple-check the equations, data, and calculations, hoping to identify a flaw and correct it. Double-checking is a very good thing to do, but when it is done more thoroughly in cases with unexpected or undesired results than in other cases, it biases the results overall to the expected/desired ones. Scientists obtaining a desired result can be too quick to publish, without

> Both group leads accepted global warming, and both therefore placed their results—opposite though they were—in the context of global warming.

adequate double-checking, whereas scientists obtaining an undesired result can be too inclined to argue away the undesired result and not publish it at all. In his 1977 overview of hail suppression attempts (see chapter 8), David Atlas addresses a similar problem, namely, how groups that fail to suppress hail can explain their failures in the face of other groups claiming to have been successful. Atlas mentions two common responses of the unsuccessful groups: "(i) to search far and wide for physical or statistical reasons to account for [the negative results] . . . or (ii) to attribute them to inadequate or inappropriate methods . . . since there is in principle an infinity of methods, negative results can almost always be discounted."[27]

In September 2007, a book review appeared in the journal *Science* that included the following statement: "It is easy to recount the history of science after Newton as something like a long series of attempts to restrain the use of the imagination and the production of overly clever explanations. In such a view, science depends on the sober consideration of data, not the creation of speculative accounts that could conceivably explain that data."[28] The author was certainly not saying that is the only interpretation of the history of science, and certainly the statement does not correctly represent the history of science in total. Nonetheless, the quote is relevant here, as it does properly identify a problem with attempted explanations. Part of the reason that Galileo and Newton steered away from explanations was that so many explanations that Aristotle and others before Galileo had given were found to be wrong and often nonsensical, a key example being the Aristotelian explanation that objects fall when dropped because they are seeking their place at the center of the universe, which was thought to be the center of the Earth. This is one explanation that Galileo could not logically hold once he accepted the Copernican theory of the solar system, with the Earth orbiting the Sun rather than vice versa. Although our explanations today tend to be more scientifically based, almost any set of data can be explained in more than one way, and which explanation is preferred depends on the preferences and larger worldview of the individual. Just as two groups can use global warming to explain a Greenland ice sheet thickening or a Greenland ice sheet thinning, respectively, two groups alternatively could explain a Greenland ice sheet thickening either in a global warming context or in a global cooling context, the first from greater evaporation and snowfall and the second from less melt.

Even today, with all our sophistication, many explanations of the results from scientific data are little more than what is often labeled as "hand waving." I have sat through research-group meetings listening to one person pontificate at great length on why we are getting a particular result, only to have me or another person in the room point out that the data actually show just the opposite, at which point the original person, almost without a pause, suddenly is pontificating a new, equally elaborate explanation but now for the entirely opposite result. Some people are simply inclined to offer explanations for every result, and although some of these people are extremely skilled at it, it often comes down to "overly clever explanations" that not only are not helping the science but actually are hindering both the science and the public's understanding of it.

Unfortunately, it has become common in at least some disciplines of Earth sciences for reviewers, program managers, and editors to insist that scientists offer explanations. Hence, even those who would prefer to hold to the "sober consideration of data" are forced to insert essentially hand-waving explanations into their publications. Many of these explanations are speculative and contorted, and they can lower the quality of the science published. In the effort to get results published, I have often discouragingly accommodated reviewers' comments and put in explanations that I would have preferred to leave out, and I am certainly not unique on that score, as even within the small group of scientists with whom I have discussed this topic, several have felt the same pressure and have also discouragingly accommodated the reviewers. Scientists have to compromise a great deal more than most people realize, as almost never does a paper get published in the peer-reviewed literature without the author having to make at least some changes, and often not all the changes are ones that the author would prefer to make.

My sense is that the overwhelming majority of scientists—whether in the minority or in the majority—very much want to get and publish correct results. However, all of us have biases of one type or another and pressures of one type or another, and these biases and pressures can affect both the research and the results obtained. It is my observation that when scientists in the majority camp get results in line with global warming, they proceed toward publication or presentation, whereas when they get results that do not line up with the expectations, they instead proceed to examine their computer code or devise explanations of why their results really fit the consensus view despite at first seeming not to. This is quite understandable, but it does create a positive feedback, with a consensus view tending to feed back to create more publications supporting and expanding on the consensus.

An entirely separate problem concerns a lack of attention to detail, with many scientists actually seeming to take pride in not being concerned about the details, highlighting instead that they are "visionary" or concerned about the "big picture." Both being visionary and being concerned about the big picture are good, but if the details are wrong, then the science is wrong, and no matter how visionary or big-picture oriented a person is, it is not of much value if the science produced is fundamentally flawed. The lack of care sometimes exercised by scientists in doing science is a difficult problem to talk about because either the conversation is just about generalities, which makes it rather vacuous, or it goes into specifics, which has the unfortunate semblance of attacking individual scientists and of course opens the speaker to the counterattack that he or she is not perfect either (which is most certainly the case of all of us). Taking an intermediate route here, I will simply state that the majority of the scientists I have worked with over the years do not exercise a level of care that would lead me to be certain of the results they present in their research papers. There are far too many instances where the text they write says something that is contradicted by the figures or by the numbers or is in error in other ways. In 2007, I discussed this with a science supervisor in the context of try-ing to explain to him my concerns about the validity of the science being published on global change issues. The supervisor was unmoved, asking whether any of the details that are in error actually affect the main conclusions of the papers and then suggesting that the review process will catch the important errors. I have a very dif-ferent perspective regarding both of his points. First, I feel that if important details are wrong, the science is flawed regardless of whether the main conclusions are affected (and, actually, sometimes they are). If scientists are lax about the details, then the chance increases that some day they will publish a paper where indeed the main conclusions are affected, even if they have been fortunate not to have done so already. Second, as should be clear from the section on peer review, I do not believe that peer review can be relied on to catch all errors that might appear in a manu-script. It is a gigantic mistake for a scientist to proceed as though the peer-review process should catch all serious errors. The correctness of the paper is always the responsibility of the authors, not the reviewers.

Doing science carefully can be extremely time consuming and difficult, and no matter how hard a person tries to be careful (and many scientists try extremely hard), errors can and often do occur. No scientist is immune, and probably none can make it through a long career without at least some errors at one point or another. Hence, even the results of the most careful scientists are not necessarily correct. Sadly, with the media and public interest and the institutional and other pressures to get results out, the likelihood of errors increases further.

MEDIA ISSUES

As global warming has emerged as an item of media and public interest, more Earth scientists are entering the public arena, which in many instances is an arena for which the individual scientist was never trained and for which he or she has no particular affinity. Perhaps partly as a result, many statements attributed to scientists that appear in the media lack scientific objectivity. Some of these are misquotes or quotes taken out of context, for instance by eliminating caveats that the scientist carefully explained but that the interviewer felt unnecessarily confused the issue. In some such cases, the misquoted scientist can become quite frustrated by the process and quite uninclined to speak with the media again. On the other hand, in other instances, statements lacking scientific objectivity are correctly quoted and are indeed the fault of the scientists. Like everyone else, scientists can make mistakes. These can arise because of the scientist's being outright wrong or for such reasons as being overly nervous, being overly desirous not to disappoint an interviewer by discussing caveats and ambiguities, or, most disturbing, wanting too much to make an attention-grabbing statement. By now, quite a few stories have been overhyped in a way that successfully gets immediate attention but in the long run impairs the public's understanding of the issues and reduces the public's confidence in the science and in the press.

> Far from this being the first time in 50 million years, as some media reports indicated, the North Pole experiences instances of open water every year.

For an example of an overhyped story, consider the summer of 2000, when the Russian icebreaker *Yamal* reached the North Pole and found the location to be free of ice, greatly surprising the many tourists and few scientists onboard.[29] Although this event had some newsworthiness in that a far greater amount of open water existed near the North Pole than would be expected—the *New York Times* reported that the ship had to travel 6 miles away from the North Pole to find an ice floe firm enough to let the passengers debark[30]—and the Northwest Passage was easily accomplished that year by another ship in a single month,[31] the press coverage went considerably overboard on the significance of the open water at the

North Pole. Most notably, the *New York Times* article that began the flurry of media attention to this incident, published on August 19, 2000, and titled "Ages-Old Ice-cap at North Pole Is Now Liquid, Scientists Find," includes the following sentences: "The thick ice that has for ages covered the Arctic Ocean at the pole has turned to water. . . . The last time scientists can be certain the pole was awash in water was more than 50 million years ago."[32] However, far from this being the first time in 50 million years, the North Pole experiences instances of open water every year. Anyone well familiar with the Arctic ice and giving a little thought to the topic realizes that, as the Arctic ice pack is a dynamic, moving mass of ice floes, open-water areas (including many roughly linear ones called "leads") are continually opening and closing within the pack, and an open-water area is bound to be situated right at the North Pole at numerous moments each year.

The author of the original *New York Times* article is not to blame. He was informed by passengers about the startling ice-free conditions, proceeded to speak with two prominent scientists who had been lecturers on the cruise (one from Harvard University and the other from the American Museum of Natural History), and wrote his article based on the information they provided. Unfortunately, neither of these scientists is a sea ice expert, so, however noted they might be in their own fields, they were not immune from making an error regarding the frequency of open water at the North Pole. Because of the startling nature of the *New York Times* report and the prominence of the *New York Times* itself, the report was quickly a major news item. Overall, I would say that the follow-up to the story was excellent, as many reporters called sea ice experts to put the event into context. I was interviewed by several print reporters and on several radio shows and in each case felt that the interviewer was conscientiously trying to get the facts corrected. Five days after the original *New York Times* article appeared, its author, John Wilford, called me and expressed dismay that the original article was flawed. We talked at length about the sea ice conditions and sea ice research, and he spoke also with other sea ice experts, resulting in a follow-up story appropriately titled "Open Water at Pole Is Not Surprising, Experts Say."[33] Science reporters have an extremely difficult job, having to become knowledgeable enough about a huge range of science topics that they can write coherently about them and often having to write under an impending deadline. I was impressed with Wilford's professionalism and with his clearly sincere desire to correct the misrepresentations appearing in the original article.

I also feel that the Harvard scientist on the ship (Jim McCarthy) did a superb job in clarifying the situation subsequent to the original article. He and I were on National Public Radio's *Morning Edition* on August 22, 2000, and on August 31, McCarthy,

Drew Rothrock, and I were the three guests on the hour-long *Diane Rehm* radio show. In both cases, McCarthy cogently explained the ice conditions that they encountered on the *Yamal* voyage and which aspects were most unusual. As is often the case, there were no "bad guys" purposely trying to mislead anyone. Quite the contrary, all the primary participants were trying to do a conscientious job. Still, the badly overhyped initial story left many people with very erroneous ideas.

For another overhyped example, consider what has been said about potential changes in the large-scale transfers of water and heat taking place in the "ocean conveyor" (chapter 1). In 2005, results were published from measurements taken in 1957, 1981, 1992, 1998, and 2004 along the Atlantic portion of the 24.5°N latitude circle. The article suggested that "the Atlantic meridional overturning circulation has slowed by about 30 per cent between 1957 and 2004."[34] The article was explicit that the slowing was in the deep ocean, not in the surface Gulf Stream flow that brings warm water northeastward across the Atlantic. Nonetheless, concern that a slowdown in part

> Each time we have a false alarm like this, it is likely that additional people become immune to further dire warnings coming from the climate community.

of the conveyor would lead to a slowdown in the rest of the conveyor resulted in media coverage declaring that northern Europe could be headed for a significant deep freeze. A year later, however, the journal *Science* published an article titled "False Alarm: Atlantic Conveyor Belt Hasn't Slowed Down After All," as new analyses indicated that changes as large as those reported among the 5 years examined in the 2005 study sometimes occur within a single year; that is, the five "snapshots" used in the 2005 study are woefully insufficient for determining a long-term trend or change.[35] Hence, within a year, the fears sparked in 2005 were shown to be essentially unfounded, as scientists agreed that the high variability in the currents makes it impossible to know from the data analyzed thus far whether any significant change in the Atlantic circulation has occurred.[36] Each time we have a false alarm like this, it is likely that additional people become immune to further dire warnings coming from the climate community. We run the serious risk of figuratively crying "wolf" or "fire" too many times, numbing people to future dire warnings that might have more substance.

Of course, inappropriate media hype also occurs with topics other than climate change, as many celebrities know painfully well. In *A Scientist's Guide to Talking with the Media*, Richard Hayes and Daniel Grossman describe a 1990 medical case in which an experimental AIDS treatment was highlighted on CNN's *Health Week*, sparking a nationwide flood of media attention. The procedure, called "hyperther-mia," involves taking blood from a patient's body, heating it, and returning it back into the body. The CNN report discussed the first human patient to undergo this procedure for the purpose of treating AIDS. The apparent success with this patient was enough to spread hope throughout the AIDS community during 2 months of media hype, even though during those 2 months, a second patient underwent treatment with no signs of improvement. Then a third patient receiving the treat-ment died, and the results from an investigation of the first patient indicated that he had not had the AIDS-associated form of cancer he was thought to have had but instead had a bacterial infection. At that point, further experimentation with the hyperthermia treatment for AIDS patients was halted.[37]

Another case of extensive media hype followed by a burst in the bubble of en-thusiasm concerned the discovery in the 1990s of microscopic organic compounds and perhaps minute fossils of microorganisms in a meteorite that had been re-trieved from Antarctica in the 1980s. The meteorite was believed to have originated on Mars, and the findings were hailed as evidence that life existed on Mars 3 billion years ago. The initial announcement of the findings was made on August 6, 1996, by NASA Administrator Dan Goldin and was quite reasonable in terms of exhibit-ing excitement while also including a good range of appropriate qualifiers. Specifi-cally, Goldin announced that "NASA has made a startling discovery that points to the possibility that a primitive form of microscopic life may have existed on Mars more than three billion years ago. . . . The evidence is exciting . . . but not conclu-sive. It is a discovery that demands further scientific investigation. . . . These are extremely small, single-cell structures that somewhat resemble bacteria on Earth. There is no evidence or suggestion that any higher life form ever existed on Mars."[38] Still, despite the reasonably phrased initial announcement, people were naturally extremely interested, and the press coverage was intense (for a scientific topic) over the next many months.[39] One of my colleagues at the time at NASA's Goddard Space Flight Center, Jerry Soffen, told me that he hardly slept for the 8 months fol-lowing the initial announcement both because of his own excitement and because of all the interviews he was giving (Soffen had been project scientist for the Viking missions that landed on Mars in 1976). Eventually, the media coverage died down, as it became clear that the evidence was not nearly as compelling as first thought.

As the public witnesses more and more instances like this, where a hope or in-terest is raised only to be shattered, or a feared disaster does not materialize, it is no wonder that people become immune to further pronouncements. Both journalists and experts would serve the public better if they would be slower to hype a story without stating (or reiterating) the reasonable qualifiers. Sadly, in today's society, that might not be feasible for reporters and researchers intent on maintaining their competitive edge.

One way to avoid contributing to overhyping a story would be to say nothing. However, that is not a viable option for scientists who feel a strong sense of responsibility to inform the public and policymakers and/or to offer suggestions. Speaking with members of the media has advantages in getting a message out and perhaps receiving favorable recognition, but it runs the risk of misinterpretations, the need for repeated clarifications, and entanglement in never-ending controversy. Hence, the decision of whether to speak with the media tends to be highly individualized. Decades ago, it was unusual for Earth scientists to have results that were of interest to the media, and consequently few media contacts were expected or encouraged. In the 1970s, the few scientists who spoke frequently with the media were often criticized by their fellow scientists for having done so. The situation now is quite different, as many scientists feel a responsibility to speak out because of the importance of global warming and related issues, and many reporters share these feelings. In addition, many scientists are finding that they enjoy the media attention and the public recognition that comes with it. At the same time, other scientists continue to shun speaking with reporters, thereby preserving more time for their science and avoiding the risk of being misquoted and the other unpleasantries associated with media coverage.

Steve Schneider, who has an extensive record of dealing with the media, starting in the 1970s when he was one of the most visible U.S. Earth scientists, from his positions at the Goddard Institute for Space Studies and the National Center for Atmospheric Research, and continuing in his later position as a professor at Stanford University, is quoted by Hayes and Grossman as saying, "In my view, staying out of the fray is not taking the 'high ground,' it is just passing the buck."[40] This is reasonable, as is the need "to simplify for the sake of effectiveness," another strategy attributed to Schneider by Hayes and Grossman.[41] Simplification is often highly desirable, sometimes essential; however, it should not override truth and honesty, which unfortunately it sometimes does. Within their discussion of Schneider (although not quoting him), Hayes and Grossman write, "Thus, the goal of honesty has the potential or, even likelihood, to undermine efficacy. . . . To be effective, you have to be concise, lucid, memorable, and, if possible, colorful. These requirements are at odds with the requirements for honesty."[42] Fortunately, they are not always at odds with honesty. When they are, I think the person should back off on the conciseness, colorfulness, and so on rather than on the honesty. Still, it is not easy no matter which route is taken. Certainly, any goal of telling "the whole truth" is bound to crumble, as more details can always be added. Trying to be honest, with all the relevant caveats, can easily lead to statements that are too long and cumbersome and that get abbreviated in the media to something that is even more misleading than what the scientist might have said with a shorter, less complete statement. Recognizing that in most cases completeness is impossible (e.g., you cannot provide the temperature trend for every location in the world), an aim should be for honest, balanced summary statements. Examples include the following: "On average, temperatures have risen" (providing the average value if it is available) and "Although some regions have experienced unusually cold conditions, the predominance of the data indicates warming over much of the Earth."

Schneider has clearly grappled with the issue of effective media presentation. In his 1989 book *Global Warming*, he refers to it as a "double ethical bind," wanting as a scientist to present the whole truth, including the caveats, but wanting as a citizen to improve the world, which often means making "simple and dramatic statements" in order to get media coverage and the public's attention.[43] Although I might not agree with each choice Schneider makes (especially not the choices in favor of dramatic statements), I probably would agree with most of them, and I very much like the fact that he is up front about the issue and has considered it at some length. The reality is that we can never be totally complete in our presentations, whether written or oral, as we neither know nor would have time to mention every piece of potentially relevant data. Choices do need to be made by all of us, no matter on which side of a debate or how firmly in the middle we might be. Further, Schneider most certainly does not give only one side of the story, instead often being careful to present some of the key uncertainties. For instance, in his book *Laboratory Earth: The Planetary Gamble We Can't Afford to Lose*, he explicitly states that although it is not likely that changes in the Sun can explain the climate changes of the past 100 years, this extremely unlikely possibility cannot be ruled out with 100% certainty. Similarly, he mentions several mismatches between the observations and the model simulations, including the magnitude of amplified polar warming, the magnitude of stratospheric cooling, and the magnitude of the contrast in the warming of the Northern and Southern hemispheres.[44]

Because of the increased media interest in Earth sciences, many organizations now offer their scientists media training of one form or another. I have attended several such sessions and can attest to the value of much of what was taught in the sessions I attended. I can also attest to the value of some written guidebooks for scientists and others regarding how to work well with the media.[45] These books emphasize the need for clarity, focus, and brevity; the need to speak without jargon or unnecessary acronyms; the need to be enthusiastic (or at least not dull); and the need to identify beforehand a small number of key messages and not get distracted away from those messages. The classes and the books are extremely useful for those of us for whom talking in front of cameras and microphones does not come easily. They are especially useful in their presentation of examples of actual cases where a message became badly twisted or outright misrepresented, as it is overwhelmingly less jarring to learn from the mistakes of others and their pain than to learn instead through making the same mistake oneself. The examples are also good in showing both cases in which a reporter has intentionally twisted a story and cases in which the reporter did not intentionally twist anything but instead was provided with an insufficiently clear message to start with. However, despite my favorable impressions overall, I also have some major concerns with what is taught, especially regarding sound bites.

Quite understandably, mass media welcome attention-grabbing sound bites, both to lure the reader or listener and to instill a central message. Many scientists are becoming attuned to creating and vocalizing sound bites effectively, and as long as the sound bites are valid, I am all in favor of short, informative, memorable statements. However, unfortunately, the sound bites are not always valid, with scientists sometimes being far from objective or reasonable in what they say. Sea ice examples mentioned earlier in this chapter include "the ice-albedo feedback is starting," "once you start melting and receding, you can't go back," and "the last time scientists can be certain the pole was awash in water was more than 50 million years ago."[46] Other examples could be provided for sea ice, and similarly examples could be provided for many other climate components. Sound-bite types that have become noteworthy in the 2005–2009 time frame are ones along the lines of "we have only ten years left to prevent disaster" and ones claiming that we are approaching a "tipping point" of one category or another.

Talk of tipping points has become increasingly popular in the mass media, a fact borne out by statistics reported in the June 15, 2006, issue of the journal *Nature*. Specifically, the number of newspaper articles mentioning a climate tipping point increased from 45 in the 12 months of 2004 to 234 in just the 5-month period January–May 2006.[47] The term has also become popular among scientists, appearing in journal publications[48] as well as frequently in conversations and presentations. In fact, "tipping point" is used so much now, in some cases where it really is not appropriate, that the term has become trivialized, which is quite unfortunate, as otherwise it could be a very colorful, meaningful, and effective term.

Another issue, comparable to that of sound bites, is the use of superlatives. In at least some media classes and literature, scientists are taught that the media favor stories with superlatives (smallest, largest, warmest, first, and so on).[49] This is fine (in fact, quite helpful) in terms of suggesting which research results are more likely to attract media interest. It is not fine, however, when scientists or their public affairs personnel proceed to twist the results to add superlatives where they do not belong. And that definitely happens, at least on occasion, in the media classes and in front of the cameras and microphones in actual interviews. If something is not the first or largest or smallest, it should not be called that just to get media attention. Years ago, I would not have imagined that the previous sentence would need to be mentioned, being as obvious as it is, but by now, I have heard too many inappropriate superlatives to ignore the existence of a problem.

AUGUST 13, 2007, NEWSWEEK COVER STORY

I am a longtime *Newsweek* subscriber and an admirer of the science writing of Sharon Begley. However, in its August 13, 2007, issue, *Newsweek* had a cover story by Begley on global warming that I feel is seriously misleading and illustrative of several problems.

First, consider the cover. The main title, centered and in large letters, is "Global Warming Is a Hoax.*" Knowing quite a bit about global warming issues and some basics about *Newsweek*, I realized immediately that *Newsweek* would not have a cover article seriously claiming that global warming is a hoax. Hence, I realized that the asterisk would disclaim the thrust of the main title, which indeed it does, in smaller lettering positioned in the lower left of the cover. Given that *Newsweek* is a well-known, reputable magazine for the general public, I feel that it should not have put as the main title on the cover a statement that it (meaning the *Newsweek* upper management) believes to be false. *Newsweek* is read by a large range of people, some of whom know little about the scientific issues surrounding global warming, and the cover is seen by far more people than those who read the article or even than those who read the explanatory asterisk information. In addition to reading *Newsweek*, another approximately weekly activity of mine is to go to the grocery store. This activity includes seeing the cover of *Newsweek* and other magazines as I wait in the checkout line. There could be many tens of thousands of people who read the "Global Warming Is a Hoax" main title, read no further, and went away in disgust, with thoughts along the lines of the following: "All those scientists and reporters claiming all these years that we should be worried about global warming, and now we find out it was all a hoax. Scientists are just as bad as politicians; we can't believe anything they say." I think *Newsweek* in this case did a disservice to scientists and science writers in general, not just those who write about global warming. I doubt that they would have presented a similar title saying that the Holocaust was a hoax with a footnote clarifying that they don't really mean it or a title saying that the time has come to return to segregation, again explained away in a footnote as not being what they really mean. Similarly, they should not have done it in the case of global warming or any other topic. As a newsmagazine, *Newsweek* should seek above all to inform and to be correct in the information it transmits. The August 13, 2007, cover headline failed horrendously on that score.

Now to the article itself, which is titled "The Truth about Denial." The article is excessively one-sided. Begley seems to accept the consensus IPCC view in full and to label anyone who fails to do so a "denier." She uses the words "the denial machine" at least 14 times and claims that it is a "well-coordinated, well-funded campaign by contrarian scientists, free-market think tanks and industry."[50] If there is such a well-funded campaign, some of the key skeptics are not in on it.[51] Begley puts down skeptic Patrick Michaels for receiving funding from industry without mentioning that some mainstream scientists receive corporate funding as well. She mentions thousands of deaths during the 2003 heat wave in western Europe without mentioning thousands of deaths from the cold. (According to Lomborg, in Europe as a whole, approximately 1,500,000 people die from cold each year, and approximately 200,000 people die from heat each year, for a very heavy weighting toward more deaths from cold than from heat.[52]) She criticizes media training proposed for naysayers without mentioning that some mainstream scientists are also getting media training. Many of her points are important and valid, and if she had presented the facts instead in a balanced and nonderisive way, it could have been an excellent article. However, the context in which everyone not solidly in the mainstream is a "denier," with the repeated mention of a "denial machine," is unfair. Begley clearly feels that there should be more effort to reduce greenhouse gas emissions into the atmosphere, and I strongly agree with her on that issue. But I do not feel that the way to get to that goal is to ridicule everyone who has any disagreements with the consensus view, which, after all, is a continually evolving view, as is likely the case with at least some views in any scientific discipline.

Much to *Newsweek*'s credit, the week after Begley's article appeared, *Newsweek* published a strong, full-page, personal-opinion criticism of the article. The criticism was by *Newsweek*'s noted columnist Robert J. Samuelson, who correctly indicated that the Begley article oversimplified a very complicated problem and "was a wonderful read, marred only by its being fundamentally misleading."[53] Samuelson emphasizes the difficulty of dealing effectively with global warming and specifically mentions that the "'denial machine' is a peripheral and highly contrived story. Newsweek implied, for example, that Exxon-Mobil used a think tank to pay academics to criticize global-warming science. Actually, this accusation was long ago discredited." Samuelson closes his superb article with the following well-stated advice: "Journalists should resist the temptation to portray global warming as a morality tale—as Newsweek did—in which anyone who questions its gravity or proposed solutions may be ridiculed as a fool, a crank or an industry stooge. Dissent is, or should be, the lifeblood of a free society."[54]

THE APRIL 2007 GALILEO'S LEGACY CONFERENCE

In April 2007, I participated in the Second Annual Galileo's Legacy Conference, held at Missouri Western State University. This conference was one of the best-conceived conferences I have ever attended. The purpose of the conference series is to bring experts on both sides of a highly contentious issue to Missouri Western to present their respective views and engage in a civil debate, with significant audience participation. The first Galileo's Legacy Conference dealt with evolution, and the second, in April 2007, dealt with global warming and the various issues surrounding it. At the April 2007 conference, I was the scientist representing a moderate form of the consensus viewpoint, Dr. Willie Soon of Harvard University was the scientist representing the skeptics, and Dr. John Nolt of the University of Tennessee, a philosopher and ethicist, represented the ethical viewpoint of a nonscientist who accepted the consensus view of the IPCC and proceeded from there to ethical consequences. Each of us presented our respective cases in presentations on the first evening of the conference, and these presentations were followed (both the same evening and the next day) by extensive discussion among the speakers and with the audience, the latter coming both from the university and from the surrounding community, including members of the media. Many people in the audience participated, several making strong and well-articulated points. I think by far the majority of us left having learned and having felt that it was a quite worthwhile and eye-opening discussion.

Of particular relevance to this book, I found Soon to be very sincere and passionate about his work and his results, which are currently far outside the mainstream of climate change research. He finds the paleoclimate evidence to be convincing that CO_2 is not a driver of global temperature, pointing out in particular that in the long-term record, the CO_2 increases tend to follow rather than precede the temperature increases. He also finds solar influences to be far more important to global temperature changes than CO_2 influences. I assume that many in the mainstream would agree in substantial part with Soon regarding the paleoclimatology of the past million years, for which the record is strong that CO_2 changes lag rather than lead the temperature changes (see chapter 2) and Milankovitch insolation-based forcing is widely accepted. However, many (including myself) disagree with him regarding the important drivers for today's climate changes. In any event, whether right or wrong, Soon presented his argument energetically and passionately, and he came across as sincere in attempting to do good science and to relay his results to the broader community. Soon specifically mentioned that he never wanted to be in the minority; being in the minority is just where his results placed him. I suspect that is true of others of the skeptics as well, as being in the minority is not in general a sought-after position.

After Soon spoke, Nolt gave a wonderful description of the ethical issues surrounding what we are currently doing to our environment and the anticipated consequences to future generations. As a philosopher, he quite reasonably accepted as his starting premise the conclusions of the consensus IPCC report. A portion of his argument can be summarized as follows: 1) This is an extraordinarily important ethical problem, as the people we are harming are our descendants, who have no regress; 2) we have benefited tremendously from our ancestors (including the founders of our country, who wrote a Constitution for "posterity"), and we in turn owe a debt to our descendants, to provide them with conditions at least as good as those that were left to us; and 3) we all have a personal stake in the future, as our descendants will ultimately be the earthly judges over the value of our lives.

CLOSING COMMENT

Currently, there are many advantages for a scientist to fit his or her results into a global warming scenario. For instance, it can help get funding, help get manuscripts accepted for publication, help avoid the wrath of the mainstream, and help generate media interest. Putting everything into one perspective, however, runs the risk of generating a very biased overall picture that appears to be much more complete than it really is. Hence, I close this chapter on social pressures with the following relevant quote from mainstream climate scientist Steve Schneider: "Consensus is a poor way to do science."[55] Schneider wrote that statement decades ago in the context of the then-current consensus that pollution could cause a cooling climate. The statement was valid when Schneider wrote it decades ago, and it remains valid today. Obtaining a consensus can be extremely valuable, in fact in some cases almost essential, especially for presenting a case to politicians and other decision makers. But obtaining a consensus should never be confused with obtaining proof. In any field of science, scientists skeptical of the mainstream consensus could have important points to bring to the table, and mainstream scientists should never ignore that fact.[56]

PART V

Avoiding Paralysis despite Uncertainty

CHAPTER 12

What Are the Alternatives?

Many suggestions have been made regarding how to ease the predicted climate crisis, including more conservation, less waste, and development and use of alternative energy sources. Although more mundane than geoengineering, many of these suggestions are far safer, more reasonable, and perhaps even more likely to succeed.

AT THIS point, it is logical to ask the following: Even in the absence of certainty, if the likelihood exists that global climate change will bring serious difficulties that are in substantial part due to human activities, then is it not imperative that we do something? And if geoengineering is not the appropriate solution, then what is? Many plausible steps toward a solution have been suggested, ranging over broad spectrums both of how easy they might be to implement and of how effective they might be in addressing the predicted problems. This chapter presents some of the safer, nongeoengineering suggestions for reducing future climate-related and broader environmental damage. Because much has already been written and discussed about such suggestions, I make no attempt to be complete but instead attempt to make it clear that viable alternatives exist, although none of them is yet proving to constitute a quick and easy solution.

Many of the suggestions in this chapter revolve around reducing emissions to the atmosphere and oceans. These suggestions have an overwhelming advantage over geoengineering in that they tackle the core of the problem and hence address the broad range of the impacts of the emissions, not just global warming, which has tended to be the focus of most of the geoengineering proposals. Most notably, geoengineering to cool the planet does not address the increased acidification of the oceans, a problem that could prove fatal to many marine species. People hearing of geoengineering and concluding that we can emit as much as we want because geoengineering can solve all the resulting problems are badly misunderstanding the range of the problems humans are causing and the potential fixes of the geoengineering schemes. By far the surest way of reducing human-induced damage is to reduce the actions causing the damage, and that is the focus of many of the suggestions presented in this chapter. Although some of the suggestions proposed here might have problems of their own, at least none of them have the fundamental flaw inherent in some of the geoengineering schemes, namely, purposefully adding to the atmosphere or oceans even more emissions, with the intent of counteracting other emissions, a scene reminiscent of slapstick comedy except for the fact that it has been proposed in all seriousness (see chapter 7).

SAMPLE PROGRESS

Fortunately, much progress has been made in recent decades toward better stewardship of the planet, both regarding the present and regarding the anticipated future. This has occurred on all fronts, in attitudes as well as accomplishments, from the very small scale to the very large scale, and by individuals, corporations, other organizations, and governmental entities. This section presents an illustrative sampling of some of the efforts and successes, almost all of which derive at least in part from an increased awareness and a concern either about one's own health and quality of life or about the health and quality of life of others, sometimes encompassing future generations and sometimes encompassing also species other than humans. Although this chapter and book are focused mostly on global warming and its attendant issues, it is relevant additionally to mention ways in which humans have come together to make progress on other environmental issues as well.

Progress on Environmental Problems
outside the Global Warming Context

Tremendous progress has been made in cleaning individual rivers and lakes and reducing air pollution in numerous locations around the world. Often this is done through a grassroots effort in which individuals in the community who want their river or lake or air cleaner proceed to act. In some locations, wonderful results have been achieved, with, for instance, lakes once again becoming suitable for swimming and for housing a healthy ecosystem of fish and other freshwater or marine life.

A regional example of at least some improvement has come in the progress made in reversing damage done to the northernmost segment of the Aral Sea (see chapter 4), named the Small Aral Sea. The Small Aral Sea, along with the rest of the Aral Sea, experienced severe shrinkage in the twentieth century due to humans siphoning water away from inflowing rivers. After the World Bank got involved in 1999, a 13-kilometer dike was completed, and in only 7 months versus a projected 10 years, the Small Aral Sea reached its target level of rising to 42 meters above the Baltic Sea, providing a very encouraging example of success coming much more quickly than expected. Former fishing villages are now not nearly as far away from the sea as they had been, a prime example being Aralsk. Like the other Aral fishing villages, Aralsk had bordered the sea in the early twentieth century and had stood by helplessly as the sea retreated in the mid- and late twentieth century. Eventually, it was 80 kilometers from the Aral shores, making it impossible for it to remain a productive Aral fishing village. With the Small Aral's revival, the sea was only 15 kilometers from Aralsk by 2006.[1] The overwhelming majority of the former Aral Sea (versus the small, northernmost segment) has not experienced such progress, but progress even in a small portion can be encouraging.

On a global level, notable success has occurred in reducing sulfur emissions to the atmosphere. Because sulfur emissions constitute a serious health and pollution problem, communities and larger governmental entities have tended to legislate sulfur emissions more aggressively than the less visibly damaging greenhouse gases at the core of the human contribution to global warming. As mentioned in chapter 4, global annual sulfur emissions, as estimated in *The Economics of Climate Change*, peaked in 1989, at 74.1 million tons, and had been reduced to 55.2 million tons by 2000.[2] The much-reduced annual amount still leaves a great deal of sulfur entering the atmosphere, but it is a marked improvement over what it had been and offers a reason for hope that environmental problems can be reduced not only on local and regional scales but on the global scale as well.

Also on the global scale is the international response, beginning in the 1980s, to the human-induced damage to the stratospheric ozone layer, especially as manifested in the Antarctic ozone hole (chapter 4). The 1987 Montreal Protocol and follow-on agreements were major accomplishments in terms of the international community coming together and acting definitively to restrict future use of chlorofluorocarbons (CFCs) and other compounds that contribute to stratospheric ozone destruction. The Antarctic ozone levels are a long way from recuperating, but the accomplishment of the Montreal Protocol and subsequent agreements has led to substantial reductions in CFC emissions, and great optimism exists that eventually, sometime in the twenty-first century, the stratospheric ozone layer will have fully recuperated.[3]

From the beginning of the international efforts, all parties involved knew that the ozone hole would not disappear immediately, as many of the CFCs poured into the atmosphere in the twentieth century will remain in the atmosphere for decades. Early estimates suggested that it would take several decades for the ozone above Antarctica to recover fully. In 2003, the World Meteorological Organization estimated that full recovery of the Antarctic ozone amounts back to 1980 levels would occur by 2050, and in June 2006, based on new model calculations, scientists from NASA, the National Oceanic and Atmospheric Administration (NOAA), and the National Center for Atmospheric Research (NCAR) revised the estimated recovery date to about 2068, marking a further delay of 18 years.[4] The NASA/NOAA/NCAR study indicated that the ozone hole was "no longer growing,"[5] although a few months later, the ozone hole set new records for both area and depth, with a record-minimum ozone column amount of 85 Dobson units recorded over East Antarctica on October 8, 2006.[6] These new results were not encouraging, but they did not invalidate what the NASA/NOAA/NCAR group had written, as, for instance, their article had warned, again from modeling results, that the reduction of the ozone hole area would likely not be statistically detectable until about 2024.[7] Further, the 2007 ozone hole returned to a size near the average for the previous 15 years.[8] There are certainly many remaining unknowns and a strong possibility that between now and the 2068 predicted recovery date, factors unrecognized today might lead to either a faster or a slower recovery.

The case of the ozone hole is one where the international community has acted and has reduced the culprit CFC emissions to the atmosphere but where it remains to be seen whether the desired response of fully restored stratospheric ozone amounts will occur. The current understanding of the chemistry of ozone creation and destruction (chapter 4) suggests that ozone recovery will come eventually, although the success so far is most evident in the international cooperation and the large-scale reductions in CFC emissions. Full recovery will depend not only on CFC levels in the stratosphere but also on the levels of any other chemicals that might enter the stratosphere and affect the maze of chemical reactions creating and destroying stratospheric ozone.

Progress Relevant to Global Warming

The human contribution to global warming derives largely from our release of greenhouse gases and heat into the atmosphere and from changes in land use that reduce the drawdown of greenhouse gases from the atmosphere. Hence, when it comes to making progress on the global warming issue, a large component of this is reducing our greenhouse gas and heat emissions to the atmosphere and changing land use patterns or doing other activities that increase drawdown of greenhouse gases. With use of fossil fuels being a major culprit in greenhouse gas emissions, a reduction in fossil fuel emissions is key. This can be done through cleaner fossil fuel usage, less energy usage altogether, or use of cleaner alternative energy sources. A primary desirable land use change for increased drawdown of CO_2 in particular is more tree coverage. At an individual level, many people have planted trees and have cut back on energy usage by turning off lights, reducing heat and air-conditioning in their homes, replacing inefficient lightbulbs and appliances by more efficient versions, recycling, traveling less in planes and personal vehicles, and otherwise lessening their personal contributions to greenhouse gas increases.

At a corporate level, much progress has occurred in individual corporations and categories of corporations. For example, many companies that cut down trees for the wood now replant routinely, which is both to their own long-term benefit, maintaining a supply of trees, and to the benefit of the world, maintaining the drawdown of atmospheric CO_2 by the trees. Many hotels now provide patrons the option of having towels and sheets washed less frequently than once a day, again saving money for the corporations as well as benefiting the world, this time by less needless use of energy. Automobile manufacturers have developed cars that achieve improved efficiencies, and these are becoming more common in many countries. Architects and builders are designing buildings with features that make them more energy efficient by at least 30% without increased construction costs, as demonstrated, for instance, through the certification program of the U.S. Green Building Council's Leadership in Energy and Environmental Design.[9] Individual businessmen also are taking important leadership roles, one example being Rupert Murdoch with his May 9, 2007, announcement to his worldwide News Corp employees of an ambitious initiative to transform the company's energy usage to become carbon neutral—not a simple feat—by 2010.[10]

By now there are numerous organizations whose central purpose revolves around one or more environmental issues, including issues of global warming and related issues, such as biodiversity. An excellent example of such an organization is the Rainforest Alliance, which, among other things, is helping to preserve the Earth's remaining rain forests and thereby retain their drawdown of atmospheric CO_2. The Rainforest Alliance's explicit mission is to work to "conserve biodiversity and ensure sustainable livelihoods by transforming land-use practices, business practices and consumer behavior" (as stated in various Rainforest Alliance publications, including on the organization's website at http://www.rainforest-alliance.org, from where it was obtained on December 10, 2007). Rainforest Alliance personnel have instructed local inhabitants in a wide range of sustainable business activities that make use of rather than destroy the rain forests, creating in at least a portion of the local population an awareness of the personal financial as well as environmental benefits of retaining the rain forests. These efforts are complemented by a quite substantial certification program. Rainforest Alliance–certified seals, picturing a frog, are now visible on products as diverse as coffee, chocolate, wood, greeting cards, bananas, furniture, and books. The certification process is neither simple nor cheap but often proves to be advantageous not only from an environmental standpoint but also from an economic standpoint for the company whose products are being certified. For a case in point, Chiquita Brands International began working toward Rainforest Alliance certification in 1992. The process took 8 years and cost approximately $20 million for improved facilities and education of the company employees, but in the end, the process resulted in significantly upgraded worker housing, improved quality of life for workers and their families, upgraded equipment, reduced waste, reduced use of pesticides, increased efficiency, improved water quality, newly instituted recycling and waste disposal programs, and reduced costs overall. Further, the certification seal has been an important factor in successful sales, especially to leading retailers in Europe. The result in this and other cases has been that the environmentally conscientious changes have also been fiscally sound business decisions.[11]

Notable progress has also been made in developing and using alternative energy technologies, including use of renewable energy sources. Each of the following has become a major energy source at least in some locations: photovoltaics, solar heating, geothermal heating, winds, hydrosystems, nuclear energy, and biofuels. The first two make use of solar energy, the first through photovoltaic cells converting solar radiation to electricity and the second through thermal panels converting solar radiation to heat. The others also rely on renewable energy sources although from within the Earth system (some driven by the Sun). Geothermal heating relies on the Earth's internal heat, which presumably will not be running out for many millions of years, and winds rely on the atmospheric circulation, powered by the Sun and the Earth's rotation. Hydrosystems use flowing water, for instance with energy captured by waterwheels along streams and dams along rivers or with larger systems capturing the energy of tides and waves. Nuclear energy is created through the conversion of mass to energy in atomic nuclei during nuclear fusion, providing an essentially limitless energy source, although accompanied by fears of accidental release of cripplingly dangerous levels of radioactivity. Biofuels are fuels produced from plant or animal organic matter, with common usage often restricting the term to cover only those fuels created from recently living plant or animal matter, that is, eliminating fossil fuels from this category. Some biofuels, especially those based on corn, have received considerable criticism, as they can be quite counterproductive if land is cleared in order to produce crop-based biofuels or if crops otherwise used for food are converted to biofuels, lessening the food supply. However, there are other biofuel possibilities that do not necessarily involve land clearing, for instance biofuels from algae and waste biomass, and these could become important energy sources.

The biofuel case is instructive and illustrates yet again the wide-ranging reality of unintended consequences (chapter 6). Biofuels had looked promising as an alternative fuel in part because of returning to the atmosphere only what they had recently received from the atmosphere, versus fossil fuels returning to the atmosphere carbon that had been taken from the atmosphere millions of years ago. However, converting cropland from food crops to fuel crops can lead to much higher food prices and, even more critically, to food shortages, and converting forests to cropland for fuels has all the disadvantages of any deforestation.

Palm oil is among the cheapest of biofuels and consequently has been pursued aggressively in some countries, to the great detriment of the local environment. In fact, 87% of the deforestation in Malaysia between 1985 and 2000 was due to the creation of palm oil plantations.[13] The net result is that although biofuels were encouraged as alternative energy sources for environmental reasons, the reality is that the environmental consequences of biofuel usage have been so bad that British syndicated columnist George Monbiot has declared unequivocally, "The decision by governments in Europe and North America to pursue the development of biofuels is, in environmental terms, the most damaging they have ever taken."[14] I feel for the respective governments and the individuals most responsible for the well-intentioned but ultimately damaging decisions.

> The decision by governments in Europe and North America to pursue the development of biofuels is, in environmental terms, the most damaging they have ever taken.
> —George Monbiot[12]

One city that has conscientiously attempted to reduce greenhouse gas emissions is Burlington, Vermont, which in 2002 launched a "10% challenge" to reduce emissions throughout the city by 10%. Although by 2006 they had not succeeded in the 10% goal (or even in decreasing overall emissions by a lesser amount), they had undertaken efforts that presumably kept their emissions below the level to which they otherwise would have risen. Among these are the following: the city buses now have bicycle racks on them; the city supports a City Market grocery store that contains considerable local produce and provides an alternative to driving to the suburbs for food purchases; the Burlington Electric Department (BED) obtains close to half its energy from renewable sources, including a wind turbine that generates electricity for 30 homes (clearly just a beginning) and a 50-megawatt power plant powered by wood chips; the BED leases energy-efficient compact fluorescent lightbulbs to customers for $0.20 per month; and a composting facility turns vegetable waste from restaurants into soil.[15]

On the state level, California has taken a lead on many fronts in the United States, being the first state to legislate strict clean-air regulations, the first to legislate energy efficiency standards for appliances, and the first to require that products containing toxic chemicals display warnings to that effect.[16]

On the country level, several countries, including Japan, have made strides in the use of photovoltaic cells. Japan progressed from obtaining 21 megawatts of energy from photovoltaic cells in 1996 to 129 megawatts in 2000 and to 602 megawatts in 2004. Other countries are conscientiously considering alternatives to traditional energy sources as they further develop their economies. For instance, as part of a program to provide electricity to rural communities, Argentina is making use of photovoltaics, wind, and small hydrosystem schemes for producing electricity.[17]

Iceland stands out as an impressive example of an entire (albeit small) country conscientiously making substantial progress toward reduced pollution and reduced greenhouse gas emissions.[18] Iceland has transformed itself from a coal-based economy in the mid-twentieth century to an economy based largely on renewable energy in the early twenty-first century. As of 2006, 70% of Iceland's energy use derived from hydroelectric and geothermal sources, with 85% of the country's electricity based on hydroelectric power, heating based mostly on geothermal energy, and transportation, the third of the big-three energy-consuming segments of the economy, beginning to make progress, with some buses and cars fueled by hydrogen. As a result of this considerable progress throughout Iceland's economy, even in the capital city of Reykjavik, where black smoke dominated in the mid-twentieth century, the skies in the early twenty-first century are usually clear.

> As of 2006, 70% of Iceland's energy use derived from hydroelectric and geothermal sources.

Although Iceland is favored with substantial geothermal energy readily available at or near the Earth's surface, the country has also taken a lead in drilling for geothermal energy, doing so by 2006 to depths as great as 3 kilometers, with plans for drilling to 5-kilometer depths.[19] Additionally, Iceland is taking a leadership role in the global community both by serving as an example and by cooperating with other countries to help increase the use of renewable energy around the world. As eloquently stated by Iceland's President Ólafur Ragnar Grímsson at a climate conference in Washington, D.C., in September 2006, the importance of some nations is not measured by their size but by "what we can contribute to the solutions of some of the challenges that face people all over the world, and what we can offer in terms of ideas."[20] I was honored to be seated at President Grímsson's table at the luncheon preceding his talk and was impressed and encouraged by his eloquence and by the successes that Iceland has achieved through making a conscious decision to reduce the negative impacts that human activities have on the environment and then proceeding to enact the policies that would do that.

Many cities and other localities have banded together to develop stronger emission reduction policies than the policies in effect through their respective national governments. One major international effort of this type is Cities for Climate Protection, which was established in 1993 and by 2006 had engaged over 670 local and city governments committed to reducing greenhouse gas emissions through systematic inventories and forecasts, establishment of emission targets, local action plans, policies to reduce emissions, and monitoring of progress. In the United States, the U.S. Congress of Mayors endorsed a Climate Protection Agreement introduced by the mayor of Seattle in February 2005, with 230 mayors signing on by May 2006. This agreement urges bipartisan action in the U.S. Congress to pass legislation regarding greenhouse gas reductions, urges federal and state governments to reduce global warming emissions to at least 7% below 1990 levels by 2012, and commits to attempting at the local level to reach at least the targets set in 1997 in the Kyoto Protocol,[21] a major but highly contentious international agreement on greenhouse gas emissions, with nowhere near the level of acclaim or support enjoyed by the Montreal Protocol on CFC emissions.

The Kyoto Protocol is one of a sequence of attempts to deal with global warming issues at the international level. Five years prior to the Kyoto meeting, there was a June 1992 Earth Summit in Rio de Janeiro following which over 150 countries signed the UN Framework Convention on Climate Change (UNFCCC), acknowledging that climate change is a concern and agreeing to develop and update national inventories of greenhouse gas emissions, to work on mitigating and adapting to climate change, and to promote education and public awareness of climate change issues.[22] The aim of the UNFCCC was to stabilize atmospheric greenhouse gas concentrations prior to their reaching a level that would cause dangerous interference with the global climate system. The UNFCCC was signed by U.S. President George H. W. Bush and approved unanimously by the U.S. Senate as well as being approved in many other countries. It went into effect in March 1994. However, by 1995, almost no countries were making progress in greenhouse gas reductions except former members of the Soviet Union, whose progress came as a side effect—this time a favorable unintended consequence—of their economies being in rapid decay. Negotiations to strengthen the UNFCCC took place in March 1995 in Berlin, in July 1996 in Geneva, and, most important, in December 1997 in Kyoto.

Emerging from the Kyoto meeting, the Kyoto Protocol was an addendum to the UNFCCC, strengthening the wording to call for mandatory commitments rather than mere goals. The protocol covers four individual greenhouse gases and two additional groups of greenhouse gases—CO_2, methane, nitrous oxide, sulfur hexafluoride, hydrofluorocarbons, and perfluorocarbons—all with their emission amounts converted to carbon dioxide equivalents. Targets vary among nations (for the United States, it would have been reductions to 7% below 1990 levels) and can be met in part by buying and selling emission "credits" and investing in "clean development" projects in other nations. The U.S. Senate firmly opposed signing the Kyoto Protocol, in substantial part because only developed countries were obligated to reduce emissions, leaving the developing nations free rein to continue to increase their emissions and hence not offering even a chance of solving the global problem. Nonetheless, enough other countries signed the Kyoto Protocol that it went into effect, for the signatory nations, on February 16, 2005. This required approval by countries responsible for at least 55% of emissions from the UNFCCC Annex I countries, a tall order when the United States, with 34% of the total, refused to sign.[23] The Kyoto Protocol is viewed by many scientists as well as others as not being the right solution,[24] but, however flawed, it was an important instance of nations coming together to attempt to address the widely perceived global warming problem.

Unfortunately, whether at the international, governmental, or corporate level, the road toward progress in addressing global warming and related issues is not one-way. An example of a setback is the planned retreat to coal usage announced in 2008 by Italy's major electricity producer, Enel, which expects to increase its reliance on coal from 15% to 33% over the period 2008–2013 through converting its massive power plant in Civitavecchia, Italy, from oil to coal. Other European countries are also returning to coal, with about 50 coal-fired plants expected to begin operating between 2008 and 2013. Economics and the availability of coal play a large role in the decisions, and the companies are conscientiously incorporating new technology that will limit the local pollution from the coal-fired plants, which is positive but perhaps not of major significance in terms of carbon emissions. The new Enel plant has many favorable aspects, including using the waste nitrous oxide to generate ammonia, which is then sold, selling the waste coal ash and gypsum to the cement industry, and using the waste heated water to heat a fish farm.[25] Thus, although the return to coal is a considerable setback, at least it is a more environmentally conscious coal usage than was typical in the past.

Alongside the issue of trying to restrict further human-induced climate change is the issue of our adapting to whatever climate changes do occur. As described in chapter 2, the history of the Earth shows decisively that climates around the globe change regardless of whether humans play a major role (or even whether humans exist). As in the past, some of the future changes will essentially force adaptations. Humans have a long (and by and large successful) history of adaptation, prime examples from early on being the use of clothing and fire to facilitate adaptation to the cold, complementing the other benefits that clothing and fire brought. Today, one adaptation that could be of increasing importance is the adaptation to sea level rise. Humans have certainly adapted to sea level rise in the past, but in many ways it becomes more difficult in the modern world, with all the infrastructure, land-ownership, and political boundaries to complicate an individual's or community's simply picking up and moving landward from the encroaching coastline.

Out of necessity, the Netherlands has dealt with the complications of rising sea levels for many decades, since so much of that country (about 25% in the early twenty-first century) is below sea level, and another substantial portion (also about 25% in the early twenty-first century) is low enough to be flooded regularly.[26] As detailed in the final section of chapter 8, the Dutch have spent billions of dollars to build thousands of miles of dikes that have indeed allowed the country to survive and thrive, although not without major devastating incidents of destruction of life and property from occasional major storm surges. In a further effort at adaptation, one of the largest Dutch construction firms, Dura Vermeer, is building buoyant roads and amphibious houses.[27] These would surely not be ideal for individuals subject to motion sickness, but many other people already living in amphibious houses have adjusted quite well, in some cases much enjoying it. In any event, humans do have a history of adaptation.

WHAT MORE SHOULD BE DONE?

Suggestions for Individuals

The previous section centers on actions that have already been taken, many of which should continue indefinitely, as should many others not mentioned. Most straightforward of the actions to encourage are those—and there are many—that have no negative side effects. Examples include the following: reduce excessive waste (save leftover food for future meals, reuse wrapping paper for gifts, refrain from throwing away a can of oil when a third remains, and even refrain from throwing away a piece of paper when only a third of it has been used), avoid excess use of energy-consuming items (take the stairs rather than an elevator, walk rather than ride when feasible, use buses and subways in preference to private cars, combine trips when a car is necessary, drive more slowly when within reason, cut back on airplane travel, do not overdo heating and air-conditioning, and turn off lights, computers, and television sets when not in use for an extended period), and reuse and recycle as much as feasible. Many of these suggestions are cheaper as well as being better for the environment, and some, like walking, are healthier in addition. Although some are trivial and do not matter much in the large scheme, they are mentioned to point out that all of us have excesses that we could easily reduce generally at no loss to ourselves and often at a financial savings. If human-driven emissions to the atmosphere are leading to as major a climate crisis as many people think they are, then we need to make some adjustments, and among them should be less waste and extravagance.

Social pressure can play an important role. High society has largely adjusted to not buying or wearing fur coats, in significant part because of social pressures regarding animal treatment. We also can and should adjust to less waste and less materialism that breeds the need for waste, in part by making excess consumption just as distasteful as wearing fur coats has become. If society as a whole starts regarding conspicuous consumption negatively, then, bit by bit, many people will stop being so wasteful. Much of this can be done without sacrifice, as individuals often find that becoming less materialistic and living more simply, with fewer impacts on the environment, not only is less expensive but also improves rather than worsens their quality of life.

Other cutbacks are more difficult to make because they entail discomfort and/or additional work. Here as elsewhere, which cutbacks are most reasonable is a highly individualized matter, dependent greatly on priorities and circumstances, but almost all of us outside of the most poverty-stricken classes can make at least some cutbacks. For instance, I live in a location where I have found it feasible (although not always pleasant) for over 25 years now to go without heat or air-conditioning at home for at least the months of April, May, June, September, and October, in recent years expanding that list to include November, December, and March. For people with large families or frequent visitors or for people living in very cold or very hot climates, such severe cutbacks on heating and air-conditioning would probably not be appropriate. However, for readers who might consider cutting back on air-conditioning, here are some suggestions: open the windows on cool, dry nights; take cold rather than hot or warm showers on hot days; and do not use the stove or oven on hot days. I change my diet considerably in summertime specifically to avoid using the oven, and indeed I have found that there are plenty of healthy, tasty, easy-to-make meals that can be prepared without an oven or stove. These include cereal, sandwiches, yogurt and fresh fruit, and, one of my favorites, a plate full of fresh spinach overlain by cottage cheese topped by applesauce. Even people wanting air-conditioning can reduce how much is needed by judicious opening and closing of windows and by limiting summertime use of the oven and stove. In winter, the need for heating can be reduced by wearing layered, cold-weather clothing, and in both summer and winter, the need for air-conditioning and heating can be reduced by keeping one's house or other dwelling place well insulated.

One cutback in energy usage that I do not do but that could be more convenient for some people than cutting back on heating and air-conditioning is to drip-dry clothes rather than use an energy-consuming clothes dryer. All of us could do this, but circumstances that would make it easier would be having a largely unused portion of a basement or garage or living in a location where weather and regulations make outdoor drip-drying feasible.

In view of the high environmental costs of meat-based versus non–meat-based diets,[28] meat-eating individuals could further help the environment by lowering their meat consumption or going further and switching to a vegetarian diet. For some people, such a switch might even be a major health benefit, although that is likely not the case for others, and I am certainly not recommending that everyone become a vegetarian. In any event, most nonvegetarians could probably at least cut back on their meat eating without health disadvantages, and if enough people were to do so, this could be a major benefit from the environmental and animal-rights perspective (although not from the meat-industry perspective).

For individuals ready to step up to actions that would require expenditures in the short term, suggestions include energy-efficient lightbulbs (relatively easy and inexpensive), energy-efficient appliances, energy-efficient cars, smaller cars, solar heating or other alternative energy usage, white roofs in locations where buildings overheat in summer, or grass-covered roofs, perhaps with vegetable gardens.

All the suggestions mentioned so far and dozens of others in a similar vein, if done by enough people, could be important. However, it is unlikely that any of them approaches in importance the following two: limit reproduction and limit plane travel. Without question, a major reason why humans are impacting the environment and climate is the large number of humans now on the planet. Hence, one of the most important things that any individual can do is to limit himself or herself to having no more than two natural offspring. Probably next in importance for most of the people reading this book and more broadly for a sizable fraction of people in the middle and upper classes would be to limit plane travel. Because of the high emissions of greenhouse gases from jet planes, two or three round-trip plane excursions a year can easily cancel out all the reductions an individual might make through other lifestyle changes throughout the entire year. The significance of plane travel is so great that I grant it a separate section.

Plane Travel

In his book *Sustainable Energy—Without the Hot Air*, University of Cambridge physics professor David MacKay estimates that the energy cost of a Boeing 747-400 flight of 14,200 kilometers (8,800 miles) is 2,400,000 kilowatt-hours, from the fuel usage alone.[29] Assuming a full load of 416 passengers, this equates to approximately 5,770 kilowatt-hours per person, and, more realistically, assuming a flight with 80% of the seats occupied, it equates to approximately 7,200 kilowatt-hours per person. Elsewhere in the book, MacKay indicates that driving 50 kilometers consumes approximately 40 kilowatt-hours.[30]

The plane-versus-car energy costs can be looked at from several perspectives, but here are two relevant ones:

1. The energy cost of each person on a 14,200-kilometer Boeing 747-400 flight that is 80% full is the same as the energy cost of that person's driving his or her car 50 kilometers a day for 180 days. Hence, for most people, it would be very difficult (and totally impossible within 180 days for those people who do not drive at least 50 kilometers a day) to compensate for their long flight by reducing their driving distances.
2. On the other hand, scaling up the 40-kilowatt-hour figure for driving 50 kilometers, the energy cost of driving a car 14,200 kilometers (an unrealistically long car trip) comes to about 11,360 kilowatt-hours. Hence, if a person has to go 14,200 kilometers and is going alone, then on the basis of energy cost alone (as well as time), it is better that he or she fly (at an energy cost of approximately 7,200 kilowatt-hours, assuming a plane at 80% passenger capacity) than drive. However, if two or more people need to take the same trip and can ride in the same car, then the energy cost is less by driving, as the

energy cost per person for the car is lowered to about 5,700 with two people in the car and is lowered further for additional people. (All these numbers are only approximate.)

The second example above illustrates that the energy cost of flying does not always exceed the energy cost of traveling to the same location by other means. The problem comes instead in that flying allows a person to reach his or her destination quickly, thereby freeing time for more flying or other energy-consuming activities.

MacKay mentions that the numbers he gives for the energy costs for flying take account of the CO_2 emissions from the plane but not the emissions of such other greenhouse gases as water vapor and ozone. Adding the other greenhouse gas emissions from planes would yield a CO_2-equivalent carbon footprint two to three times larger than the values MacKay gives for CO_2 alone, adding considerably further to the environmental costs of flying.[31] Others also have noted the additional environmental costs of plane travel, and, in line with but more specific than MacKay, the Intergovernmental Panel on Climate Change (IPCC) states that the total warming effect from planes is 2.7 times the effect from their CO_2 emissions alone.[32]

Another relevant approximation given by MacKay is that the energy cost of home heating comes to about 24 kilowatt-hours per day.[33] This figure clearly can be only extremely approximate, as home heating varies a great deal depending on the house, the location, the energy source, the outside weather, and the setting of the thermostat. However, using Mackay's figure as a very rough approximation, the approximate 7,200 kilowatt-hour per person energy cost of traveling on one 14,200 kilometer flight that is 80% full equates to the energy cost of heating one's house for 300 days. So once again, compensating for that one flight by reducing energy usage elsewhere (e.g., in home heating) becomes quite difficult: no heat for 300 days—and that even without factoring in that the greenhouse gas emissions from planes are two to three times greater than the CO_2 emissions that were used to obtain the 7,200-kilowatt-hour figure for the energy cost of the plane ride.

Instead of focusing on a flight's energy cost in kilowatt-hours, George Monbiot focuses on the actual amount of CO_2 emitted. The United Kingdom Department of Transport estimates 110 grams of CO_2 emissions per passenger per kilometer of a typical long-distance flight. Accepting this estimate, the CO_2 emission per passenger on a flight between New York and London (distance of 5,585 kilometers) totals 614,350 grams each way, or approximately 1.2 tons per passenger for a round-trip.[34] This 1.2 tons of CO_2 emissions comes to what should be an individual's total fair-share yearly allotment of CO_2 as of 2030 according to Monbiot[35] (more about the 2030 fair-share amount later in this chapter). However, this low yearly allotment is only after a severe cutback in CO_2 emissions.

For a more realistic per person fair-share amount of CO_2 emissions in 2009, divide the total CO_2 worldwide emissions of about 26 billion tons by the global population of 6.8 billion, to get a per person value of about 3.8 tons. This means that a person uses more than his or her fair share of CO_2 by flying a total distance exceeding 35,100 kilometers in an individual year even if he or she does absolutely nothing else throughout the entire year that contributes to CO_2 emissions. This would allow a round-trip between New York and Melbourne (round-trip distance of 33,344 kilometers) but would not additionally allow a round-trip between New York and Chicago or any other pair of locations separated by more than 900 kilometers.

For many people—including many of the most vocal scientists, journalists, and environmentalists raising concerns about global warming—cutting their flying distance down to no more than 35,100 kilometers a year would be a severe cutback. While I can accept the impracticality for many of cutting back to no more than 35,100 kilometers, it does seem that those who are building their careers and reputations on sounding the alarm about the dangers of greenhouse gas emissions should be a great deal more conscientious than many of them seem to be about how much flying they do. Keep in mind also that the numbers given here are based only on CO_2 emissions; considering the other emissions from planes makes the situation even worse.

An article in the journal *Science* in October 2007 highlighted the environmental costs of the travel engaged in by scientists attending major scientific conferences.[36] A telling statistic in the first paragraph of the article is that the estimated share of CO_2 aircraft emissions attributed to the 9,500 participants (including many scientists who repeatedly express concern for the environment) at the December 2002 annual fall meeting of the American Geophysical Union because of their travel to and from the meeting totaled approximately 11,000 metric tons, "roughly the same as 2250 Honda Civics during a year's worth of normal driving."[37] It seems axiomatic that individuals and organizations calling for less fossil fuel usage should consider cutting back on organizing and participating in large in-person meetings that are contributing significantly to the problem through the air travel of the participants.

Fortunately, through modern technology, there are now several alternatives to attending meetings in person. Replacing face-to-face meetings by meetings where participants interact through videoconferencing, Internet conferencing, or some other alternative would limit carbon emissions, save money for the participants and the organizations, and save time for all the participants who no longer would have to travel beyond their home base. I have participated in many teleconferences that have been highly productive and entailed no need for anyone to be displaced from his or her workplace and no need for the time, expense, or other inconvenience of additional travel. There are also multiple options for videoconferencing and Internet conferencing and for making audio, video, or written statements that can, through the Internet, reach a vastly larger audience than would be possible at a traditional conference.

A widespread complaint against all the alternatives to having conference participants physically in the same rooms and corridors is that the participants would miss out on what many people find to be the most valuable aspect of scientific conferences, that is, the random meetings and discussions in the hallways, at meals, and elsewhere outside the scheduled program. This indeed is an important factor, although the latest computer technology is facilitating the random discussions as well. I do not advocate ending all in-person scientific conferences with people coming from afar, as the cross-fertilization of ideas that takes place at these meetings can be a major boost to the scientific effort, and, additionally, the occasional change of scenery and routine can be refreshing and stimulating for the participants. However, the number of large meetings, at least in the Earth science and climate change arenas, has gotten way out of hand, and cutting back on the number could be of benefit for the scientists as well as for the environment. Many scientists express dismay at how many conferences are now taking place and how much time this is taking away from the time available to do the science, yet many of those same scientists continue not just to attend the conferences but to organize sessions for them and recruit others to attend as well. Perhaps the issue of carbon emissions from plane travel can be the additional leverage point to increase the number of scientists conscientiously saying no to participation in some of these activities.

Suggestions beyond the Individual Level

Communities can scale up the various suggestions made for individuals to incorporate these in the community spaces and workforce and can further help by encouraging conveniently located farmers' markets and creating and maintaining sidewalks and bicycle paths. The farmers' markets, sidewalks, and bicycle paths all facilitate reduced automobile usage, and the farmers' markets additionally reduce the packaging and transport needed for getting food distributed. Beneficial community regulations can include incentives for environmentally conscientious behaviors and disincentives for wasteful and otherwise damaging behaviors, including severe restrictions against the most damaging behaviors, such as building coal-fired power plants that fail to include carbon capture and sequestration. Communities and larger governmental units can also place restrictions on the size and capacities of airports.

Educators, school systems, and universities all have a major role to play, as educating the youth of the world from an early age to understand the finiteness of the Earth system and its resources and the impacts that an individual's actions have on the environment and climate can go a long ways toward reducing the acceleration of the current problems. One person's restraint will not make a dent in any but the most local of problems, but if everyone were to exercise restraint in consumption, carbon-emitting activities, and reproduction, then collectively there could be tremendous progress not just locally but around the globe.

Entrepreneurs also can play major roles, for instance by making environmentally desirable changes more palatable, convenient, and even outright advantageous. One suggestion along these lines, made by Alan Storkey and advocated by George Monbiot, centers on major upscaling of coach (bus) travel, with dedicated high-speed lanes on major highways, routes that are direct and convenient, and vastly improved conditions inside the coaches. The improved conditions would include more leg room, improved seat quality, the addition of work stations, and perhaps films to watch on long trips. The resulting upscaled coaches would be closer to large limousines than to what we normally think of as buses and could conceivably become a vehicle of choice for many excursions.[38] Luring people away from private vehicles could take the best of entrepreneurial skills, although reliable Internet connectivity on the coaches could be a major draw and even necessitate restraint so as not to lure people into additional trips rather than simply luring them from car to coach on trips that they would take anyway.

Employers, whether in the corporate, academic, or governmental realm, can, like communities, scale up the suggestions for individuals and can also encourage telecommuting by dependable employees and encourage and facilitate teleconferencing, videoconferencing, and Internet conferencing in lieu of sending employees on travel. Ironically, many of the organizations at the forefront of declaring concern about global warming have large annual or more frequent meetings to which thousands of people fly in by plane, with all the greenhouse gas emissions that flying entails. One such organization well aware of the conflict between message and action is the American Meteorological Society (AMS), which serves as a good example of an organization attempting to tackle the dilemmas involved.

Like others, the AMS is considering changes to lessen the carbon impact of its meetings, although is not ready to eliminate its productive, stimulating, face-to-face annual meeting. Instead, the AMS is taking measures to: ensure less paper waste at its meetings, purchase locally produced products where reasonable, select environmentally conscious hotels and convention centers, and provide attendees with information on ways in which they can offset their transportation-related carbon emissions.[39] Individual AMS members are advocating stronger measures, with, for instance, Thomas Hamill advocating that the AMS become fully carbon neutral, conserving where it can and offsetting the remaining carbon emissions by paying other organizations to fund projects that will achieve emission reductions equal to the AMS's remaining emissions. Regarding conferences, Hamill recommends that the AMS facilitate remote participation, broadcast presentations live over the Internet, establish a robust videoconferencing capability, and add a mandatory average carbon-offset charge to the conference registration fee. Hamill's rough calculation indicates that this fee might be only $13.20,[40] which should not be a major burden on any of the participants. A fee of this amount would have provided a total of $62,040 for carbon offsets from the 2005 AMS annual meeting, with 4,700 attendees, and $55,440 from the 2006 AMS annual meeting, with 4,200 attendees.

Although paying carbon offsets is, in concept, beneficial in helping to fund activities that provide carbon savings, care must be taken to ensure that the funded activities are indeed providing the desired carbon savings. Recall from chapter 7 the case of commercial companies looking toward the possibility of making money by seeding portions of the ocean with iron and selling carbon credits for doing so, even though such iron fertilization might do more harm than good. With money on the line, well-intentioned companies can step into presumed carbon-saving activities too aggressively and as a result bring about very damaging unintended consequences. When dealing with carbon offsets and carbon credits, which activities qualify as carbon saving needs to be considered extremely carefully.

More broadly, "carbon trading" allows companies, governments, or other entities to compensate for failing to meet their required or self-imposed greenhouse gas emission reductions by purchasing carbon credits. Specifically, "cap-and-trade" schemes set limits on the amount of emissions of one form or another that an entity is allowed to produce (the cap) but provide the option that this entity can buy credits from another entity to cover emissions that exceed the cap. Companies/entities can sell credits up to the amount of how much lower their emissions are than the cap set for them. A major benefit of a cap-and-trade scheme is that it provides financial incentives to keep emissions even lower than the cap level. In addition, by putting a price on the continued excess emission of greenhouse gases, carbon offsets, carbon taxes, and carbon trading all provide financial incentives for energy efficiency, renewable energy, and otherwise reducing greenhouse gas emissions. Thus, if regulated appropriately, they can help considerably by making the excesses more expensive and hence less profitable and less desirable.

However, even if all the carbon-offset and carbon-credit payments were to fund appropriate activities that do indeed reduce carbon emissions, making these payments can never be the total solution to the greenhouse gas problem and should not be viewed by the individuals or organizations as a route by which they can absolve themselves of further responsibilities. Carbon offsets and carbon trading essentially allow the wealthy to continue their carbon emissions by paying someone else to offset those emissions. In the long run, we as a global community cannot buy our way out of our excesses. Straight across the line from individuals to corporations to government entities, no one should be thinking that the world needs to reduce excess but that "we" can continue our excesses just by paying others to reduce theirs.

Sources for Additional Suggestions

Hundreds of additional suggestions of how individuals, organizations, and governments can take action to help reduce human impacts on climate are readily available. Al Gore at the end of his book *An Inconvenient Truth* (a companion to the movie of the same name) includes a 17-page list of actions that individuals can take to help solve the predicted climate crisis or at least reduce their contribution to it. This list overlaps significantly with suggestions in the previous sections and includes eating more locally grown produce, replacing regular incandescent lightbulbs with compact fluorescent lights, setting the household hot water temperature at no more than 120°F (49°C), taking showers rather than baths, driving less and walking and biking more, reducing air travel, purchasing less and giving preference to items that use recycled packaging, composting organic waste rather than sending it to landfills, and shopping predominantly at stores that are at least attempting to reduce emissions.[41]

Beyond the individual level, the IPCC's 851-page volume subtitled *Mitigation of Climate Change* is replete with mitigation possibilities and details about them.[42] Dozens of these possibilities involve technologies that are already available, such as use of available alternative energies, more fuel-efficient vehicles, more efficient lighting, improved stoves, improved insulation, more recovery of excess heat generation during industrial processes, improved land management, reduced and more effective use of fertilizer, improved forest management, recovery of emitted methane at landfills, and increased composting of organic wastes. The IPCC volume also details financial incentives and disincentives that governments can provide, such as taxes and other charges for carbon emissions, and subsidies and tax credits for emission reductions.

Many additional sources also include wide-ranging suggestions for reducing our global warming impacts.[43] One source of particular value is David MacKay's *Sustainable Energy—Without the Hot Air*, mentioned earlier in the context of the energy costs of plane travel. MacKay provides the approximate energy costs of a very large range of activities, giving an extremely convenient means of identifying which conservation or other measures are most likely to make a difference and which are essentially irrelevant. For instance, turning your home thermostat down by 1 degree in winter is overwhelmingly more meaningful than unplugging your cell phone charger.[44]

CAN ENOUGH BE DONE?

Even among those most concerned about global warming and other aspects of climate change, there are varying estimates of exactly how much must be done to keep the climate from major destabilization due to human activities. Most of these estimates suggest that quite considerable reductions in greenhouse gas emissions will be needed by the middle of the twenty-first century. This recognition of the need for major reductions is reflected in the commitments made by the Group of Eight (G8) at their annual summit held in L'Aquila, Italy, on July 8–10, 2009, when the G8 countries (Canada, France, Germany, Italy, Japan, Russia, the United Kingdom, and the United States) and the European Union committed to reduce their greenhouse gas emissions by at least 80% by 2050,[45] meaning that each country committed to getting its emissions down to no more than 20% of what they had been.

Several studies have examined in some depth whether major reductions in emissions are feasible. Here I summarize portions of two of these studies to give the reader a feel for what might be involved. The two studies are quite different, and the first even assumes that it would be acceptable to hold emissions steady until 2054 and have the emission reductions begin after that. In sharp contrast, the second study is based on the perceived need for even greater and faster emission reductions than the G8 commitments. Despite the vast contrast in their aims, both studies conclude that it will be difficult but possible to achieve the respective required emission levels.

Stabilization Wedges

The first study, carried out by Stephen Pacala and Robert Socolow of Princeton University, begins with the presumed need—stated repeatedly in the early twenty-first century—to keep atmospheric CO_2 concentrations below a doubling of the preindustrial CO_2 levels of approximately 280 parts per million. Somewhat more stringently, Pacala and Socolow focus on keeping the CO_2 levels below 500 ± 50 parts per million, and they do so by breaking the task into two parts: first, hold the CO_2 emissions during the 50 years from 2004 (the year of their publication) through 2054 at 7 billion tons (i.e., 7 gigatons) of carbon per year, using existent technology, and, second, implement more advanced mitigation strategies to reduce the emission rates below this level after 2054.[46] The basic concept is that 1) another 50 years at current emission levels will not raise the atmospheric CO_2 concentrations to the doubled-CO_2 danger point, and 2) 50 years should provide adequate time for developing viable means for emission reductions without severely damaging the global economy. They then concentrate on how to keep carbon emissions at 7 billion tons per year for the next 50 years.

Pacala and Socolow estimate, very roughly, that if the world continues "business as usual," emissions will linearly rise from 7 billion to 14 billion tons of carbon per year by 2054. Plotted on a graph with time (2004–2054) along the x-axis and carbon emissions per year along the y-axis, the desired level for the 2004–2054 time frame is a straight horizontal line at 7 billion tons of carbon per year, and the projected level is a straight tilted line going from 7 billion tons of carbon per year at 2004 to 14 billion tons of carbon per year at 2054. Those two lines plus the vertical line connecting them at year 2054 make up the three lines of what Pacala and Socolow term the "stabilization triangle." The goal for the 50 years 2004–2054 is to eliminate all the carbon emissions represented within that stabilization triangle, while coming up with the technologies or other means of tackling the more difficult problem of significantly reducing emissions well below 7 billion tons of carbon per year after 2054.

Eliminating all the projected carbon emissions within the stabilization triangle, rising to 7 billion tons of additional carbon per year by 2054, is not feasible with any single CO_2-reduction strategy. However, Pacala, Socolow, and their colleagues feel that it becomes feasible by employing multiple reduction strategies.[47] They detail this by dividing the stabilization triangle into seven wedges, each representing additional carbon emissions rising uniformly from 0 billion tons of extra carbon per year in 2004 to 1 billion tons of extra carbon per year in 2054. If seven different strategies each eliminated the carbon represented by one of the seven wedges, then the goal of eliminating the carbon from the full stabilization triangle would be accomplished.

Pacala, Socolow, and their colleagues identify 15 carbon-reduction strategies, each of which could be among the seven needed to cover the seven stabilization wedges, that is, each preventing 1 billion tons of additional carbon emissions per year by 2054. Each of the 15 strategies is based on known and functioning technologies, although none of them could easily be scaled up to the level of a full 1-billion-ton-carbon-reduction wedge. The 15 strategies are not meant to be comprehensive but instead to show that the potential for eliminating the carbon represented by the full stabilization triangle exists.[48]

The 15 strategies are grouped by Socolow and colleagues into five categories: energy conservation, renewable energy, nuclear energy, enhanced natural sinks, and fossil-carbon management. Among energy-conservation measures they include increased efficiency in cars and trucks, driving less, greater use of mass transit, making new buildings more energy efficient, and replacing incandescent lightbulbs with fluorescent bulbs. Renewable energy includes hydropower, wind, solar photovoltaic electricity, geothermal energy, and renewable fuels such as ethanol from either sugarcane or corn, with many complications, or hydrogen fuel. Nuclear energy could be considered renewable also although is separated out into its own category. Enhanced natural sinks of carbon include new forestation, reduced deforestation, and expanded conservation tillage. Fossil-carbon management includes altering the mix of fossil fuel usage and engaging in carbon capture at the point of emission, followed by transport and storage of the carbon, rather than allowing it to accumulate in the atmosphere.[49]

For each of the 15 strategies, covering a full wedge of carbon reductions would be possible but challenging. For instance, to get to a full wedge from increased use of wind power—which is already widely used in some countries—the early twenty-first-century level of use would need to be multiplied by 50, with the deployment of 2 million 1-megawatt wind turbines (or the equivalent) and a total land requirement on the order of 30 million hectares, roughly the area of Italy or the Philippines.[50] A wedge from photovoltaic electricity would require less land, about 2 million hectares by 2054, but a considerably larger scaling up, to about 700 times the 2004 usage of photovoltaic electricity. Nonetheless, there are 15 strategies, each of which could conceivably cover a full wedge, and only seven are needed. This leaves considerable leeway, with, for instance, a wedge being potentially covered by two strategies each accomplishing half the goal. By slicing the stabilization triangle and focusing on strategies to cover individual wedges, Pacala and Socolow conclude that stabilization through 2054 is feasible.[51] Further reductions, not just stabilization, would be needed after 2054, but there should be sufficient time by 2054 for people to come up with appropriate means.

A Single-Country Example: The United Kingdom

For many people, the concept that we can hold off until 2054 before making severe emission reductions is hopelessly inadequate. As already mentioned, the major industrialized G8 countries have agreed to reductions of at least 80% by 2050, and some people, including George Monbiot, feel that even that is not enough.

In his 2007 book *Heat: How to Stop the Planet from Burning*, Monbiot proceeds on the premise that humans need to reduce our annual global carbon emissions from about 7 billion tons in 2007 to 2.7 billion tons by 2030, for a global reduction of about 60% by that time. Estimating a global population of about 8.2 billion by 2030, this averages out to 0.33 tons of carbon (or 1.2 tons of CO_2 equivalent) per person. Each country's fair allotment by 2030 then becomes 0.33 tons of carbon multiplied by the country's anticipated 2030 population. Rich countries currently emitting well above their fair share of carbon hence will need to cut back considerably more than the global average, and some developing countries will quite reasonably be allowed to emit more than they are currently emitting in their unindustrialized state. Under this scheme and with realistic estimates for future population numbers, the United States, Canada, and Australia will each need to reduce their emissions by 94%, Germany by 88%, and the United Kingdom by 87%.[52]

Rather than tackle the problem of the globe as a whole, Monbiot narrows down to the one industrialized country that he knows best, the United Kingdom, to examine how feasible it might be for that country to achieve as dramatic an energy cut as the industrialized nations are being called on to achieve, recognizing that other industrialized nations would have similar issues to handle although with differences dependent on the individual geographic, population, and other circumstances.

One by one, Monbiot examines several major sectors of the British economy and concludes in each case that a 90% cut could be achieved although not without difficulty.[53] Some of the suggested changes are unlikely to be readily accepted by a society grown used to our current lifestyles. Not unexpectedly, in view of the high global warming contribution from airplanes, key among these cutbacks would be a reduction by about 87% in the total number of airplane flights. In view of this need, Monbiot suggests that governments limit the capacity of airports and place a moratorium on new runways. However, recognizing the widespread desire for rapid long-distance travel, he proceeds to examine alternatives to airline travel as it currently exists, looking at trains, ships, alternative airplane designs, and alternative fuels. Sadly, in the end, he concludes that—with current technology—to travel long distances quickly requires large emissions into the atmosphere no matter what means of travel are employed; that is, neither high-speed trains nor anything else is going to solve the problem of the high climatic costs of rapid travel. Airships held aloft by hydrogen gas offer some environmental advantages, but they are not without environmental consequences, and their top speeds would require about 43 hours for a flight from London to New York. Among the conclusions are that individuals should no longer vacation in distant places unless they use environmentally conscientious means to get there, which will take a very long time, and that meetings among people from widely separated locations should occur over the Internet or through videoconferencing or other such means, not through having the participants travel to the same physical location.

Throughout his book, Monbiot offers valuable analyses of different energy sources and different possibilities for cutting back on energy usage. Heat pumps penetrating into the ground could take advantage of the near constant ground temperatures that exist below about 1.5 meters. Solar energy capture is not ideal year-round at the latitudes of the United Kingdom but could be useful there in summer and elsewhere for a larger portion of the year. Wind turbines attached to a house can cause structural damage, and the noise from windmills can enrage people living nearby, although wind power from offshore winds could be an important energy source. Micro–power generators located where the energy is used have the advantage of avoiding the approximately 7.5% of electricity loss typical in the transmission from a large generator to the individual users, and microgenerators, such as solar panels for home heating, can be viable in some circumstances.

One interesting concept that Monbiot describes for the housing sector is the *passivhaus* (passive house), first developed in Germany in the late 1980s. The passivhauses have such efficient ventilation systems and high-quality windows and insulation that even with no active heating or cooling systems, they can apparently maintain comfortable inside temperatures throughout the year, even at the latitudes of the United Kingdom. Converting current houses to passivhauses would require expensive renovations, but when building new houses, the added expense of the efficient ventilation system and quality windows and insulation would be canceled by the cost savings of not needing to put in heating and cooling systems.[54]

Another sector for which Monbiot goes well beyond the standard energy-conservation suggestions is the retail food sector. He certainly joins others in encouraging selling locally grown produce to avoid the need for long-distance transport of the food. But he also recommends replacing many modern grocery stores with food warehouses. This would avoid the considerable excess heating/cooling costs in a grocery store where frozen foods are kept cooled only inches away from where the air is at normal room temperature for the comfort of the customers. It would also result in a net reduction in driving needed to get the food to the customers' homes, as instead of customers separately driving to and from the store, delivery trucks would combine multiple trips. Another savings could come in the reduced need for fancy packaging.[55]

Although Monbiot encourages less use of cars (and more walking, bicycling, riding in coaches/buses, and telecommuting from home), he recognizes that cars will remain important for the foreseeable future. Reducing the weight and drag of cars will help somewhat, but to reach the goal of 90% emission reductions, he advocates electric cars. These could become feasible for a mass market after solving the problem of having to recharge an electric car's battery for hours following 100 to 300 miles of driving. Dave Andrews has proposed a solution: have filling stations stock charged batteries, one of which the customer would take (for a fee) when he pulls in, leaving the battery that needs recharging. The filling station would then proceed to recharge the used battery, for instance using wind power from offshore wind farms, and add it to the collection for future customers.[56] This would be a far more viable solution in the near term than hydrogen cars, which are still quite a ways from being ready for mass usage.[57]

By the end of his book, Monbiot expresses confidence that he has identified feasible methods of reducing carbon emissions in the United Kingdom by about 90% in each of the sectors he examined, with the single exception of aviation, where he realizes that cutting the number of flights by anywhere close to 90% is not feasible, given the expectations of modern society. On the other hand, all those flights are not essential to civilization, so Monbiot does feel that even under the very stringent premise that annual global carbon emissions need to be reduced to no more than 2.7 billion tons by 2030, he has demonstrated that "it is possible to save the biosphere."[58] Although he confined his analysis to the United Kingdom, it has much broader relevance. Basically, if the United Kingdom is able to reduce its emissions by 90%, the other industrialized countries should be able to do so also, and the developing countries should be able to restrict their increased emissions enough so that by 2030, they are still not exceeding the per capita emissions of the currently industrialized countries.

REASONS FOR HOPE

Part of the reason that I am not as concerned as some people about a possible coming climate crisis is that I believe that solutions will be found (even if the situation is as bad as some people think). With so much attention being given to the problem, surely some very bright and innovative inventors will turn their efforts to finding solutions and will come up with economically viable energy alternatives that do not emit massive volumes of greenhouse gases into the atmosphere. Not just will these inventors have the satisfaction of making a major contribution to the welfare of their and future generations, but they will almost certainly have a chance to reap financial rewards from their innovations as well in view of the pervasive worldwide need for the innovative solutions.

> Surely some very bright and innovative inventors will come up with economically viable energy alternatives.

Perhaps the solutions will involve capturing energy from the Sun that would otherwise have gone to undesired further heating of the Earth system and funneling it instead to desired energy uses. Perhaps the solutions will involve capturing the CO_2 and other emissions from fossil fuel usage and making valuable use of these emissions, allowing fossil fuel usage to continue. Already, valuable use is being made of methane emitted from some landfills. For instance, NASA's Goddard Space Flight Center harnesses methane from a landfill nearby its Greenbelt, Maryland, facility and uses this methane to fire boilers that produce the steam that heats 31 buildings at the center. This not only has helped the environment by lessening the methane emissions to the atmosphere but also has come at a cost savings to NASA. More activities like this should produce win–win situations. Perhaps additionally, the amazing growth rates of algae—some populations can double in mass every day—can be put to use in a controlled setting for biofuel generation. In that case, the amazing growth rates could become a major benefit rather than the considerable nuisance they become in an out-of-control situation in a lake or waterway. With an estimated 10,000 types of algae, there is an encouraging chance that one or more of them could become a major alternative energy source.

I am not certain what the solutions will be, but surely greenhouse-gas-emitting fossil fuel usage will eventually go the way of horse-drawn carriages, computer punch cards, and numerous other extremely useful innovations that were eventually super-seded. Fossil fuels have been important ingredients in helping to fuel the industrial-ization of the past two centuries, but they are not the only energy sources available. They were preceded by a multitude of other sources and will eventually be superseded. In the meantime, if it becomes necessary to reduce our energy usage to stabilize our greenhouse gas emissions, we just might find that adjusting to less excess and extrava-gance is not so bad. (Even drastic cutbacks in the amount of flying we do could have its disadvantages outweighed by the advantages of more time at home with family and friends and more time for thinking, reading, writing, and other activities.) Within only a few decades of when overpopulation was condemned in the mid-twentieth century as the greatest problem the globe faced in the long run, societies in some developed nations have not just adjusted to smaller families but have reached a point where a great many households have no desire for more than two children, even leading to concerns in several countries (including Japan, Italy, and Russia) about population declines. We can also adjust to less fossil fuel usage and perhaps can do so with no major hardship involved.

CHAPTER 13

Closing Plea

Humans have often failed to be good stewards of the magnificent planet that we have inherited. However, despite the considerable damage we have caused over the centuries, we have also reversed some of that damage and have many plans to reverse more of it and to limit future damage. In our enthusiasm to right past wrongs, we need to be careful not to make matters even worse by implementing one or more inadequately thought-through massive geoengineering schemes that could be replete with horrifying unintended consequences.

As OVER the course of time we have looked outward into the universe with more sophisticated and varied instruments, we have seen spectacular wonders, from exploding stars to interacting galaxies to close-up details of our orbiting Moon and our neighboring red planet Mars. But of all the wonders we have seen, nothing else comes close to our home planet Earth. Not to downplay any of the other wonders throughout the universe, the Earth does stand out as something extraordinarily special versus everything else that we have seen so far. (There might be many other places like Earth out there, but so far we have not seen them.) Earth is indeed a jewel in the vast universe.

Earth's magnificence is captured in many ways, but one is its multicolored gem-like beauty when viewed from afar against the stark blackness of space. The Apollo astronauts voyaging to and from the Moon in the 1960s and 1970s took amazingly meaningful photographs of our finite, isolated planet with its white clouds, white ice, deep blue oceans, and brown and green land areas, all surrounded by an extremely thin atmosphere. This sliver of an atmosphere, with a thickness less than 1% of the

Earth's diameter, is a vital protective covering that has evolved in concert with explosive volcanic eruptions and with the multitude of life forms at and near the Earth's surface. Without that thin protective sliver, neither humans nor countless other life forms could survive.

Elsewhere in the universe there might be many other places just as magnificent and just as suited for life as planet Earth. But if so, those others are very far away and will remain unreachable by humans for a very long time. Earth is what we have inherited, and we are tremendously fortunate for how well preserved and favorable for life this inheritance is. We have an atmosphere with an abundance of oxygen for comfortable breathing but not so much oxygen as to produce constant fire hazards. We have liquid water and a water cycle that freshens and transports the water. We have a surface temperature range that for many locations is comfortable for humans for much of the time. We have minerals, soils, oil, and a multitude of other resources. And we have an abundance of life forms, with an untold number of intricately evolved mutually beneficial symbiotic relationships, from very visible instances such as honeybees collecting nectar from plants and in turn pollinating those plants to microscopic instances such as the billions of bacteria and fungi that reside within each human gut, making their home there and in turn performing vital functions that keep us alive and, usually, well. These microbes, comfortable in the dark, moist, and oxygen-free confines of our individual intestines, help us considerably by breaking down toxins, manufacturing vitamins and amino acids, and assisting our personal immune systems.

Of course, not everything in the natural world is mutually supportive. There are predator/prey relationships that play out with tremendous acts of violence where captured prey are ripped to death in ways that appear orders of magnitude worse than a quick rifle shot to the heart or brain. Still, despite the brutality, in the natural world predator and prey generally each have at least a fighting chance in any individual encounter. Humans have totally altered that equation, as livestock led to a slaughterhouse have no chance, nor do fish in a lake that is suddenly being massively polluted with a deadly influx of industrial waste, nor do many other animals, plants, and other life forms have a chance against human activities of one type or another.

Humans have too often failed to act in a way that recognizes the finiteness of the Earth system and its offerings. We have hunted numerous animal species to extinction and others to near extinction, acting as though the populations will continue to thrive and be available to us despite what we might do to them. Similarly, we have depleted the ground of substances such as coal and oil and water as though those supplies either are limitless or will continue to replenish indefinitely. And we have treated the atmosphere and oceans all too often as though they were infinite sinks, pouring waste products into them with abandon. They are not infinite sinks, and the waste products are having an impact.

Among the multitude of undeniable human impacts on the Earth system are the following: vegetated land surfaces have been replaced by amalgamations of buildings, pavements, and other artificial surfaces; dams and causeways have been constructed that quite significantly alter natural water flow; and the chemicals we pour into the atmosphere are well recorded in changed atmospheric chemical composition. What is less certain is exactly how damaging these impacts are, and here there are plenty of disagreements, ranging from issues of small-scale personal preferences to large-scale considerations of the survival of civilization as we know it or even the survival of the Earth itself. For two personal-preference examples on the relatively unimportant small scale, some people prefer manicured gardens over the former wild vegetation, while others prefer the wild vegetation, and some prefer a massive rock face sculpted into familiar figures, while others feel that the sculpting is a degradation. On the large scale, some think that our input of additional CO_2 to the atmosphere has net favorable aspects for plant growth and other processes, while others regard it as a deciding factor precipitating an upcoming major climate crisis. Furthermore, some people predict and fear that human activities are leading toward widespread destruction of all coastal communities from sea level rise, widespread poleward expansion of tropical diseases, and massive species extinctions, while others neither predict nor fear any of those eventualities.

Personal preferences and predictions aside, by now there is widespread recognition that among the many human activities, some have caused definitive damage, at least on local and regional scales and for individual species. Fortunately, in many instances this recognition has come before it was too late to reverse the damage, and effective action has been taken. As a result, many lakes, rivers, and other sites have been cleaned up, and many seriously threatened species have revived, at least temporarily, with whales, cod, and bald eagles being notable examples. Reviving the species does little good for the individuals that died before the revival, but it does a great deal of good for the species as a whole.

At the global level, it often becomes far more difficult to identify what change has occurred, to establish whether the change was caused by humans, and to assess whether the change is favorable or unfavorable. The Earth is large and varied, and within its expanse some regions cool while others warm, some regions become wetter while others become drier, some glaciers advance while others retreat, and, on down the litany of climate variables, some regions experience one change while others experience the opposite change. All this provides ample basis for confusion, complicated additionally by the realities that consistent global data sets are not available for any period prior to the advent of satellite technology in the second half of the twentieth century and that even with satellite technology there are many difficulties and uncertainties in interpreting the data and in consistently calibrating the data records through time. (Most satellite instruments relay information only about radiation. Scientists and programmers convert those data into information about temperatures, humidities, ice, oceans, vegetation, greenhouse gases, and a host of other variables, all by way of sets of equations that are meant to approximate the relationships in the Earth system but are never totally inclusive of all the intricacies of that system and consequently could yield significant errors.) Hence, documenting a global change is complicated both by the fact that changes can vary greatly from place to place (and time to time) and by the fact that the data records are generally quite incomplete and imperfect.

Difficult as it is to document past and ongoing changes, those difficulties pale in comparison to the difficulty of predicting what will happen in the future. The global system involves an intricate and incompletely understood interplay between ocean, atmosphere, land, ice, and life, all considerably affected by the energy input from the Sun. If ocean or atmospheric circulations change, then regional climates can change markedly, some toward warmer conditions because of increased flow from lower latitudes and/or decreased flow from higher latitudes and others toward colder conditions because of flow changes in the opposite direction. Furthermore, increased moisture flow over high-latitude regions could result in increased rainfall and vegetation if the temperatures are sufficiently warm or in increased snowfall and glaciation if the temperatures are cold. If the simulated high-latitude temperatures are in error by a few degrees, this could mean the difference between simulating healthy vegetative conditions versus simulating a major glaciation.

However, despite all the uncertainties and complications in documenting past changes and predicting future changes, quite a bit is known with some certainty. By now a sizable accumulation of evidence suggests that the globe, on average, warmed over the course of the twentieth and early twenty-first centuries and did so with the unwitting assistance of humans because of our voluminous emission of greenhouse gases into the atmosphere. With less certainty, a strong consensus exists among climate scientists that the Earth system will continue to warm, on average over multidecadal time scales, again in part because of the emission of greenhouse gases as a result of human activities. It is by no means certain that the warming will not reverse in the coming decades—perhaps a reversal has already begun—but at this point the greatest likelihood is that warming, overall, will continue.

We know that humans are pouring CO_2 and other greenhouse gases into the atmosphere. We know that greenhouse gases help to retain heat within the Earth/atmosphere system and hence contribute to warming the system. However, we also know that we are pouring other emissions into the atmosphere, including many aerosols, and that some of these other emissions encourage cooling rather than warming. We think, although do not know, that the warming impacts will dominate over the cooling impacts, whether human caused or natural, at least through the current century. Looking at the distant past, we also know that Earth's climate experienced quite dramatic changes well before there was any possibility of a global human impact, and that fact makes us aware that natural changes could throw a significant curveball into any of our predictions. Hence, although continued warming is expected, with considerable scientific basis, it is not preordained and might not happen. Assuming that it does happen, continued warming will almost certainly have both beneficial and detrimental consequences. Of crucial significance, many people, including many scientists, are quite convinced that, overall, continued warming will be detrimental, at least from the human perspective.

The projected impacts of the predicted warming include more frequent and severe heat waves, continued reductions in the global ice cover, further sea level rise (destroying coastal properties and communities), more of the most intense and damaging hurricanes, more infrastructure damage from thawing permafrost, enhanced spreading of assorted diseases (especially tropical diseases), and accelerated extinctions of species, among numerous others (see chapter 5 for more of the damaging impacts and also for some favorable impacts). In total, our impacts could have devastating consequences, adding to the devastating consequences that natural changes, including volcanic eruptions, earthquakes, and naturally occurring hurricanes and other storm systems, might have.

Most devastating presumably would be changes that are both rapid and large. Paleoclimate evidence unveiled especially from the 1990s on has revealed that the Earth system is fully capable of quite abrupt and widespread climate changes. Furthermore, satellite imagery in the early twenty-first century has shown recent ice shelf shattering followed by increased outward flow of upstream glaciers along the Antarctic Peninsula, satellite altimetry has shown ice losses in the Thwaites and Pine Island Glacier region, and people on the Greenland ice sheet have witnessed meltwater pouring in torrents downward through the ice sheet through moulins, perhaps reaching and lubricating the base. All of this has combined to increase the fear that a collapse of either the Greenland or the West Antarctic ice sheet might be possible. A major collapse is not thought to be imminent, but even though it is not likely, rapid loss of considerable ice mass from Greenland or Antarctica, with a corresponding rise in sea level and flooding of coastal communities, is now conceivable in a way that would not have been conceivable for most scientists even as recently as the end of the twentieth century.

Even though neither sizable sea level rise nor other projected changes are certain to take place, some of the projections are sufficiently alarming that scientists, policymakers, and others recognize the need to examine how we might lessen the anticipated damage, either by adjusting our future damage-producing activities or by counteracting them. With many disagreements on how damaging our impacts are and on which activities are the most objectionable, and with tremendous economic costs involved in significantly reducing our impacts, it is not surprising that part of the discussion has become highly contentious and polarized, to the great detriment of finding appropriate solutions. While we can all agree that it would be favorable to reduce unnecessary increases in atmospheric particulate matter that causes serious health problems, we cannot all agree on the need to reduce invisible greenhouse gases that are being injected into the atmosphere through activities that help fuel our economy and modern lifestyles and might or might not be leading to detrimentally high levels of global warming.

Despite my belief that the consensus view on global warming is most likely correct in its broad outline, I am not absolutely certain. I realize the following:

1. We are nowhere close to understanding fully the Earth's past and ongoing climate changes. Scientists continue to unearth new data, continue to reinterpret old data (sometimes with significant revisions), and continue to offer new explanations of recorded climate changes both during the era of major human impacts and prior to that era (chapters 2 to 4).

2. The models simulating future warming are full of imperfections, and at best their simulations are only an approximation to what might happen in the physical world. However sophisticated, the models include only the variables and processes that scientists and programmers have inserted into them, consciously or unconsciously. The model results do not prove anything about the physical world, only about the simulated world, and there is considerable leeway for a skilled scientist/programmer to fine-tune the results from a model, including to simulate a specific temperature rise or polar amplification in one hemisphere but not the other. We should regard the predictions

from state-of-the-art models with due respect, as at this point these models are by far the best predictive tools we have, but we should not regard the predictions as unfailingly correct (chapter 10).

3. Many social pressures have helped to consolidate the consensus view and perhaps make it appear more solid than it is. Most poignantly, with the tensions as high as they are on this topic, any deviation by a climate scientist from the consensus risks alienation and condemnation from the scientific community (chapter 11). This provides a gigantic positive feedback toward consolidating the consensus view.

4. History is replete with examples that even in circumstances with a large majority consensus among experts, the experts are sometimes wrong (chapter 9). This is a reality in all fields, and not only is science not immune, but the phenomenon is common enough in science that a name is given to it: when the consensus view is overturned, it is termed a "paradigm shift" or, in the case of particularly significant changes, a "scientific revolution." Mainstream textbooks and scientists are far too inclined to present the current scientific understanding as the final answer despite all the evidence that science is an evolving process. The absurdity of continuing to present the latest result as "truth" is increasingly apparent as the changes in our "truth" now come so rapidly, with numerous accounts every month of one science concept or another being revised. Unlikely though it might be, the majority viewpoint on the reality of a coming climate crisis could conceivably be wrong.

Other people are far less equivocal and overwhelmingly more dogmatic than I am on the topic of upcoming climate changes. In listening to Al Gore's moving Nobel lecture (http://nobelprize.org/nobel_prizes/peace/laureates/2007/gore-lecture .html), given on December 10, 2007, in connection with his and the Intergovernmental Panel on Climate Change's receipt of the Nobel Peace Prize, I was reminded of how much I would prefer to be as certain as he and others are, which would be far easier psychologically, would make it far easier for me to write and speak about the topic, and additionally would even allow me to attempt to help lead the charge. Being certain has wonderful advantages, a point that I was coincidentally also reminded of earlier in 2007 in an entirely different context, specifically, at a Martin Luther King Day celebration centered on a panel discussion regarding civil rights activities in Vermont and Mississippi in the 1960s. When as a high school student in central Vermont in the mid-1960s I stood in defense of civil rights for everyone, regardless of race, I knew that I was right, and knowing that I was right made it easy for me to stand up for this cause despite the local unpopularity of that stance and the resulting criticisms and threats. Being on the 2007 Martin Luther King Day panel not only was an honor but also reminded me of the fabulous feeling of being certain. With certainty, I had no difficulty as a high school student and would have no difficulty as an adult being strong and steadfast and unequivocal. With certainty, I could advocate a compelling case, whether popular or unpopular, and could do so without the hindrance of nagging doubt. But sadly for me, on the issue of global warming—surely one of the most important issues I have to face in my career as a climate scientist—I do not have certainty, for all the reasons detailed throughout this book.

> We do not have the final answers, but this does not mean that we should stand paralyzed against any action, as we need to make the best decisions we can with the incomplete knowledge we have.

Although I understand when Gore, as a politician, exudes confidence in what he writes and says, it is more troublesome to me that some of my fellow scientists exude a comparable level of confidence. I regard such confidence as unwarranted both in view of the limitations of our understanding and in view of the frequency with which current understandings are revised. We as a scientific community did not have the final answers in the 1970s when some among us feared a coming ice age. We did not have the final answers in the early 1990s when we still thought climate change was necessarily a very gradual process, before the Greenland ice cores and an abundance of additional evidence told us otherwise. And we do not have the final answers now. Nor will we have them a year from now, 10 years from now, or 100 years from now. This does not mean that we should stand paralyzed against any action, as we need to make the best decisions we can with the incomplete knowledge we have, but it does mean that our pronouncements should be tempered by our uncertainties even as we advocate actions based on our best judgments at the time.

In listening to congressional testimony by scientists (and being one of those giving testimony in one instance), I am struck by the frequency with which senators and representatives make comments suggesting that they hold scientists in awe. Having worked within the science community for decades now, I am well aware that scientists as a whole are neither as objective nor as careful in their research as many people think they are. Both scientists and the science they do fall far short of the ideals many of us grew up assuming. Being treated with awe by politicians, reporters, and others has perhaps contributed to some scientists becoming overly confident and too personally tied to their pronouncements.

Still, despite my many uncertainties, I agree with Gore that we should be doing something proactively. In view of the predicted propagating consequences of continued global warming, Gore and others are adamant regarding the need for humans to cut back on our greenhouse gas emissions. Although I am not fully convinced that our situation is as dire as Gore believes, I wholeheartedly support reducing our emissions, where feasible, as this is a responsible course of action given the uncertainties. In fact, I agree with almost all the suggestions I have heard regarding cutting back on our fossil fuel usage, seeking and developing alternative energy sources, recycling, reusing, and in general cutting back on our extravagances (chapter 12). Moreover, I have added to the list small additional cutbacks in my personal energy usage, and, partly in respect for the millions who have made their own small sacrifices, I challenge each person who is adamant and vocal on the issue of human culpability in a coming climate crisis to commit to cutting back his or her own carbon footprint at least to the point where that personal footprint is no greater than the average personal footprint in his or her country of residence. For some of the most adamant and vocal, this would almost certainly require a major cutting back on airplane travel and probably quite a bit else as well. Frankly, it seems somewhat hypocritical when highly vocal global warming alarmists do so much plane traveling to get their message out that they become among the very worst of the world's offenders in terms of their personal carbon footprints.

I challenge each person who is adamant and vocal on the issue of human culpability in a coming climate crisis to commit to cutting back his or her own carbon footprint at least to the point where that personal footprint is no greater than the average personal footprint in his or her country of residence. Frankly, it seems somewhat hypocritical when highly vocal global warming alarmists do so much plane traveling to get their message out that they become among the very worst of the world's offenders in terms of their personal carbon footprints.

Hypocrisy aside, not everyone shares a sense of urgency about the climate situation. This is quite understandable given the uncertainties. However, some go so far as to argue that the world economy should continue "business as usual" while scientists engage in further studies, essentially awaiting definitive conclusions. Although this scenario has the admirable intent of allowing society to avoid what might later be determined to be unnecessary expenditures, it also has a fundamental flaw, for we will never have definitive conclusions about what the future climate will be. With the warnings as dire as they are and with many noted scientists fully behind these warnings, it is only prudent that we attempt to restrict further damage. To the extent feasible, we should cut back on all emissions to the atmosphere and oceans, some for health reasons, some for reasons related to climate change, and some simply because of the uncertainties regarding the damage we might be doing.

Because of the contrasts around the globe in emissions, lifestyles, and wealth, the emission cutbacks cannot be uniformly distributed either individually or nationally. For me and the other middle-class and upper-class residents of developed nations, cutting back can be done without even coming close to sinking into abject conditions, as our lifestyles routinely include excess and waste. Quite in contrast, hundreds of millions of other people not only have not indulged in great waste but have suffered severe deprivations throughout their lives. These people have a right to improve their circumstances even though it might mean significantly increasing their energy usage. If indeed we are facing a human-caused climate crisis, then the industrialized nations are the ones that, by and large, created the crisis. It is neither fair nor reasonable to expect people in the less developed nations to forgo industrialization in order not to increase their current relatively low emission levels.

Fortunately, with the knowledge gained from the past 200 years and with new, more efficient technologies, new industrialization need not produce the same level of damage that the industrialization of the nineteenth and twentieth centuries produced. Nations developing their economies today have a chance to do so in a way that is kinder to the environment and future generations than the examples set by the nations that are already developed. They can benefit from much that has been learned in the past 200 years and can avoid some of the most damaging steps along the circuitous routes taken toward industrialization by the already industrialized nations. Still, it would be unrealistic to expect that there will not be significant increases in the emissions to the atmosphere by nations emerging from undeveloped to developed status. In principle, at least as a rough approximation, I favor George Monbiot's suggestion, discussed in the previous chapter, that, if allotments are to be made, each country should have approximately the same per capita carbon-emission allotment (probably not immediately but perhaps by 2030), although complemented by a separate incentive to keep population growth in check. To do this and still lower total global emissions would require severe cutbacks in emissions by developed nations while still allowing developing nations the chance to develop.

With increased emissions by developing nations and all the uncertainties in model predictions, the possibility is very real that global temperatures will rise even faster than the mainstream predictions. On the other hand, the difficult financial times experienced by many nations around the globe in the 2008–2009 time frame could produce an overall reduction in emissions. Further, the attention given in recent years to environmental issues and to the potential dangers from continued greenhouse gas emissions might have created enough awareness of the problems to lead to a reduction in emissions even regardless of financial conditions. I am tremendously encouraged that many successful companies are finding economic as well as social advantages to adjusting their business practices in environmentally conscientious ways.

My greatest hope, however, and where I think we have a solid chance of truly solving the problem of human-induced climate change is that outstanding innovative inventors, probably from among today's youth, will develop clean alternative energy sources and make them affordable and available. Doing so could well be a creative modern inventor's best chance for fame, fortune, a Nobel Peace Prize, and the satisfaction of having done something really meaningful in life. In the second half of the twentieth century, clever people putting their minds to miniaturizing various computer components and writing software to make the machines user friendly succeeded beyond all initial expectations in creating computers for the masses, with such computers now pervasive in one form or another throughout modern society. Clever people putting their minds to developing clean energy sources will almost certainly end up succeeding there as well. The solutions might come from algae-based biofuels, a more effective capturing of solar radiation, or continued use of fossil fuels but accompanied by capturing and using the fossil fuel emissions rather than releasing them into the atmosphere, or they might come from some other scheme, perhaps one not even hinted at yet. Just as horses and oxen were superseded as major sources of energy for human societies and camphene and whale oil and lard oil were superseded, so too will our use of fossil fuels be superseded. If there were no possibility of geoengineering looming on the horizon, I would be optimistic that we can make appropriate adjustments to limit our climate impacts, for instance by cleaning up our use of fossil fuels, developing appropriate alternative energy sources, and appropriately modifying our lifestyles.

However, the possibility of geoengineering is looming. My fear is that before the solutions to our fossil fuel emissions are in place, the rising recognition of the need for action on the climate change front and the eagerness of people to charge forward could lead to a very serious compounding of our climate problems. Small

> Just as horses and oxen were superseded as major sources of energy for human societies, so too will our use of fossil fuels be superseded. If there were no possibility of geoengineering looming on the horizon, I would be optimistic that we can make appropriate adjustments to limit our climate impacts.

but growing numbers of people are proceeding beyond contemplating how we might lessen our impacts on climate change to contemplating how we might quite purposefully affect climate change through massive geoengineering schemes aimed at counteracting the anticipated future changes (chapter 7).

Massive geoengineering could be exciting and sensible in a fully understood, idealized world, like those an author can create in a science-fiction novel. But the actual world is not a fully understood world, and, most fundamentally, although thinking about geoengineering schemes in the abstract could have value, implementing the most dramatic of them at this point would not be even close to reasonable. We simply do not know enough. We do not know enough about how and why global climate changes even in the absence of human impacts (e.g., chapters 2 and 3), we do not know enough about how climate will continue to evolve, as we do not have models that can adequately predict the future (chapter 10), and we do not know enough about how well the geoengineering schemes might work (recall chapter 8) or about what the unintended consequences might be (chapters 6 and 7).

People advocating geoengineering do so with good intentions, but even the best of intentions can lead to very undesired consequences (chapter 6), especially given the highly interconnected and incompletely understood Earth system. Using geoengineering to remove some of what humans have inserted into the atmosphere (such as carbon dioxide, other greenhouse gases, and particulate matter) could be very favorable. However, some of the potential consequences of others of the proposed geoengineering schemes are terrifying (chapter 7), scaling far above and beyond the damage that we have already done to our planet.

Most of what I have read and heard so far about geoengineering has been couched in appropriate statements warning of the potential dangers and the need for great caution before implementing any of the major geoengineering proposals. However, here and there statements are made about massive geoengineering without any cautions attached, fueling my fear that the general cautionary approach might be abandoned. This fear derives in part from observing what has happened over the past several years as the interplay among media interest, peer pressure, and individual egos and goals has led quite a few scientists and others to abandon a balanced, objective approach to the issue of global warming (chapter 11). These scientists have become way too sure of themselves, and in some cases their skill development has shifted from enhancing their skill in technical aspects of the science to developing skill in creating effective sound bites for media attention. Over the years, the sound bites about climate have tended to become more extreme, as the scary or shocking statements that made for effective sound bites a decade ago have become routine, necessitating scarier or more shocking statements to continue to gain comparable media attention. Having watched this develop, and recognizing that geoengineering could be the next hot topic, with scientists and reporters scrambling to be in the forefront in their respective fields, I have no confidence that objectivity and levelheadedness will necessarily prevail. That is a primary reason why I have gone to the considerable effort of writing this book, even though I realize that it is likely to anger many people whom I would much prefer not to anger.

I am far less concerned about the possibility of 1 or even 2 degrees of temperature rise than I am about the possibility that humans will cause much greater damage in an ill-advised effort to solve the anticipated temperature problem through a massive geoengineering effort that has been inadequately thought through but gains a bandwagon of support from scientists, journalists, politicians, and others, all anxious to be leading the pack in advocating or reporting on a grand solution to the presumed upcoming planetary climate crisis. Our technological prowess is fully adequate for implementing some of these grand schemes, but this prowess far exceeds our understanding of the potential consequences. We would be far wiser, at least in the near term, to forgo massive geoengineering with its uncertain consequences and instead attack the human-induced changes to global and regional climate at their source, that is, at the human activities themselves. As detailed in chapter 12, there are far safer and better methods of improving our stewardship of our home planet than to undertake potentially disastrous geoengineering efforts. (Here I am referring to geoengineering proposals that involve such aggressive actions as pouring additional material into the atmosphere or oceans or that cover vast areas of the land or oceans for purposes of albedo adjustment. I am not referring to efforts to limit the amount of CO_2 that enters the atmosphere or efforts to scrub CO_2 out of the atmosphere. The latter efforts are referred to as geoengineering by some people but not by others, as the term continues to evolve.)

> It hardly matters to the Earth exactly where sea level is; but to humans, who cannot readily live underwater, a few meters rise in sea level can matter a great deal.

Fortunately, whatever we do, the Earth itself is likely to survive just fine, just as it has survived earthquakes and explosive volcanic eruptions from within, repeated impacts from extraterrestrial objects, mountain-building episodes as continental plates collide, cycling of glacial and interglacial periods, and all sorts of other changes over the past 4.6 billion years. Whatever our impacts, they are likely to be most relevant not to the Earth, which should be able to withstand them, but to ourselves and future generations and to the Earth's ecosystems. After all, for one telling example, it hardly matters to the Earth exactly where sea level is; but to humans, who cannot readily live underwater, a few meters rise in sea level can matter a great deal, as it would destroy coastal environments throughout the world, wiping out some entire cities, islands, and even nations.

Despite my uncertainties regarding future climate, I am certain about some of the surrounding issues: I am certain that our understanding of climate is incomplete, I am certain that we should not be demonizing the skeptics, and I am certain that we should be extremely cautious before engaging in large-scale geoengineering. I do not have all the answers, and I recognize that in the long term it may be that one or another large-scale geoengineering effort will become the best route to take. My plea is that this not be done based on faulty assumptions about the soundness of our climate predictions. However authoritative some scientists may sound, no scientist is all-knowing, and none can predict with certainty the future course of climate change or the full consequences of any of the proposed geoengineering schemes. Good intentions do not guarantee favorable results (recall chapter 6). If society opts to undertake geoengineering, then we should do so with great care and great caution, in full recognition of our ignorance and the potential seriousness of unintended consequences. For those with offspring, consider your descendants, and for everyone, consider our responsibility to future generations and to the planet as a whole and its many ecosystems. Humans have done much damage to portions of our inherited environment and sometimes have not succeeded when trying to reverse that damage. Please be careful about implementing a geoengineering corrective action that could backfire to become a greater disaster than the one we are trying to correct.

Abbreviations

%	percent
AAAS	American Association for the Advancement of Science
ACIA	Arctic Climate Impact Assessment
A.D.	anno Domini, Latin for "in the year of the Lord"
AIDS	acquired immunodeficiency syndrome
AMIP	Atmospheric Model Intercomparison Project
AMS	American Meteorological Society
Ar	argon
AR4	Fourth Assessment Report of the IPCC
AVHRR	Advanced Very High Resolution Radiometer
B.C.	before Christ
BED	Burlington Electric Department
c.	circa
°C	degrees Centigrade
$CaCO_3$	calcium carbonate
CDC	Centers for Disease Control and Prevention
CERES	Clouds and the Earth's Radiant Energy System (a satellite instrument)
CFC	chlorofluorocarbon
CH_4	methane
$C_6H_{12}O_6$	glucose
$C_{12}H_{22}O_{11}$	sucrose
CH_3SCH_3	dimethylsulfide (also DMS)
Cl	chlorine
ClO	chlorine monoxide
CNN	Cable News Network

CO	carbon monoxide
CO_2	carbon dioxide
CO_3^{2-}	carbonate ions
COS	carbonyl sulfide
CS_2	carbon disulfide
D.C.	District of Columbia
DDT	dichloro-diphenyl-trichloroethane
DMS	dimethylsulfide
DNA	deoxyribonucleic acid
DOD	Department of Defense
DOE	Department of Energy
doi	digital object identifier
°E	degrees east longitude
e.g.	for example (abbreviated from Latin, "exempli gratia")
ENSO	El Niño/Southern Oscillation
EPA	Environmental Protection Agency
et al.	and others (abbreviation for the Latin "et alia")
etc.	et cetera
°F	degrees Fahrenheit
FACE	Florida Area Cumulus Experiment
G8	Group of Eight, consisting of the following eight major industrialized nations: Canada, France, Germany, Italy, Japan, Russia, the United Kingdom, and the United States
GCM	global climate model (alternatively: general circulation model)
GISP2	Greenland Ice Sheet Project 2
GRACE	Gravity Recovery and Climate Experiment
GRIP	Greenland Ice Core Project
H	hydrogen (atomic)
H^+	hydrogen ions
H_2	hydrogen (molecular)
HadAM3	Hadley Centre atmospheric model version 3
HBr	hydrogen bromide
HCFC	hydrochlorofluorocarbon
HCl	hydrogen chloride
HCO_3^-	bicarbonate ions
H_2CO_3	carbonic acid
He	helium

HF	hydrogen fluoride
HFC	hydrofluorocarbon
H_2O	water; water vapor
HOCl	hypochlorous acid
H_2S	hydrogen sulfide
ICS	International Commission on Stratigraphy
i.e.	that is (abbreviated from Latin, "id est")
IPCC	Intergovernmental Panel on Climate Change
ITCZ	Intertropical Convergence Zone
Jr.	Junior
K	Kelvin (a unit of temperature)
Kr	krypton
LEED	Leadership in Energy and Environmental Design
m	atmospheric attenuation coefficient
MIT	Massachusetts Institute of Technology
m.y.a.	million years ago
°N	degrees north latitude
NAS	National Academy of Sciences
NASA	National Aeronautics and Space Administration
NCAR	National Center for Atmospheric Research
Ne	neon
NGRIP	North Greenland Ice Core Project
NH_3	ammonia
NHRE	National Hail Research Experiment
no.	number
NO_2	nitrogen dioxide
N_2O	nitrous oxide
NOAA	National Oceanic and Atmospheric Administration
NSF	National Science Foundation
NSIDC	National Snow and Ice Data Center
O	oxygen (atomic)
O_2	oxygen (molecular)
O_3	ozone
O16	oxygen with an atomic weight of 16 (by far the most prevalent oxygen)
O18	oxygen isotope with an atomic weight of 18
p.	page
PEP	Precipitation Enhancement Project

pH	a measure of the concentration of hydrogen ions in (or the acidity of) a solution
PhD	Doctor of Philosophy
pp.	pages
R & D	research and development
RNA	ribonucleic acid
°S	degrees south latitude
S_2	sulfur (molecular)
SAR	Second Assessment Report of the IPCC
SF_6	sulfur hexafluoride
SO_2	sulfur dioxide
SUV	sport utility vehicle
TAR	Third Assessment Report of the IPCC
UCLA	University of California, Los Angeles
UFO	Unidentified Flying Object
UK	United Kingdom
UN	United Nations
UNEP	United Nations Environment Programme
UNFCCC	United Nations Framework Convention on Climate Change
U.S.	United States
USA	United States of America (= U.S.)
U.S.S.R.	Union of Soviet Socialist Republics
vol.	volume
°W	degrees west longitude
W/m^2	Watts per square meter
WMO	World Meteorological Organization

Notes

CHAPTER 1. INTRODUCTION

1. Loeb and Wong 2007.
2. Francis 1993; Officer and Page 1993; Stothers 1984.
3. Francis 1993.
4. Hansen et al. 1992.
5. Hansen et al. 1996; Soden et al. 2002.
6. Fu and Johanson 2007.
7. Broecker 1991, 1995a.
8. Linden 2006.
9. For instance, Hughes 1973; Shepherd and Wingham 2007.
10. Obasi and Tolba 1990.
11. Solomon et al. 2007b.
12. Intergovernmental Panel on Climate Change 2007.
13. Solomon et al. 2007a.
14. Parry et al. 2007.
15. Metz et al. 2007.
16. Meehl et al. 2007.
17. Christensen et al. 2007.
18. Announcement video available at http://nobelprize.org/nobel_prizes/peace/laureates/2007/announcement.html.

CHAPTER 2. 4.6 BILLION YEARS OF GLOBAL CHANGE

1. Textor et al. 2004.
2. Segrè 2002.
3. Alley 2000, p. 85.
4. Schneider 1997.
5. Wilson 1992, p. 186.
6. Segrè 2002.
7. Bradbury 1998.
8. Sepkoski 1993.
9. Segrè 2002, p. 164.
10. Margulis 1998.
11. Bradbury 1998.
12. Lovelock 1988.
13. Anbar et al. 2007.
14. Sepkoski 1993; Wilson 1992.
15. Narbonne 2005.
16. Bradbury 1998; Margulis 1998.
17. Margulis 1998.
18. Morrison et al. 2007.

19. Keeling 2007.
20. Wilson 1992.
21. Segrè 2002.
22. Macdougall 2004.
23. Kirschvink et al. 2000.
24. Hoffman et al. 1998.
25. http://www.stratigraphy.org/chus.pdf (as of January 20, 2008).
26. Harland and Rudwick 1964.
27. Macdougall 2004.
28. Kirschvink et al. 2000.
29. Segrè 2002.
30. Caldeira and Kasting 1992.
31. Hoffman and Schrag 2000; Kirschvink et al. 2000.
32. Hoffman et al. 1998.
33. Hoffman and Schrag 2000; Hoffman et al. 1998.
34. Segrè 2002.
35. Narbonne 2005.
36. Knoll et al. 2006.
37. Knoll et al. 2006; Narbonne 2005.
38. Shen et al. 2008.
39. Narbonne 2005, p. 421.
40. Narbonne 2005, p. 421.
41. Narbonne 2005.
42. Knoll et al. 2006.
43. Narbonne 2005.
44. Shen et al. 2008.
45. Marshall 2006.
46. Marshall 2006.
47. Wray et al. 1996.
48. Marshall 2006, p. 356.
49. Marshall 2006; Sepkoski 1993.
50. Benton 1993c.
51. Sepkoski 1993; Wilson 1992.
52. Sepkoski 1993.
53. Wilson 1992.
54. Benton 1993c.
55. Bradbury 1998.
56. Stokstad 2006.
57. Benton 1993b, p. 83.
58. Benton 1993b; Wilson 1992.
59. Wilson 1992.
60. Macdougall 2004.
61. Macdougall 2004.
62. Benton 1993b; Macdougall 2004.
63. For example, Jansen et al. 2007.
64. Berner 1993.
65. Benton 2008.
66. Benton 1993b.
67. Benton 1993b.
68. Bambach 2006.
69. Bambach 2006.
70. Bambach 2006.
71. Benton 2008; Sepkoski 1993.
72. Bambach 2006.
73. Becker et al. 2001; Xu et al. 1985.
74. Renne and Basu 1991.
75. Benton 2008.
76. Campbell et al. 1992.
77. Benton 2008.
78. Renne and Basu 1991.
79. Benton 2008.
80. Baksi and Farrar 1991; Renne and Basu 1991.
81. Renne et al. 1995.
82. For example, Benton 2008.
83. Bambach 2006.
84. Becker et al. 2001; Erwin 1994.
85. Benton 1993b.
86. Benton 2008.
87. Benton 2008.
88. Benton 2008; Eshet et al. 1995.
89. Benton 2008.
90. Bradbury 1998.
91. Benton 1993b.
92. Bradbury 1998.
93. Benton 1993a.
94. Benton 1993b.
95. Bradbury 1998.
96. Benton 1993b.

97. Benton 2008.
98. Bambach 2006.
99. Benton 1993a.
100. Tudge 1997.
101. Benton 1993a.
102. Bradbury 1998.
103. Benton 1993a.
104. Prentice et al. 2001.
105. Benton 1993a.
106. Tudge 1997.
107. For example, Benton 2008; Segrè 2002.
108. Alvarez et al. 1980.
109. Benton 2008.
110. Hildebrand et al. 1991; Pope et al. 1993.
111. Bradbury 1998.
112. Hildebrand et al. 1991.
113. Benton 2008.
114. Benton 2008.
115. Benton 1993a.
116. Bambach 2006.
117. Jansen et al. 2007.
118. Moran et al. 2006.
119. Pearce 2007, p. 90.
120. Hansen et al. 2007; Pearce 2007.
121. Segrè 2002.
122. Jansen et al. 2007.
123. Pearce 2007.
124. Jansen et al. 2007.
125. Whaley 2007.
126. Macdougall 2004.
127. Tudge 1997.
128. DeConto and Pollard 2003.
129. DeConto and Pollard 2003.
130. Kennett 1977; Scher and Martin 2006.
131. Bradbury 1998.
132. Gibbons 2006.
133. Holman 1995.
134. Gibbons 2006.
135. Gibbons 2006.
136. Bradbury 1998.
137. Gibbons 2006.
138. Jansen et al. 2007.
139. Jansen et al. 2007.
140. For example, Raymo et al. 2006.
141. Pearce 2007.
142. Macdougall 2004.
143. Ruddiman 2005.
144. Tudge 1997; Bradbury 1998.
145. Macdougall 2004.
146. Macdougall 2004.
147. Diamond 2005b.
148. Wood and Collard 1999.
149. Wood and Collard 1999.
150. Macdougall 2004.
151. Spoor et al. 2007; Wood and Collard 1999.
152. Spoor et al. 2007.
153. Gibbons 2006.
154. Dennell and Roebroeks 2005.
155. Gibbons 2007.
156. Segrè 2002.
157. Dennell and Roebroeks 2005.
158. Diamond 2005b.
159. Macdougall 2004.
160. Tudge 1997. Ruddiman 2005, p. 23, indicates that it was sometime between 150,000 years ago and 100,000 years ago that "nearly modern people evolved" and that this occurred in Africa.
161. Dennell and Roebroeks 2005.
162. Tudge 1997.
163. For example, Dennell and Roebroeks 2005.
164. Macdougall 2004.
165. Tudge 1997; Wood and Collard 1999.
166. Jansen et al. 2007.
167. Petit et al. 1999; Vimeux et al. 2002.
168. Hansen et al. 2007.
169. Holman 1995.

170. Holman 1995.
171. Jansen et al. 2007.
172. Forster et al. 2007.
173. Petit et al. 1999.
174. Hansen et al. 2007.
175. Detailed in Walter et al. 2007a.
176. Walter et al. 2007a.
177. Martin et al. 1990; Pearce 2007.
178. Jansen et al. 2007, from the Intergovernmental Panel on Climate Change. However, like so much else in Earth history, the timing of when natural forcings would lead to a new ice age is disputed. Ruddiman (2005) suggests that without human impacts the ice age would already be upon us.
179. Anklin et al. 1993.
180. Jansen et al. 2007.
181. Mercer 1978.
182. De Vernal and Hillaire-Marcel 2008; Emiliani 1969.
183. De Vernal and Hillaire-Marcel 2008.
184. Jansen et al. 2007.
185. Thompson et al. 1997.
186. Alley 2000.
187. Thompson 2000.
188. Pearce 2007.
189. Segrè 2002.
190. Goebel et al. 2008.
191. Diamond 2005b.
192. Fagan 2008.
193. Officer and Page 1993.
194. Macdougall 2004.
195. Imbrie and Imbrie 1979.
196. Jansen et al. 2007.
197. Church et al. 2001.
198. Jansen et al. 2007.
199. Pearce 2007.
200. Broecker and Kunzig 2008; Pearce 2007.
201. Fagan 2004.
202. Pearce 2007.
203. For example, Fagan 2004, 2008; Linden 2006.
204. Pearce 2007.
205. Thompson et al. 2000.
206. Grove 1988.
207. Macdougall 2004.
208. Hansen et al. 2006; Trenberth et al. 2007.

CHAPTER 3. ABRUPT CLIMATE CHANGE

1. Stommel 1961, p. 228. A few scientists had mentioned abrupt climate change earlier, although generally the time frame of what was considered "abrupt" was much longer than what is now considered "abrupt." For example, in 1960 Wally Broecker, Maurice Ewing, and Bruce Heezen published a paper titled "Evidence for an Abrupt Change in Climate Close to 11,000 Years Ago," but in that case "abrupt" referred to changes that "occurred in less than 1000 years" (Broecker et al. 1960, p. 429).
2. Rooth 1982, pp. 131 and 135.
3. Rooth 1982.
4. Dansgaard et al. 1969.
5. Dansgaard et al. 1982.
6. Oeschger et al. 1984.
7. For instance, Broecker 1998; Broecker et al. 1985, 1990.
8. For instance, Bond et al. 1993; Broecker et al. 1988, 1990.
9. Broecker et al. 1985.
10. Johnsen et al. 1992; Mayewski and White 2002.
11. Dansgaard et al. 1993.
12. NGRIP Members 2004.

13. Mayewski and White 2002.
14. Alley 2000; Mayewski and White 2002.
15. NGRIP Members 2004.
16. Rasmussen et al. 2006.
17. Andersen et al. 2007.
18. Johnsen et al. 1992, p. 313.
19. Dansgaard et al. 1993, p. 218.
20. Anklin et al. 1993.
21. NGRIP Members 2004.
22. Overpeck and Cole 2006.
23. For example, Dansgaard et al. 1982; Oeschger et al. 1984.
24. Ganopolski and Rahmstorf 2001; Rahmstorf 2002.
25. Jansen et al. 2007.
26. Alley 2007.
27. Braun et al. 2005; Clemens 2005; Overpeck and Cole 2006.
28. Alley 2007.
29. Bond et al. 1993.
30. Heinrich 1988.
31. Hemming 2004.
32. Bond et al. 1992; Hulbe et al. 2004.
33. Broecker and Hemming 2001.
34. Pearce 2007.
35. MacAyeal 1993.
36. Johnson and Lauritzen 1995.
37. Hulbe 1997; Hulbe et al. 2004.
38. Stocker and Marchal 2000.
39. Overpeck and Cole 2006.
40. Alley 2007.
41. Macdougall 2004; Mayewski and White 2002.
42. Alley 2000, p. 111.
43. For example, Broecker 1995a; Teller et al. 2002.
44. Lowell et al. 2005.
45. For example, Broecker 2006.
46. Overpeck and Cole 2006.
47. Mayewski and White 2002.
48. Alley 2000.

49. Tarasov and Peltier 2005.
50. Thompson et al. 1997.
51. Broecker 1998.
52. Barrows et al. 2007.
53. Thompson et al. 2000.
54. Firestone et al. 2007.
55. Kennett et al. 2009.
56. Firestone et al. 2007; Kennett et al. 2009.
57. Teller et al. 2002.
58. Barber et al. 1999; Jansen et al. 2007; Kleiven et al. 2008.
59. Alley et al. 1997.
60. Pearce 2007.
61. Pearce 2007.
62. Jansen et al. 2007.
63. Thompson et al. 2002.
64. Kolbert 2006.
65. Overpeck and Cole 2006.
66. LeGrande et al. 2006.
67. For example, Jansen et al. 2007.
68. Thompson et al. 2006.
69. Overpeck and Cole 2006.
70. Overpeck and Cole 2006.
71. Thompson et al. 2000.
72. Thompson et al. 2006.
73. For example, Thompson 2000.
74. For example, Thompson et al. 2000.
75. Steffensen et al. 2008.
76. Rahmstorf 2002.
77. Blunier et al. 1998.
78. Stott et al. 2007.
79. Thompson et al. 1997.
80. Thompson et al. 2002.
81. For example, Alley 2000; Dansgaard et al. 1993.
82. Petit et al. 1999.
83. Overpeck and Cole 2006; Rahmstorf 2002.
84. Ganopolski and Rahmstorf 2001.
85. Visible imagery is available at, for instance, http://nsidc.org/iceshelves/

larsenb2002/animation.html or http://
earthobservatory.nasa.gov/Newsroom/
NewImages/images.php3?img_id=8257.

86. Rott et al. 1996; Vaughan and Doake
1996.

87. For example, Scambos et al. 2000,
2004.

88. Hughes 1977; Mercer 1978.

89. Hindmarsh and LeMeur 2001.

90. Rignot et al. 2004; Rott et al. 2002;
Scambos et al. 2004.

91. For example, Bindschadler and Vornberger 1990.

92. For example, Shabtaie and Bentley
1987.

93. Bindschadler and Bentley 2002;
Joughin and Tulaczyk 2002.

94. Hughes 1981.

95. Shepherd et al. 2001.

96. Bentley 1997.

97. Rignot 1998; Shepherd et al. 2001.

98. Zwally et al. 2005.

CHAPTER 4. A SHORT HISTORY
OF HUMAN IMPACTS

1. Goebel et al. 2008; Kolbert 2006; Meyer
1996.

2. Diamond 2005b.

3. Holman 1995.

4. Benton 2008.

5. Wilson 1992.

6. Balter 2007.

7. Dubcovsky and Dvorak 2007.

8. Diamond 2005b.

9. Fagan 2004.

10. Diamond 2005b; Dubcovsky and
Dvorak 2007.

11. Balter 2007; Dillehay et al. 2007.

12. Balter 2007.

13. Diamond 2005b.

14. Diamond 2005b.

15. Dillehay et al. 2007.

16. Ruddiman 2005.

17. Diamond 2005b; Driscoll et al. 2007.

18. Ruddiman 2005.

19. Klein Goldewijk 2001.

20. Numbers are updated every few seconds on the U.S. Census Bureau website
at http://www.census.gov.

21. Durant 1957.

22. Evelyn 1661.

23. Officer and Page 1993.

24. Christianson 1999.

25. Bergreen 2007.

26. Officer and Page 1993.

27. Gordon 2002.

28. Gordon 2004.

29. Gordon 2002.

30. Gordon 2004.

31. Standage 1998.

32. Gordon 2004.

33. Standage 1998.

34. Gordon 2002.

35. Gordon 2002.

36. Officer and Page 1993.

37. Christianson 1999.

38. Officer and Page 1993.

39. Officer and Page 1993.

40. Stern 2007.

41. Crutzen 2006; Prentice et al. 2001.

42. Forster et al. 2007.

43. For example, Sabine et al. 2004.

44. Feely et al. 2004.

45. Royal Society 2005.

46. Determined from data listed in Keeling
et al. 2009.

47. Hansen et al. 2007.

48. Stern 2007.
49. Lelieveld 2006.
50. Archer 2007.
51. Dow and Downing 2006.
52. Forster et al. 2007.
53. Arguez et al. 2007.
54. Hansen et al. 2007.
55. Bousquet et al. 2006; Lelieveld 2006.
56. Stern 2007.
57. Denman et al. 2007.
58. Dow and Downing 2006.
59. Arguez et al. 2007.
60. Denman et al. 2007.
61. Arguez et al. 2007.
62. Forster et al. 2007.
63. Christianson 1999; Meyer 1996.
64. Christianson 1999.
65. Stern 2005.
66. Stern 2005.
67. Ohmura 2006.
68. Andreae et al. 2005.
69. Forster et al. 2007.
70. Arguez et al. 2007.
71. Thompson 2000.
72. Wan 2007.
73. Arguez et al. 2007.
74. Macdougall 2004.
75. Arguez et al. 2007.
76. Trenberth et al. 2007.
77. Arguez et al. 2007.
78. Mann et al. 1999.
79. For example, McIntyre and McKitrick 2003, 2005; Soon et al. 2003.
80. McIntyre and McKitrick 2003, 2005.
81. McIntyre and McKitrick 2003.
82. Jansen et al. 2007.
83. Trenberth et al. 2007.
84. Bindoff et al. 2007.
85. Pearce 2007.
86. Dow and Downing 2006.
87. Stern 2007.
88. Abdalati et al. 2004; Arendt et al. 2002; Williams and Ferrigno 2002.
89. Thompson et al. 2003; Williams and Ferrigno 1998.
90. Allison and Peterson 1988; Dyurgerov and Meier 2000; Thompson et al. 2003.
91. Hastenrath and Kruss 1992; Kaser et al. 2004; Thompson et al. 2002.
92. Cook et al. 2005.
93. Dyurgerov and Meier 1997.
94. Williams and Ferrigno 1993.
95. For example, Lemke et al. 2007.
96. For more complete reviews, see Lemke et al. 2007 and Parkinson 2006.
97. Rayner et al. 2003; Walsh and Chapman 2001.
98. Rothrock et al. 1999; Yu et al. 2004.
99. Parkinson and Cavalieri 2008.
100. Arguez et al. 2007.
101. Comiso et al. 2008. Late-summer Arctic sea ice rebounded somewhat in 2008 and 2009 from the record low values in September 2007.
102. Cavalieri and Parkinson 2008.
103. Derocher 2008; Stirling and Parkinson 2006.
104. Dow and Downing 2006.
105. Parkinson 2006; Walsh et al. 2005.
106. Stern 2007; Walter et al. 2007b; Zimov et al. 2006.
107. Royal Society 2005.
108. Orr et al. 2005; Royal Society 2005.
109. Royal Society 2005.
110. Hoegh-Guldberg 2005.
111. Christianson 1999.
112. Royal Society 2005.
113. Kolbert 2006.
114. Dow and Downing 2006.
115. Kolbert 2006.
116. Root et al. 2003.

117. Parmesan 2006.
118. For example, Pounds et al. 1999.
119. Parmesan 2006, p. 639.
120. Parmesan 2006, p. 637.
121. Pounds et al. 2006.
122. Pounds et al. 2006.
123. Kolbert 2006.
124. Royal Society 2005.
125. Wilson 1992.
126. Wilson 1992.
127. Wilson 1992.
128. Wilson 1992.
129. Wilcove 2008.
130. Wilcove 2008.
131. Micklin 2007.

132. Pala 2005.
133. Micklin 2007; Pala 2005.
134. Johnson 1997.
135. Farman et al. 1985.
136. Solomon 1990; Stolarski 1988.
137. Molina and Rowland 1974.
138. For example, Solomon 1990.
139. Solomon 1990.
140. Solomon 1990.
141. Newman et al. 2004.
142. Solomon 1990.
143. Rowlands 1993.
144. Gore 2006, p. 295.
145. Arguez et al. 2007.
146. Dow and Downing 2006.

CHAPTER 5. THE FUTURE: WHY SOME PEOPLE ARE SO CONCERNED WHILE OTHERS AREN'T

1. Jansen et al. 2007.
2. Jansen et al. 2007.
3. Jansen et al. 2007; Stern 2005.
4. Cubasch et al. 2001.
5. IPCC 2007.
6. Jansen et al. 2007.
7. Arguez et al. 2007.
8. Pearce 2007; Stern 2007.
9. Stern 2007, p. 151.
10. Barnett et al. 2008.
11. Pearce 2007.
12. Parry et al. 2007.
13. Stern 2007.
14. Randall et al. 2007; Stern 2007.
15. Archer 2007.
16. For example, Dessler et al. 2008.
17. Pearce 2007.
18. Pearce 2007.
19. Andreae et al. 2005.
20. Dow and Downing 2006.
21. Pearce 2007.
22. Pearce 2007.
23. Bindoff et al. 2007.

24. Meehl et al. 2007.
25. Das et al. 2008; Zwally et al. 2002.
26. Hansen et al. 2007.
27. Joughin et al. 2008.
28. Parizek and Alley 2004.
29. Pfeffer et al. 2008.
30. Monbiot 2007.
31. Meehl et al. 2007.
32. For example, Hughes 1973; Shepherd and Wingham 2007.
33. For example, Helsen et al. 2008; Shepherd and Wingham 2007; Zwally et al. 2005.
34. Shepherd and Wingham 2007.
35. IPCC 2007.
36. Hansen et al. 2007.
37. For example, Pfeffer et al. 2008.
38. Rahmstorf 2007.
39. Chao et al. 2008.
40. Barnett et al. 2005.
41. Davis and Topping 2008; Wilbanks et al. 2007.
42. Royal Society 2005.

43. Royal Society 2005.
44. Feely et al. 2004.
45. Royal Society 2005.
46. Royal Society 2005.
47. Hoegh-Guldberg 2005.
48. Royal Society 2005.
49. Royal Society 2005.
50. Ingermann et al. 2003.
51. Pounds et al. 2006.
52. Thomas et al. 2004, p. 145.
53. Thomas et al. 2004, p. 147.
54. Wilson 1992.
55. Wilson 1992.
56. Dow and Downing 2006.
57. Dow and Downing 2006.
58. Lindgren and Gustafson 2001.
59. Dow and Downing 2006.
60. Manabe and Stouffer 1994.
61. Bryden et al. 2005.
62. Bryden et al. 2005.
63. Kerr 2006.
64. Alley 2007.
65. Randall et al. 2007.
66. For example, Pielke et al. 2005.
67. Smith 1999.
68. Emanuel 2005; Webster et al. 2005.
69. Emanuel et al. 2008.
70. Knutson et al. 2008.
71. Webster et al. 2005.
72. Emanuel 2005.
73. Emanuel 2005; Hoyos et al. 2006.
74. Elsner 2008.

75. Leatherman and Williams 2008.
76. Pielke et al. 2005.
77. For example, Lomborg 2007.
78. Lomborg 2007.
79. Lomborg 2007.
80. Lomborg 2007.
81. Gosnell 2005.
82. For example, Buzin et al. 2007.
83. ACIA 2005.
84. Pearce 2007.
85. Pearce 2007.
86. Lindzen et al. 2001.
87. Lindzen et al. 2001; see also Spencer et al. 2007.
88. Chambers et al. 2002; Lin et al. 2002.
89. For example, Spencer 2008.
90. Spencer 2008.
91. Spencer 2008, p. 8.
92. Spencer 2008, pp. ix and 15.
93. Solomon et al. 2007a, p. 60.
94. Spencer 2008.
95. Spencer 2008, p. 67.
96. Spencer 2008.
97. Spencer 2008, p. 138.
98. Spencer 2008.
99. Lomborg 2007.
100. The 1973 agreement on the conservation of polar bears is reviewed in depth by Prestrud and Stirling 1994.
101. Lomborg 2007.
102. Pittock 2008.

CHAPTER 6. GOOD INTENTIONS GONE AWRY

1. Ward 1989.
2. Ward 1989.
3. Alexander 2001; Percival 1890.
4. Diamond 2005a; Tudge 1997.
5. Diamond 2005a.
6. Freeman 1992.
7. Freeman 1992.

8. Freeman 1992.
9. Clarke et al. 1984.
10. Clarke et al. 1984; Wilson 1992.
11. Clarke et al. 1984.
12. Wilson 1992.
13. Civeyrel and Simberloff 1996.

14. Wilson 1992; see also the American Museum of Natural History website at http://www.amnh.org/nationalcenter/Endangered/perch.html, September 13, 2007.
15. Kitchell et al. 1997.
16. Gutro and Cole 2006.
17. Beschta and Ripple 2006; Ripple and Beschta 2007a.
18. Beschta and Ripple 2006.
19. Ripple and Beschta 2007b.
20. Diamond 2005a.
21. Schneider 1989.
22. Schneider 1989.
23. Pearce 2007.
24. Page et al. 2002.
25. McDonald et al. 2007.
26. Apt et al. 1996.
27. Fagan 2008.
28. Diamond 2005a.
29. Diamond 2005a, p. 118.
30. Schneider with Mesirow 1976, p. 222.
31. Schneider with Mesirow 1976.
32. Schneider with Mesirow 1976, p. 222.
33. Apt et al. 1996.
34. Stone 2008.
35. Normile 2007.
36. Normile 2007.
37. Leslie 1980; Midgley 1937.
38. Christianson 1999.
39. Leslie 1980; Midgley 1937; Midgley and Henne 1930.
40. Midgley 1937.
41. Midgley and Henne 1930.
42. Christianson 1999; Giunta 2006.
43. Christianson 1999.
44. Kolbert 2006.
45. Dow and Downing 2006.
46. Segrè 2002.
47. Molina and Rowland 1974.
48. Christianson 1999.
49. For example, Solomon 1990.
50. Betts 2000.
51. Kaiser 2005.
52. Available at http://www.sti.nasa.gov/tto.

CHAPTER 7. GEOENGINEERING SCHEMES

1. Keith 2000, p. 245.
2. Marchetti 1977.
3. For example, Keith 2000.
4. For example, Schneider 2001.
5. Marchetti 1977.
6. Marchetti 1977, p. 62.
7. Rochelle 2009.
8. Keith 2009.
9. Orr 2009.
10. Schrag 2009.
11. Orr 2009.
12. Chu 2009; Haszeldine 2009.
13. Budyko 1977.
14. Crutzen 2006.
15. Andreae et al. 2005.
16. McCormick et al. 1995; Soden et al. 2002.
17. Bluth et al. 1992.
18. Keith 2000.
19. National Academy of Sciences 1992.
20. Keith 2000, p. 263.
21. Crutzen 2006.
22. Crutzen 2006, p. 212.
23. Rasch et al. 2008.
24. MacCracken 2006; Murphy 2009.
25. As reported in Nel 2005.

26. Crutzen 2006.
27. Keith 2000.
28. For example, Tilmes et al. 2008.
29. Tilmes et al. 2008.
30. Keith 2000.
31. Monbiot 2007, p. 209.
32. For example, Early 1989.
33. Govindasamy and Caldeira 2000.
34. Govindasamy and Caldeira 2000, p. 2143.
35. Seifritz 1989.
36. Keith 2000.
37. For example, Calder 1976.
38. Gaskill 2007.
39. Gaskill 2007.
40. Gaskill 2007.
41. Keith 2000.
42. MacCracken 2006, p. 237.
43. For example, MacCracken 2006; Schneider with Mesirow 1976.
44. Calder 1976; Kellogg and Schneider 1974.
45. Keith 2000.
46. Schneider with Mesirow 1976.
47. Fleming 1998.
48. Wexler 1958.
49. Wexler 1958.
50. Kellogg and Schneider 1974.
51. Wexler 1958, p. 1063.
52. Schneider with Mesirow 1976.
53. Maykut and Untersteiner 1971.
54. Schneider with Mesirow 1976.
55. Aagaard and Coachman 1975.
56. Micklin 1986.
57. Aagaard and Coachman 1975.
58. Budyko 1972.
59. Micklin 1981.
60. Micklin 1981.
61. Keith 2000.
62. Johnson 1997.
63. Johnson 1997.

64. Johnson 1997.
65. Martin 1990.
66. For example, Martin and Gordon 1988; Martin et al. 1989, 1990.
67. Martin et al. 1990.
68. Martin et al. 1990.
69. Martin 1990.
70. For example, Broecker 1990.
71. Kintisch 2007.
72. Broecker 1990.
73. Kintisch 2007.
74. Monbiot 2007.
75. Boyd et al. 2007.
76. Buesseler et al. 2008.
77. Jerch 2008; Kintisch 2007.
78. Jerch 2008.
79. Keith 2000.
80. Cicerone 2006.
81. Cicerone et al. 1991.
82. Cicerone et al. 1991.
83. Elliott et al. 1994.
84. Cicerone et al. 1991.
85. Korycansky et al. 2001.
86. Korycansky et al. 2001.
87. Korycansky et al. 2001.
88. Keith 2000.
89. Keith and Dowlatabadi 1992.
90. Fleming 2007.
91. Victor 2008.
92. Robock 2008a.
93. Barker et al. 2007.
94. Barker et al. 2007, p. 624.
95. Ekholm 1901.
96. Kellogg and Schneider 1974.
97. Kellogg and Schneider 1974.
98. Wigley 2006, p. 452.
99. Tilmes et al. 2008; Wigley 2006.
100. Gaskill 2007.
101. For example, Robock 2008b; Schneider 2001.

CHAPTER 8. THE RECORD ON SMALLER-SCALE ATTEMPTED MODIFICATIONS

1. Keith 2000.
2. Keith 2000.
3. Schaefer 1946.
4. Schaefer 1949.
5. Fleming 2004, p. 177.
6. Schaefer 1949.
7. Vonnegut 1947.
8. Fleming 2004, p. 175.
9. Fleming 2004.
10. Fleming 2004, p. 175.
11. Cotton and Pielke 2007.
12. Fleming 2004.
13. Fleming 2004.
14. Keith 2000.
15. Fleming 2004.
16. Fleming 2004.
17. Fleming 2004, p. 189.
18. Fleming 2004.
19. Fleming 2004.
20. Fleming 2004.
21. The reader is referred to Battan 1969 and Fleming 2004 for more on the Soviet efforts at weather modification and to end notes 50 to 52 in Fleming 2004 for additional relevant references.
22. Battan 1969.
23. Gagin and Neumann 1981.
24. Rangno and Hobbs 1995, p. 1169.
25. Cotton and Pielke 2007.
26. Chappell et al. 1971; Mielke et al. 1981.
27. Rangno and Hobbs 1993.
28. Ryan and King 1997.
29. Ryan and King 1997.
30. Silverman 2001.
31. Silverman 2001.
32. Silverman 2001.
33. Woodley and Rosenfeld 2004.
34. Cotton and Pielke 2007.
35. Changnon and Ivens 1981.
36. Changnon and Ivens 1981.
37. Changnon and Ivens 1981, p. 372.
38. Changnon and Ivens 1981.
39. Changnon and Ivens 1981.
40. Schaefer 1949, pp. 318–319.
41. Cotton and Pielke 2007.
42. Battan 1969.
43. Cotton and Pielke 2007.
44. Calder 1976.
45. Cotton and Pielke 2007.
46. Atlas 1977.
47. Atlas 1977.
48. Atlas 1977.
49. Changnon and Ivens 1981.
50. Cotton and Pielke 2007.
51. Gray 2003; Simpson and Malkus 1964.
52. Gray 2003.
53. Gray 2003.
54. Gray 2003.
55. Cotton and Pielke 2007.
56. Sheets 2003.
57. Calder 1976; Simpson and Malkus 1964.
58. Simpson and Malkus 1964.
59. Sheets 2003.
60. Cotton and Pielke 2007.
61. Calder 1976.
62. http://www.rapidcitylibrary.org/lib_info/1972Flood/index.asp, September 10, 2008.
63. Schneider with Mesirow 1976.
64. Farhar 1974.
65. Schneider with Mesirow 1976.

66. Jay Trobec, http://www.trobec.net/ update3/rapid-city-flood.html, September 10, 2008.

67. Farhar 1974.

68. Schneider with Mesirow 1976.

69. Kellogg and Schneider 1974.

70. Kellogg and Schneider 1974.

71. Ward 1989.

72. Scheffel and Wernert 1980; Ward 1989.

73. Ward 1989.

74. Kolbert 2006.

CHAPTER 9. THE POSSIBLE FALLIBILITY OF EVEN A STRONG CONSENSUS

1. Kuhn 1970. This is the book's second edition; the first edition was published in 1962.

2. Kuhn 1970, p. 6.

3. Segrè 2002.

4. Parkinson 1985; Segrè 2002.

5. Imbrie and Imbrie 1979.

6. Imbrie and Imbrie 1979; Parkinson 1985, 1996.

7. For example, Hays et al. 1976; Milankovitch 1920, 1941.

8. Emiliani 1955.

9. Panofsky 1956.

10. Broecker et al. 1968, p. 300.

11. Broecker et al. 1968.

12. Imbrie and Imbrie 1979.

13. Budyko 1972.

14. Hays et al. 1976.

15. Broecker et al. 1968; Emiliani 1955; Hays et al. 1976. The latter reference lists 10 publications published within the previous 10 years that provide evidence for the importance of orbital forcings.

16. Paillard 2006.

17. Raymo et al. 2006.

18. Raymo et al. 2006.

19. Raymo et al. 2006.

20. Huybers 2006.

21. Hönisch et al. 2009.

22. Ruddiman 2005.

23. Paillard 2006.

24. Johnson 1997.

25. Johnson 1997.

26. Arrhenius 1896.

27. Imbrie and Imbrie 1979.

28. Fleming 1998.

29. Tyndall 1861.

30. Arrhenius 1896.

31. Arrhenius 1908.

32. Meehl et al. 2007.

33. Arrhenius 1908; Fleming 1998.

34. Fleming 1998.

35. Anderson and Gainor 2006.

36. Fleming 1998.

37. Trenberth et al. 2007.

38. Callendar 1938.

39. Fleming 1998.

40. Ewing and Donn 1956.

41. Friedan 1958.

42. Panofsky 1956.

43. Revelle and Suess 1957.

44. Revelle and Suess 1957, p. 19.

45. Mitchell 1961.

46. Lomborg 2007.

47. Anderson and Gainor 2006.

48. Douglas 1975, p. 139.

49. Anderson and Gainor 2006.

50. Fleming 1998, p. 133.

51. Fleming 1998.

52. Bryson 1974.

53. Gribbin 1977.
54. Rasool and Schneider 1971, p. 138.
55. Rasool and Schneider 1971, p. 138. Schneider has been criticized for having suggested a coming ice age in this 1971 paper and then later becoming one of the most prominent scientists warning of global warming. I most certainly do not share in this criticism. Quite the contrary, I would far prefer for a scientist to change his or her mind when the facts warrant it than to hold firm to a position that has become unjustified. In fact, I feel the far greater concern is that scientists and others become too bound to a position and are not willing to change despite new information.
56. Broecker and Kunzig 2008, p. 63.
57. Gribbin 1977, p. 68.
58. Gribbin 1977, p. 137.
59. Gribbin 1977.
60. Peterson et al. 2008.
61. Sellers 1969, p. 399.
62. Sellers 1969, p. 398.
63. Budyko 1972.
64. Broecker 1975, p. 460.
65. Kellogg and Schneider 1974.
66. Schneider with Mesirow 1976.
67. Schneider 1989.
68. Kerr 1989.
69. Manabe and Wetherald 1967.
70. Manabe and Wetherald 1975.
71. Hansen et al. 1984.
72. Meehl et al. 2007.
73. For example, Alexeev et al. 2005; Hall 2004; Parkinson 2004.
74. Parkinson and Kellogg 1979.
75. Parkinson and Bindschadler 1984.
76. For instance, William Ruddiman argues that the greenhouse gases already released into the atmosphere through human activities have prevented the Earth system from sinking into a new Northern Hemisphere ice age. In fact, he suggests that the renewed glaciation would have arisen many centuries ago if human activities had not interfered. Further, he relates the cooling during the Little Ice Age to reduced human population (and the resultant reduced greenhouse gas emissions) during the major pandemics of the Middle Ages. Still, he, like many others, is concerned about continued global warming (Ruddiman 2005).

CHAPTER 10. THE UNKNOWN FUTURE: MODEL LIMITATIONS

1. Detailed overviews of climate modeling can be found in Trenberth 1992 and in Washington and Parkinson 2005.
2. For example, Jansen et al. 2007; Randall et al. 2007.
3. Cotton and Pielke 2007, p. 247.
4. Randall et al. 2007.
5. Randall et al. 2007.
6. Cess et al. 1989.
7. Gates 1992.
8. Randall et al. 2007.
9. Randall et al. 2007; the quoted segments are specifically from pp. 623 and 625. See also Guilyardi et al. 2009 regarding progress and continuing difficulties in ENSO simulations.
10. Randall et al. 2007.
11. Cerveny et al. 2007.
12. Randall et al. 2007, p. 600.
13. Reichler and Kim 2008.

14. Randall et al. 2007, p. 600.
15. Maykut and Untersteiner 1971.
16. Parkinson 1983.
17. Martinson et al. 1981.
18. Borino 1983, p. 4.
19. Parkinson 1983.
20. Manabe and Wetherald 1975.
21. Walter et al. 2007a.
22. Running 2008.
23. Murphy et al. 2004.
24. Ohfuchi et al. 2007; Spencer et al. 2007.
25. Cotton and Pielke 2007.
26. Winckler et al. 2008.
27. Winckler et al. 2008.
28. Ohfuchi et al. 2007.
29. Elliott et al. 1994.
30. Parkinson 1996.
31. For example, Manabe and Stouffer 1980; Rind et al. 1995; Senior and Mitchell 1993.
32. Rind et al. 1995.
33. For example, Murphy and Mitchell 1995; Washington and Meehl 1989.
34. Pollard and Thompson 1994.
35. Flato and Boer 2001.
36. For example, Walsh 1991.
37. Cess et al. 1991.
38. Santee 2007.
39. Farman et al. 1985.
40. Shindell et al. 2001.
41. Shindell et al. 1999.
42. Parizek and Alley 2004.
43. Parizek and Alley 2004.
44. Pfeffer et al. 2008.
45. Walter et al. 2007b.
46. Walter et al. 2007b.
47. Walter et al. 2006, 2007a.
48. Keppler et al. 2006.
49. Spencer 2008.
50. Wentz et al. 2007.
51. Walsh et al. 2002.
52. Spencer 2008.
53. Stroeve et al. 2007.
54. Allan and Soden 2007.
55. Zhang et al. 2007.
56. Yu and Weller 2007.
57. Allan and Soden 2008.
58. Rahmstorf 2007.
59. Meehl et al. 2007.
60. Murphy et al. 2004.
61. Stainforth et al. 2005.
62. Roe and Baker 2007.
63. For example, Brohan et al. 2006; Hansen and Lebedeff 1987; Hansen et al. 1999, 2006; Jones et al. 1986, 1999; Rayner et al. 2003.
64. For example, Pielke et al. 2007.
65. Mann et al. 1999.
66. For example, Parkinson et al. 1987; Zwally et al. 1983a.
67. Wielicki et al. 2002.
68. Xie 2004.
69. Running 2008.
70. Deckert and Dameris 2008; Rosenlof and Reid 2008.
71. Rind et al. 1995.
72. Running 2008.

CHAPTER 11. COMPOUNDING SOCIAL PRESSURES

1. For example, Turner et al. 2009.
2. Kukla and Gavin 1981.
3. Parkinson and Cavalieri 2008.
4. Zwally et al. 1983b.
5. Walker 2006, p. 802.
6. Serreze and Francis 2006, p. 69.
7. Broecker and Kunzig 2008, p. 58.
8. Monbiot 2007, chap. 2.
9. Spencer 2008, p. 176.
10. Spencer 2008.

11. Lindzen 2006, p. A14.
12. Lindzen 2006, p. A14.
13. Lindzen 2006.
14. Lindzen 2006, p. A14.
15. Lindzen 2006, p. A14.
16. Lomborg 2007.
17. Lomborg 2007, p. x.
18. Rothrock et al. 1999.
19. Yu et al. 2004, p. 11.
20. For example, Morison et al. 2007.
21. Rothrock et al. 1999; Yu et al. 2004.
22. Comiso et al. 2008.
23. Zwally et al. 1989.
24. Douglas et al. 1990.
25. Zwally 1989.
26. For example, Kuhn 1970.
27. Atlas 1977, p. 143.
28. Jones 2007, p. 1327.
29. Wilford 2000a.
30. Wilford 2000a.
31. Brooke 2000.
32. Wilford 2000a, p. A1.
33. Wilford 2000b.
34. Bryden et al. 2005, p. 655.
35. Kerr 2006.
36. Kerr 2006.
37. Hayes and Grossman 2006.
38. Goldin 1996.
39. Hayes and Grossman 2006.
40. Hayes and Grossman 2006, p. 168.
41. Hayes and Grossman 2006, p. 171.
42. Hayes and Grossman 2006, p. 170.

43. Schneider 1989, p. xi.
44. Schneider 1997, p. 86.
45. For example, Hayes and Grossman 2006; Merlis 2003.
46. Quotes from Serreze and Francis 2006, p. 69; Walker 2006, p. 802; and Wilford 2000a, p. A1, respectively.
47. Walker 2006.
48. For example, Foley 2005; Lindsay and Zhang 2005.
49. For example, Hayes and Grossman 2006.
50. Begley 2007, p. 22.
51. For example, see Spencer 2008.
52. Lomborg 2007.
53. Samuelson 2007, p. 47.
54. Samuelson 2007, p. 47.
55. Schneider with Mesirow 1976, p. 136.
56. This book was written, peer-reviewed, and revised well before the release in November 2009 of the University of East Anglia e-mails that form the core of what soon came to be known as "Climategate." However unfortunate the circumstances were surrounding the release of the e-mails, these released e-mails from the mainstream climate community confirm several of the points made in chapter 11 regarding the treatment of the skeptics and the added difficulties faced by them in the peer-review process.

CHAPTER 12. WHAT ARE THE ALTERNATIVES?

1. Pala 2006.
2. Stern 2005.
3. For example, Newman et al. 2006.
4. Newman et al. 2006.
5. Newman et al. 2006, p. 1.

6. Arguez et al. 2007.
7. Newman et al. 2006.
8. Newman 2008.
9. Dow and Downing 2006.
10. Murdoch 2007.

11. From an interview with George Jaksch of Chiquita International, as recorded in Rainforest Alliance 2007.

12. Monbiot 2007, p. 161.

13. Monbiot 2007.

14. Monbiot 2007, p. 161.

15. Kolbert 2006.

16. Klebnikov 2006.

17. Dow and Downing 2006.

18. For example, Diamond 2005a.

19. Grímsson 2008.

20. Grímsson 2008, p. 236.

21. Dow and Downing 2006.

22. The full text of the UNFCCC is available at http://unfccc.int.

23. Kolbert 2006.

24. For example, Cotton and Pielke 2007.

25. Rosenthal 2008.

26. Kolbert 2006.

27. Kolbert 2006.

28. For example, Eshel and Martin 2006.

29. MacKay 2009.

30. MacKay 2009.

31. MacKay 2009.

32. Monbiot 2007.

33. MacKay 2009.

34. Monbiot 2007.

35. Monbiot 2007.

36. Lester 2007.

37. Lester 2007, p. 36.

38. Monbiot 2007.

39. Seitter 2007.

40. Hamill 2007.

41. Gore 2006, pp. 305–321.

42. Metz et al. 2007.

43. Among the many additional sources containing suggestions for reducing our global warming impacts are Dow and Downing 2006, parts 5 and 6, and numerous informative websites that can be located by using Internet search engines on topics such as "carbon footprint reduction" and "alternative energy."

44. MacKay 2009.

45. The July 2009 G8 statement is available at the G8 website at http://www.g8italia2009.it.

46. Pacala and Socolow 2004.

47. Pacala and Socolow 2004: Socolow et al. 2004.

48. Pacala and Socolow 2004; Socolow et al. 2004.

49. Socolow et al. 2004.

50. Socolow et al. 2004.

51. Pacala and Socolow 2004.

52. Monbiot 2007.

53. Monbiot 2007.

54. Monbiot 2007.

55. Monbiot 2007.

56. Monbiot 2007.

57. Monbiot 2007; Service 2009.

58. Monbiot 2007, p. 203.

References

Aagaard, K., and L. K. Coachman, 1975: "Toward an ice-free Arctic Ocean," *Eos, Transactions, American Geophysical Union*, vol. 56, no. 7, pp. 484–486.

Abdalati, W., W. Krabill, E. Frederick, S. Manizade, C. Martin, J. Sonntag, R. Swift, R. Thomas, J. Yungel, and R. Koerner, 2004: "Elevation changes of ice caps in the Canadian Arctic Archipelago," *Journal of Geophysical Research*, vol. 109, F04007, doi:10.1029/2003JF000045.

ACIA, 2005: *Arctic Climate Impact Assessment*. Cambridge: Cambridge University Press, 1,042 pp.

Alexander, W. H., 2001: *Runaway Pond: The Complete Story, A Compilation of Resources*. Glover, VT: Glover Historical Society, 90 pp.

Alexeev, V. A., P. L. Langen, and J. R. Bates, 2005: "Polar amplification of surface warming on an aquaplanet in 'ghost forcing' experiments without sea ice feedbacks," *Climate Dynamics*, vol. 24, no. 7, pp. 655–666.

Allan, R. P., and B. J. Soden, 2007: "Large discrepancy between observed and simulated precipitation trends in the ascending and descending branches of the tropical circulation," *Geophysical Research Letters*, vol. 34, L18705, doi:10.1029/2007GL031460, 6 pp.

———, 2008: "Atmospheric warming and the amplification of precipitation extremes," *Science*, vol. 321, no. 5895, pp. 1481–1484.

Alley, R. B., 2000: *The Two-Mile Time Machine: Ice Cores, Abrupt Climate Change, and Our Future*. Princeton, NJ: Princeton University Press, 229 pp.

———, 2007: "Wally was right: Predictive ability of the North Atlantic 'conveyor belt' hypothesis for abrupt climate change," *Annual Review of Earth and Planetary Sciences*, vol. 35, pp. 241–272, doi:10.1146/annurev.earth.35.081006.131524.

Alley, R. B., P. A. Mayewski, T. Sowers, M. Stuiver, K. C. Taylor, and P. U. Clark, 1997: "Holocene climatic instability: A prominent, widespread event 8200 yr ago," *Geology*, vol. 25, no. 6, pp. 483–486.

Allison, I., and J. A. Peterson, 1988: "Glaciers of Irian Jaya, Indonesia." In: Williams, R. S., Jr., and J. G. Ferrigno, editors, *Satellite Image Atlas of Glaciers of the World: Glaciers of Irian Jaya, Indonesia, and New Zealand.* Washington, DC: U.S. Geological Survey, pp. H1–H23.

Alvarez, L. W., W. Alvarez, F. Asaro, and H. V. Michel, 1980: "Extraterrestrial cause for the Cretaceous-Tertiary extinction," *Science,* vol. 208, no. 4448, pp. 1095–1108.

Anbar, A. D., Y. Duan, T. W. Lyons, G. L. Arnold, B. Kendall, R. A. Creaser, A. J. Kaufman, G. W. Gordon, C. Scott, J. Garvin, and R. Buick, 2007: "A whiff of oxygen before the great oxidation event?" *Science,* vol. 317, no. 5846, pp. 1903–1906.

Andersen, K. K., M. Bigler, H. B. Clausen, D. Dahl-Jensen, S. J. Johnsen, S. O. Rasmussen, I. Seierstad, J. P. Steffensen, A. Svensson, B. M. Vinther, S. M. Davies, R. Muscheler, F. Parrenin, and R. Röthlisberger, 2007: "A 60 000 year Greenland stratigraphic ice core chronology," *Climate of the Past Discussions,* vol. 3, pp. 1235–1260.

Anderson, R. W., and D. Gainor, 2006: "Fire and ice: Journalists have warned of climate change for 100 years," *Business and Media Institute,* May 17, 2006, 18 pp., downloaded December 28, 2007, from http://downloads.heartland.org/20560.pdf.

Andreae, M. O., C. D. Jones, and P. M. Cox, 2005: "Strong present-day aerosol cooling implies a hot future," *Nature,* vol. 435, no. 7046, pp. 1187–1190, doi:10.1038/nature03671.

Anklin, M., J. M. Barnola, J. Beer, T. Blunier, J. Chappellaz, H. B. Clausen, D. Dahl-Jensen, W. Dansgaard, M. DeAngelis, R. J. Delmas, P. Duval, M. Fratta, A. Fuchs, K. Fuhrer, N. Gundestrup, C. Hammer, P. Iversen, S. Johnsen, J. Jouzel, J. Kipfstuhl, M. Legrand, C. Lorius, V. Maggi, H. Miller, J. C. Moore, H. Oeschger, G. Orombelli, D. A. Peel, G. Raisbeck, D. Raynaud, C. Schott-Hvidberg, J. Schwander, H. Shoji, R. Souchez, B. Stauffer, J. P. Steffensen, M. Stievenard, A. Sveinbjörnsdottir, T. Thorsteinsson, and E. W. Wolff, 1993: "Climate instability during the last interglacial period recorded in the GRIP ice core," *Nature,* vol. 364, no. 6434, pp. 203–207.

Apt, J., M. Helfert, and J. Wilkinson, 1996: *Orbit: NASA Astronauts Photograph the Earth.* Washington, DC: National Geographic Society, 224 pp.

Archer, D., 2007: "Methane hydrate stability and anthropogenic climate change," *Biogeosciences,* vol. 4, pp. 521–544.

Arendt, A. A., K. A. Echelmeyer, W. D. Harrison, C. S. Lingle, and V. B. Valentine, 2002: "Rapid wastage of Alaska glaciers and their contribution to rising sea level," *Science,* vol. 297, no. 5580, pp. 382–386.

Arguez, A., H. J. Diamond, F. Fetterer, A. Horvitz, and J. M. Levy, editors, 2007: "State of the climate in 2006," *Bulletin of the American Meteorological Society* (supplement), vol. 88, no. 6, pp. S1–S135.

Arrhenius, S., 1896: "On the influence of carbonic acid in the air upon the temperature of the ground," *The London, Edinburgh, and Dublin Philosophical Magazine and Journal of Science,* fifth series, vol. 41, no. 251, pp. 237–276.

———, 1908: *Worlds in the Making: The Evolution of the Universe,* as translated by Dr. H. Borns. New York and London: Harper & Brothers, 230 pp.

Atlas, D., 1977: "The paradox of hail suppression," *Science*, vol. 195, no. 4274, pp. 139–145.

Baksi, A. K., and E. Farrar, 1991: "$^{40}Ar/^{39}Ar$ dating of the Siberian Traps, USSR: Evaluation of the ages of the two major extinction events relative to episodes of flood-basalt volcanism in the USSR and the Deccan Traps, India," *Geology*, vol. 19, no. 5, pp. 461–464.

Balter, M., 2007: "Seeking agriculture's ancient roots," *Science*, vol. 316, no. 5833, pp. 1830–1835.

Bambach, R. K., 2006: "Phanerozoic biodiversity mass extinctions," *Annual Review of Earth and Planetary Science*, vol. 34, pp. 127–155, doi:10.1146/annurev.earth.33.092203.122654.

Barber, D. C., A. Dyke, C. Hillaire-Marcel, A. E. Jennings, J. T. Andrews, M. W. Kerwin, G. Bilodeau, R. McNeely, J. Southon, M. D. Morehead, and J.-M. Gagnon, 1999: "Forcing of the cold event of 8,200 years ago by catastrophic drainage of Laurentide lakes," *Nature*, vol. 400, no. 6742, pp. 344–348.

Barker, T., I. Bashmakov, A. Alharthi, M. Amann, L. Cifuentes, J. Drexhage, M. Duan, O. Edenhofer, B. Flannery, M. Grubb, M. Hoogwijk, F. I. Ibitoye, C. J. Jepma, W. A. Pizer, and K. Yamaji, 2007: "Mitigation from a cross-sectoral perspective." In: Metz, B., O. Davidson, P. Bosch, R. Dave, and L. Meyer, editors, *Climate Change 2007: Mitigation of Climate Change. Contribution of Working Group III to the Fourth Assessment Report of the Intergovernmental Panel on Climate Change.* Cambridge: Cambridge University Press, pp. 619–690.

Barnett, T. P., J. C. Adam, and D. P. Lettenmaier, 2005: "Potential impacts of a warming climate on water availability in snow-dominated regions," *Nature*, vol. 438, no. 7066, pp. 303–309.

Barnett, T. P., D. W. Pierce, H. G. Hidalgo, C. Bonfils, B. D. Santer, T. Das, G. Bala, A. W. Wood, T. Nozawa, A. A. Mirin, D. R. Cayan, and M. D. Dettinger, 2008: "Human-induced changes in the hydrology of the western United States," *Science*, vol. 319, no. 5866, pp. 1080–1083.

Barrows, T. T., S. J. Lehman, L. K. Fifield, and P. De Deckker, 2007: "Absence of cooling in New Zealand and the adjacent ocean during the Younger Dryas chronozone," *Science*, vol. 318, no. 5847, pp. 86–89.

Battan, L. J., 1969: "Weather modification in the U.S.S.R.—1969," *Bulletin of the American Meteorological Society*, vol. 50, no. 12, pp. 924–945.

Becker, L., R. J. Poreda, A. G. Hunt, T. E. Bunch, and M. Rampino, 2001: "Impact event at the Permian-Triassic boundary: Evidence from extraterrestrial noble gases in fullerenes," *Science*, vol. 291, no. 5508, pp. 1530–1533.

Begley, S., 2007: "The truth about denial," *Newsweek*, August 13, 2007, pp. 20–29.

Bentley, C. R., 1997: "Rapid sea-level rise soon from West Antarctic ice sheet collapse?" *Science*, vol. 275, no. 5303, pp. 1077–1078.

Benton, M., 1993a: "Dinosaur summer." In: Gould, S. J., editor, *The Book of Life*. New York: W. W. Norton, pp. 126–167.

———, 1993b: "Four feet on the ground." In: Gould, S. J., editor, *The Book of Life*. New York: W. W. Norton, pp. 78–125.

——, 1993c: "The rise of the fishes." In: Gould, S. J., editor, *The Book of Life*. New York: W. W. Norton, pp. 64–77.

——, 2008: *When Life Nearly Died: The Greatest Mass Extinction of All Time*. London: Thames & Hudson, 336 pp.

Bergreen, L., 2007: *Marco Polo: From Venice to Xanadu*. New York: Alfred A. Knopf, 415 pp.

Berner, R. A., 1993: "Paleozoic atmospheric CO_2: Importance of solar radiation and plant evolution," *Science*, vol. 261, no. 5117, pp. 68–70.

Beschta, R. L., and W. J. Ripple, 2006: "River channel dynamics following extirpation of wolves in northwestern Yellowstone National Park, USA," *Earth Surface Processes and Landforms*, vol. 31, pp. 1525–1539.

Betts, R. A., 2000: "Offset of the potential carbon sink from boreal forestation by decreases in surface albedo," *Nature*, vol. 408, no. 6809, pp. 187–190.

Bindoff, N. L., J. Willebrand, V. Artale, A. Cazenave, J. M. Gregory, S. Gulev, K. Hanawa, C. Le Quéré, S. Levitus, Y. Nojiri, C. K. Shum, L. D. Talley, and A. S. Unnikrishnan, 2007: "Observations: Oceanic climate change and sea level." In: Solomon, S., D. Qin, M. Manning, Z. Chen, M. Marquis, K. B. Averyt, M. Tignor, and H. L. Miller, editors, *Climate Change 2007: The Physical Science Basis. Contribution of Working Group I to the Fourth Assessment Report of the Intergovernmental Panel on Climate Change*. Cambridge: Cambridge University Press, pp. 385–432.

Bindschadler, R. A., and C. R. Bentley, 2002: "On thin ice?" *Scientific American*, vol. 287, no. 6, pp. 98–105.

Bindschadler, R. A., and P. L. Vornberger, 1990: "AVHRR imagery reveals Antarctic ice dynamics," *Eos, Transactions, American Geophysical Union*, vol. 71, no. 23, pp. 741–742.

Blunier, T., J. Chappellaz, J. Schwander, A. Dällenbach, B. Stauffer, T. F. Stocker, D. Raynaud, J. Jouzel, H. B. Clausen, C. U. Hammer, and S. J. Johnsen, 1998: "Asynchrony of Antarctic and Greenland climate change during the last glacial period," *Nature*, vol. 394, no. 6695, pp. 739–743.

Bluth, G. J. S., S. D. Doiron, C. C. Schnetzler, A. J. Krueger, and L. S. Walter, 1992: "Global tracking of the SO_2 clouds from the June, 1991 Mount Pinatubo eruptions," *Geophysical Research Letters*, vol. 19, no. 2, pp. 151–154.

Bond, G., W. Broecker, S. Johnsen, J. McManus, L. Labeyrie, J. Jouzel, and G. Bonani, 1993: "Correlations between climate records from North Atlantic sediments and Greenland ice," *Nature*, vol. 365, no. 6442, pp. 143–147.

Bond, G., H. Heinrich, W. Broecker, L. Labeyrie, J. McManus, J. Andrews, S. Huon, R. Jantschik, S. Clasen, C. Simet, K. Tedesco, M. Klas, G. Bonani, and S. Ivy, 1992: "Evidence for massive discharges of icebergs into the North Atlantic Ocean during the last glacial period," *Nature*, vol. 360, no. 6401, pp. 245–249.

Borino, B., 1983: "UFO base found under Antarctic," *Globe*, vol. 30, no. 3, p. 4.

Bousquet, P., P. Ciais, J. B. Miller, E. J. Dlugokencky, D. A. Hauglustaine, C. Prigent, G. R. Van der Werf, P. Peylin, E.-G. Brunke, C. Carouge, R. L. Langenfelds, J. Lathière, F. Papa,

M. Ramonet, M. Schmidt, L. P. Steele, S. C. Tyler, and J. White, 2006: "Contribution of anthropogenic and natural sources to atmospheric methane variability," *Nature*, vol. 443, no. 7110, pp. 439–443.

Bowen, M., 1998: "Thompson's ice corps," *Natural History*, vol. 107, no. 1, pp. 28–41.

———, 2005: *Thin Ice: Unlocking the Secrets of Climate in the World's Highest Mountains*. New York: Henry Holt and Company, 463 pp.

Boyd, P. W., T. Jickells, C. S. Law, S. Blain, E. A. Boyle, K. O. Buesseler, K. H. Coale, J. J. Cullen, H. J. W. de Baar, M. Follows, M. Harvey, C. Lancelot, M. Levasseur, N. P. J. Owens, R. Pollard, R. B. Rivkin, J. Sarmiento, V. Schoemann, V. Smetacek, S. Takeda, A. Tsuda, S. Turner, and A. J. Watson, 2007: "Mesoscale iron enrichment experiments 1993–2005: Synthesis and future directions," *Science*, vol. 315, no. 5812, pp. 612–617.

Bradbury, I. K., 1998: *The Biosphere*, second edition. Chichester: John Wiley & Sons, 254 pp.

Braun, H., M. Christl, S. Rahmstorf, A. Ganopolski, A. Mangini, C. Kubatzki, K. Roth, and B. Kromer, 2005: "Possible solar origin of the 1,470-year glacial climate cycle demonstrated in a coupled model," *Nature*, vol. 438, no. 7065, pp. 208–211.

Broecker, W. S., 1975: "Climatic change: Are we on the brink of a pronounced global warming?" *Science*, vol. 189, no. 4201, pp. 460–463.

———, 1990: "Comment on 'Iron deficiency limits phytoplankton growth in Antarctic waters' by John H. Martin et al.," *Global Biogeochemical Cycles*, vol. 4, no. 1, pp. 3–4.

———, 1991: "The great ocean conveyor," *Oceanography*, vol. 4, no. 2, pp. 79–89.

———, 1995a: "Chaotic climate," *Scientific American*, vol. 273, no. 5, pp. 62–68.

———, 1995b: "Cooling the tropics," *Nature*, vol. 376, no. 6537, pp. 212–213.

———, 1998: "Paleocean circulation during the last deglaciation: A bipolar seesaw?" *Paleoceanography*, vol. 13, no. 2, pp. 119–121.

———, 2006: "Was the Younger Dryas triggered by a flood?" *Science*, vol. 312, no. 5777, pp. 1146–1148.

Broecker, W. S., M. Andree, G. Bonani, W. Wolfli, H. Oeschger, and M. Klas, 1988, "Can the Greenland climatic jumps be identified in records from ocean and land?" *Quaternary Research*, vol. 30, no. 1, pp. 1–6.

Broecker, W. S., G. Bond, M. Klas, G. Bonani, and W. Wolfli, 1990: "A salt oscillator in the glacial Atlantic? 1. The concept," *Paleoceanography*, vol. 5, no. 4, pp. 469–477.

Broecker, W. S., M. Ewing, and B. C. Heezen, 1960: "Evidence for an abrupt change in climate close to 11,000 years ago," *American Journal of Science*, vol. 258, no. 6, pp. 429–448.

Broecker, W. S., and S. Hemming, 2001: "Climate swings come into focus," *Science*, vol. 294, no. 5550, pp. 2308–2309.

Broecker, W. S., and R. Kunzig, 2008: *Fixing Climate: What Past Climate Changes Reveal About the Current Threat—and How to Counter It*. New York: Hill and Wang, 253 pp.

Broecker, W. S., D. M. Peteet, and D. Rind, 1985: "Does the ocean-atmosphere system have more than one stable mode of operation?" *Nature*, vol. 315, no. 6014, pp. 21–26.

Broecker, W. S., D. L. Thurber, J. Goddard, T.-L. Ku, R. K. Matthews, and K. J. Mesolella, 1968: "Milankovitch hypothesis supported by precise dating of coral reefs and deep-sea sediments," *Science*, vol. 159, no. 3812, pp. 297–300.

Brohan, P., J. J. Kennedy, I. Harris, S. F. B. Tett, and P. D. Jones, 2006: "Uncertainty estimates in regional and global observed temperature changes: A new data set from 1850," *Journal of Geophysical Research*, vol. 111, D12106, doi:10.1029/2005JD006548.

Brooke, J., 2000: "Through Northwest Passage in a month, ice-free," *New York Times*, International section, September 5, 2000, p. A3.

Bryden, H. L., H. R. Longworth, and S. A. Cunningham, 2005: "Slowing of the Atlantic meridional overturning circulation at 25°N," *Nature*, vol. 438, no. 7068, pp. 655–657.

Bryson, R. A., 1974: "A perspective on climatic change," *Science*, vol. 184, no. 4138, pp. 753–760.

Budyko, M. I., 1972: "The future climate," *Eos, Transactions, American Geophysical Union*, vol. 53, no. 10, pp. 868–874.

———, 1977: *Climatic Changes*. Washington, DC: American Geophysical Union, 261 pp.

Buesseler, K. O., S. C. Doney, D. M. Karl, P. W. Boyd, K. Caldeira, F. Chai, K. H. Coale, H. J. W. de Baar, P. G. Falkowski, K. S. Johnson, R. S. Lampitt, A. F. Michaels, S. W. A. Naqvi, V. Smetacek, S. Takeda, and A. J. Watson, 2008: "Ocean iron fertilization—Moving forward in a sea of uncertainty," *Science*, vol. 319, no. 5860, p. 162.

Buzin, V. A., A. B. Klaven, and Z. D. Kopaliani, 2007: "Laboratory modelling of ice jam floods on the Lena River." In: Vasiliev, O. F., P. H. A. J. M. van Gelder, E. J. Plate, and M. V. Bolgov, editors, *Extreme Hydrological Events: New Concepts for Security*. Dordrecht: Springer, pp. 269–277.

Caldeira, K., and J. F. Kasting, 1992: "Susceptibility of the early Earth to irreversible glaciation caused by carbon dioxide clouds," *Nature*, vol. 359, no. 6392, pp. 226–228.

Calder, N., 1976: *The Weather Machine: How Our Weather Works and Why It Is Changing*. New York: Viking, 143 pp.

Callendar, G. S., 1938: "The artificial production of carbon dioxide and its influence on temperature," *The Quarterly Journal of the Royal Meteorological Society*, vol. 64, no. 275, pp. 223–240.

Campbell, I. H., G. K. Czamanske, V. A. Fedorenko, R. I. Hill, and V. Stepanov, 1992: "Synchronism of the Siberian Traps and the Permian-Triassic boundary," *Science*, vol. 258, no. 5089, pp. 1760–1763.

Cavalieri, D. J., and C. L. Parkinson, 2008: "Antarctic sea ice variability and trends, 1979–2006," *Journal of Geophysical Research—Oceans*, vol. 113, C07004, doi:10.1029/2007JC004564, 19 pp.

Cerveny, R. S., J. Lawrimore, R. Edwards, and C. Landsea, 2007: "Extreme weather records," *Bulletin of the American Meteorological Society*, vol. 88, no. 6, pp. 853–860.

Cess, R. D., G. L. Potter, J. P. Blanchet, G. J. Boer, S. J. Ghan, J. T. Kiehl, H. Le Treut, Z.-X. Li, X.-Z. Liang, J. F. B. Mitchell, J.-J. Morcrette, D. A. Randall, M. R. Riches, E. Roeckner,

U. Schlese, A. Slingo, K. E. Taylor, W. M. Washington, R. T. Wetherald, and I. Yagai, 1989: "Interpretation of cloud-climate feedback as produced by 14 atmospheric general circulation models," *Science*, vol. 425, no. 4917, pp. 513–516.

Cess, R. D., G. L. Potter, M.-H. Zhang, J.-P. Blanchet, S. Chalita, R. Colman, D. A. Dazlich, A. D. Del Genio, V. Dymnikov, V. Galin, D. Jerrett, E. Keup, A. A. Lacis, H. Le Treut, X.-Z. Liang, J.-F. Mahfouf, B. J. McAvaney, V. P. Meleshko, J. F. B. Mitchell, J.-J. Morcrette, P. M. Norris, D. A. Randall, L. Rikus, E. Roeckner, J.-F. Royer, U. Schlese, D. A. Sheinin, J. M. Slingo, A. P. Sokolov, K. E. Taylor, W. M. Washington, R. T. Wetherald, and I. Yagai, 1991: "Interpretation of snow-climate feedback as produced by 17 general circulation models," *Science*, vol. 253, no. 5022, pp. 888–892.

Chambers, L. H., B. Lin, and D. F. Young, 2002: "Examination of new CERES data for evidence of tropical iris feedback," *Journal of Climate*, vol. 15, no. 24, pp. 3719–3726.

Changnon, S. A., Jr., and J. L. Ivens, 1981: "History repeated: The forgotten hail cannons of Europe," *Bulletin of the American Meteorological Society*, vol. 62, no. 3, pp. 368–375.

Chao, B. F., Y. H. Wu, and Y. S. Li, 2008: "Impact of artificial reservoir water impoundment on global sea level," *Science*, vol. 320, no. 5873, pp. 212–214.

Chappell, C. F., L. O. Grant, and P. W. Mielke Jr., 1971: "Cloud seeding effects on precipitation intensity and duration of wintertime orographic clouds," *Journal of Applied Meteorology*, vol. 10, no. 5, pp. 1006–1010.

Charmantier, A., R. H. McCleery, L. R. Cole, C. Perrins, L. E. B. Kruuk, and B. C. Sheldon, 2008: "Adaptive phenotypic plasticity in response to climate change in a wild bird population," *Science*, vol. 320, no. 5877, pp. 800–803.

Christensen, J. H., B. Hewitson, A. Busuioc, A. Chen, X. Gao, I. Held, R. Jones, R. K. Kolli, W.-T. Kwon, R. Laprise, V. Magaña Rueda, L. Mearns, C. G. Menéndez, J. Räisänen, A. Rinke, A. Sarr, and P. Whetton, 2007: "Regional climate projections." In: Solomon, S., D. Qin, M. Manning, Z. Chen, M. Marquis, K. B. Averyt, M. Tignor, and H. L. Miller, editors, *Climate Change 2007: The Physical Science Basis. Contribution of Working Group I to the Fourth Assessment Report of the Intergovernmental Panel on Climate Change.* Cambridge: Cambridge University Press, pp. 847–940.

Christianson, G. E., 1999: *Greenhouse: The 200-Year Story of Global Warming.* New York: Walker and Company, 305 pp.

Chu, S., 2009: "Carbon capture and sequestration," *Science*, vol. 325, no. 5948, p. 1599.

Church, J. A., J. M. Gregory, P. Huybrechts, M. Kuhn, K. Lambeck, M. T. Nhuan, D. Qin, and P. L. Woodworth, 2001: "Changes in sea level." In: Houghton, J. T., Y. Ding, D. J. Griggs, M. Noguer, P. J. van der Linden, X. Dai, K. Maskell, and C. A. Johnson, editors, *Climate Change 2001: The Scientific Basis. Contribution of Working Group I to the Third Assessment Report of the Intergovernmental Panel on Climate Change.* Cambridge: Cambridge University Press, pp. 639–693.

Cicerone, R. J., 2006: "Geoengineering: Encouraging research and overseeing implementation: An editorial comment," *Climatic Change*, vol. 77, pp. 221–226.

Cicerone, R. J., S. Elliott, and R. P. Turco, 1991: "Reduced Antarctic ozone depletions in a model with hydrocarbon injections," *Science*, vol. 254, no. 5035, pp. 1191–1194.

Civeyrel, L., and D. Simberloff, 1996: "A tale of two snails: Is the cure worse than the disease?" *Biodiversity and Conservation*, vol. 5, no. 10, pp. 1231–1252.

Clarke, B., J. Murray, and M. S. Johnson, 1984: "The extinction of endemic species by a program of biological control," *Pacific Science*, vol. 38, no. 2, pp. 97–104.

Clemens, S. C., 2005: "Millennial-band climate spectrum resolved and linked to centennial-scale solar cycles," *Quaternary Science Reviews*, vol. 24, no. 5, pp. 521–531.

Comiso, J. C., C. L. Parkinson, R. Gersten, and L. Stock, 2008: "Accelerated decline in the Arctic sea ice cover," *Geophysical Research Letters*, vol. 35, L01703, doi:10.1029/2007GL031972.

Cook, A. J., A. J. Fox, D. G. Vaughan, and J. G. Ferrigno, 2005: "Retreating glacier fronts on the Antarctic Peninsula over the past half-century," *Science*, vol. 308, no. 5721, pp. 541–544.

Cotton, W. R., and R. A. Pielke Sr., 2007: *Human Impacts on Weather and Climate*, second edition. Cambridge: Cambridge University Press, 308 pp.

Crutzen, P. J., 2006: "Albedo enhancement by stratospheric sulfur injections: A contribution to resolve a policy dilemma?" *Climatic Change*, vol. 77, pp. 211–219.

Cubasch, U., G. A. Meehl, G. J. Boer, R. J. Stouffer, M. Dix, A. Noda, C. A. Senior, S. Raper, and K. S. Yap, 2001: "Projections of future climate change." In: Houghton, J. T., Y. Ding, D. J. Griggs, M. Noguer, P. J. van der Linden, X. Dai, K. Maskell, and C. A. Johnson, editors, *Climate Change 2001: The Scientific Basis. Contribution of Working Group I to the Third Assessment Report of the Intergovernmental Panel on Climate Change*. Cambridge: Cambridge University Press, pp. 525–582.

Dansgaard, W., H. B. Clausen, N. Gundestrup, C. U. Hammer, S. F. Johnsen, P. M. Kristinsdottir, and N. Reeh, 1982: "A new Greenland deep ice core," *Science*, vol. 218, no. 4579, pp. 1273–1277.

Dansgaard, W., S. J. Johnsen, H. B. Clausen, D. Dahl-Jensen, N. S. Gundestrup, C. U. Hammer, C. S. Hvidberg, J. P. Steffensen, A. E. Sveinbjörnsdottir, J. Jouzel, and G. Bond, 1993: "Evidence for general instability of past climate from a 250-kyr ice-core record," *Nature*, vol. 364, no. 6434, pp. 218–220.

Dansgaard, W., S. J. Johnsen, J. Moller, and C.C. Langway Jr., 1969: "One thousand centuries of climatic record from Camp Century on the Greenland ice sheet," *Science*, vol. 166, no. 3903, pp. 377–380.

Das, S. B., I. Joughin, M. D. Behn, I. M. Howat, M. A. King, D. Lizarralde, and M. P. Bhatia, 2008: "Fracture propagation to the base of the Greenland ice sheet during supraglacial lake drainage," *Science*, vol. 320, no. 5877, pp. 778–781.

Davis, D. L., and J. C. Topping Jr., 2008: "Potential effects of weather extremes and climate change on human health." In: MacCracken, M. C., F. Moore, and J. C. Topping Jr., editors, *Sudden and Disruptive Climate Change: Exploring the Real Risks and How We Can Avoid Them*. London: Earthscan, pp. 39–43.

Deckert, R., and M. Dameris, 2008: "From ocean to stratosphere," *Science*, vol. 322, no. 5898, pp. 53–55.

DeConto, R. M., and D. Pollard, 2003: "Rapid Cenozoic glaciation of Antarctica induced by declining atmospheric CO_2," *Nature*, vol. 421, no. 6920, pp. 245–249.

D'Elia, T., R. Veerapaneni, and S. O. Rogers, 2008: "Isolation of microbes from Lake Vostok accretion ice," *Applied and Environmental Microbiology*, vol. 74, no. 15, pp. 4962–4965.

Denman, K. L., G. Brasseur, A Chidthaisong, P. Ciais, P. M. Cox, R. E. Dickinson, D. Hauglustaine, C. Heinze, E. Holland, D. Jacob, U. Lohmann, S. Ramachandran, P. L. da Silva Dias, S. C. Wofsy, and X. Zhang, 2007: "Couplings between changes in the climate system and biogeochemistry." In: Solomon, S., D. Qin, M. Manning, Z. Chen, M. Marquis, K. B. Averyt, M. Tignor, and H. L. Miller, editors, *Climate Change 2007: The Physical Science Basis. Contribution of Working Group I to the Fourth Assessment Report of the Intergovernmental Panel on Climate Change*. Cambridge: Cambridge University Press, pp. 499–587.

Dennell, R., and W. Roebroeks, 2005: "An Asian perspective on early human dispersal from Africa," *Nature*, vol. 438, no. 7071, pp. 1099–1104, doi:10.1038/nature04259.

Derocher, A. E., 2008: "Polar bears in a warming Arctic." In: MacCracken, M. C., F. Moore, and J. C. Topping Jr., editors, *Sudden and Disruptive Climate Change: Exploring the Real Risks and How We Can Avoid Them*. London: Earthscan, pp. 193–204.

Dessler, A. E., Z. Zhang, and P. Yang, 2008: "Water-vapor climate feedback inferred from climate fluctuations, 2003–2008," *Geophysical Research Letters*, vol. 35, L20704, doi:10.1029/2008GL035333, 4 pp.

De Vernal, A., and C. Hillaire-Marcel, 2008: "Natural variability of Greenland climate, vegetation, and ice volume during the past million years," *Science*, vol. 320, no. 5883, pp. 1622–1625.

Diamond, J., 2005a: *Collapse: How Societies Choose to Fail or Succeed*. New York: Viking, 575 pp.

———, 2005b: *Guns, Germs, and Steel*. New York: W. W. Norton, 518 pp.

Dillehay, T. D., J. Rossen, T. C. Andres, and D. E. Williams, 2007: "Preceramic adoption of peanut, squash, and cotton in northern Peru," *Science*, vol. 316, no. 5833, pp. 1890–1893.

Douglas, B. C., R. E. Cheney, L. Miller, R. W. Agreen, W. E. Carter, and D. S. Robertson, 1990: "Greenland ice sheet: Is it growing or shrinking?" *Science*, vol. 248, no. 4953, p. 288.

Douglas, J. H., 1975: "Climate change: Chilling possibilities," *Science News*, vol. 107, no. 9, pp. 138–140.

Dow, K., and T. E. Downing, 2006: *The Atlas of Climate Change: Mapping the World's Greatest Challenge*. Berkeley: University of California Press, 112 pp.

Driscoll, C. A., M. Menotti-Raymond, A. L. Roca, K. Hupe, W. E. Johnson, E. Geffen, E. H. Harley, M. Delibes, D. Pontier, A. C. Kitchener, N. Yamaguchi, S. J. O'Brien, and D. W. Macdonald, 2007: "The Near Eastern origin of cat domestication," *Science*, vol. 317, no. 5837, pp. 519–523.

Dubcovsky, J., and J. Dvorak, 2007: "Genome plasticity a key factor in the success of polyploid wheat under domestication," *Science*, vol. 316, no. 5833, pp. 1862–1866.

Durant, W., 1957: *The Story of Civilization. Volume 6. The Reformation: A History of European Civilization from Wyclif to Calvin: 1300–1564*. New York: Simon & Schuster, 1,025 pp.

Dyurgerov, M. B., and M. F. Meier, 1997: "Year-to-year fluctuations of global mass balance of small glaciers and their contribution to sea-level changes," *Arctic and Alpine Research*, vol. 29, no. 4, pp. 392–402.

———, 2000: "Twentieth century climate change: Evidence from small glaciers," *Proceedings of the National Academy of Sciences of the United States of America*, vol. 97, no. 4, pp. 1406–1411.

Early, J. T., 1989: "Space-based solar shield to offset greenhouse effect," *Journal of the British Interplanetary Society*, vol. 42, pp. 567–569.

Ekholm, N., 1901: "On the variations of the climate of the geological and historical past and their causes," *Quarterly Journal of the Royal Meteorological Society*, vol. 27, no. 117, pp. 1–62.

Elliott, S., R. J. Cicerone, R. P. Turco, K. Drdla, and A. Tabazadeh, 1994: "Influence of the heterogeneous reaction $HCl + HOCl$ on an ozone hole model with hydrocarbon additions," *Journal of Geophysical Research*, vol. 99, no. D2, pp. 3497–3508.

Elsner, J. B., 2008: "Hurricanes and climate change," *Bulletin of the American Meteorological Society*, vol. 89, no. 5, pp. 677–679.

Emanuel, K., 2005: "Increasing destructiveness of tropical cyclones over the past 30 years," *Nature*, vol. 436, pp. 686–688.

Emanuel, K., R. Sundararajan, and J. Williams, 2008: "Hurricanes and global warming: Results from downscaling IPCC AR4 simulations," *Bulletin of the American Meteorological Society*, vol. 89, no. 3, pp. 347–367.

Emiliani, C., 1955: "Pleistocene temperatures," *Journal of Geology*, vol. 63, pp. 538–578.

———, 1969: "Interglacial high sea levels and the control of Greenland ice by the precession of the equinoxes," *Science*, vol. 166, no. 3912, pp. 1503–1504.

Erwin, D. H., 1994: "The Permo-Triassic extinction," *Nature*, vol. 367, no. 6460, pp. 231–236.

Eshel, G., and P. A. Martin, 2006: "Diet, energy, and global warming," *Earth Interactions*, vol. 10, no. 9, pp. 1–17.

Eshet, Y., M. R. Rampino, and H. Visscher, 1995: "Fungal event and palynological record of ecological crisis and recovery across the Permian-Triassic boundary," *Geology*, vol. 23, no. 11, pp. 967–970.

Evelyn, J., 1661: *Fumifugium, or The Inconvenience of the Air and Smoke of London Dissipated*, extract available at http://www.cf.ac.uk/encap/skilton/nonfic/evelyn01.html, Cardiff University.

Ewing, M., and W. L. Donn, 1956: "A theory of ice ages," *Science*, vol. 123, no. 3207, pp. 1061–1066.

Fagan, B., 2004: *The Long Summer: How Climate Changed Civilization*. New York: Basic Books, 284 pp.

———, 2008: *The Great Warming: Climate Change and the Rise and Fall of Civilizations*. New York: Bloomsbury Press, 282 pp.

Farhar, B. C., 1974: "The impact of the Rapid City flood on public opinion about weather modification," *Bulletin of the American Meteorological Society*, vol. 55, no. 7, pp. 759–764.

Farman, J. C., B. G. Gardiner, and J. D. Shanklin, 1985: "Large losses of total ozone in Antarctica reveal seasonal ClO_x/NO_x interaction," *Nature*, vol. 315, no. 6016, pp. 207–210.

Federer, B., A. Waldvogel, W. Schmid, H. H. Schiesser, F. Hampel, M. Schweingruber, W. Stahel, J. Bader, J. F. Mezeix, N. Doras, G. D'Aubigny, G. DerMegreditchian, and D. Vento, 1986: "Main results of Grossversuch IV," *Journal of Climate and Applied Meteorology*, vol. 25, no. 7, pp. 917–957.

Feely, R. A., C. L. Sabine, K. Lee, W. Berelson, J. Kleypas, V. J. Fabry, and F. J. Millero, 2004: "Impact of anthropogenic CO_2 on the $CaCO_3$ system in the oceans," *Science*, vol. 305, no. 5682, pp. 362–366.

Firestone, R. B., A. West, J. P. Kennett, L. Becker, T. E. Bunch, Z. S. Revay, P. H. Schultz, T. Belgya, D. J. Kennett, J. M. Erlandson, O. J. Dickenson, A. C. Goodyear, R. S. Harris, G. A. Howard, J. B. Kloosterman, P. Lechler, P. A. Mayewski, J. Montgomery, R. Poreda, T. Darrah, S. S. Que Hee, A. R. Smith, A. Stich, W. Topping, J. H. Wittke, and W. S. Wolbach, 2007: "Evidence for an extraterrestrial impact 12,900 years ago that contributed to the megafaunal extinctions and the Younger Dryas cooling," *Proceedings of the National Academy of Sciences of the United States of America*, vol. 104, no. 41, pp. 16016–16021.

Flato, G. M., and G. J. Boer, 2001: "Warming asymmetry in climate change simulations," *Geophysical Research Letters*, vol. 28, no. 1, pp. 195–198.

Fleming, J. R., 1998: *Historical Perspectives on Climate Change*. New York: Oxford University Press, 194 pp.

———, 2004: "Fixing the weather and climate: Military and civilian schemes for cloud seeding and climate engineering." In: Rosner, L., editor, *The Technological Fix: How People Use Technology to Create and Solve Problems*. New York: Routledge, pp. 175–200.

———, 2007: "The climate engineers," *The Wilson Quarterly*, Spring 2007, pp. 46–60.

Fogg, G. E., 1992: *A History of Antarctic Science*. Cambridge: Cambridge University Press, 483 pp.

Foley, J. A., 2005: "Tipping points in the tundra," *Science*, vol. 310, no. 5748, pp. 627–628.

Forster, P., V. Ramaswamy, P. Artaxo, T. Berntsen, R. Betts, D. W. Fahey, J. Haywood, J. Lean, D. C. Lowe, G. Myhre, J. Nganga, R. Prinn, G. Raga, M. Schulz, and R. Van Dorland, 2007: "Changes in atmospheric constituents and in radiative forcing." In: Solomon, S., D. Qin, M. Manning, Z. Chen, M. Marquis, K. B. Averyt, M. Tignor, and H. L. Miller, editors, *Climate Change 2007: The Physical Science Basis. Contribution of Working Group I to the Fourth Assessment Report of the Intergovernmental Panel on Climate Change*. Cambridge: Cambridge University Press, pp. 129–234.

Francis, P., 1993: *Volcanoes: A Planetary Perspective*. New York: Oxford University Press, 443 pp.

Freeman, D. B., 1992: "Prickly pear menace in eastern Australia 1880–1940," *Geographical Review*, vol. 82, no. 4, pp. 413–429.

Friedan, B., 1958: "The coming ice age: A true scientific detective story," *Harper's Magazine*, September 1958, pp. 39–45.

Fu, Q., and C. M. Johanson, 2007: "Warming and cooling of the atmosphere." In: King, M. D., C. L. Parkinson, K. C. Partington, and R. G. Williams, editors, *Our Changing Planet: The View from Space*. Cambridge: Cambridge University Press, pp. 53–55.

Gagin, A., and J. Neumann, 1981: "The Second Israeli Randomized Cloud Seeding Experiment: Evaluation of the results," *Journal of Applied Meteorology*, vol. 20, no. 11, pp. 1301–1311.

Ganopolski, A., and S. Rahmstorf, 2001: "Rapid changes of glacial climate simulated in a coupled climate model," *Nature*, vol. 409, no. 6817, pp. 153–158.

Gaskill, A., 2007: "Global Albedo Enhancement Project," downloaded November 27, 2007, from http://www.global-warming-geo-engineering.org/Albedo-Enhancement/Introduction/ag1.html.

Gates, W. L., 1992: "AMIP: The Atmospheric Model Intercomparison Project," *Bulletin of the American Meteorological Society*, vol. 73, no. 12, pp. 1962–1970.

Gibbons, A., 2006: *The First Human: The Race to Discover Our Earliest Ancestors*. New York: Doubleday, 307 pp.

———, 2007: "Food for thought: Did the first cooked meals help fuel the dramatic evolutionary expansion of the human brain?" *Science*, vol. 316, no. 5831, pp. 1558–1560.

Giunta, C. J., 2006: "Thomas Midgley, Jr., and the invention of chlorofluorocarbon refrigerants: It ain't necessarily so," *Bulletin for the History of Chemistry*, vol. 31, no. 2, pp. 66–74.

Goebel, T., M. R. Waters, and D. H. O'Rourke, 2008: "The late Pleistocene dispersal of modern humans in the Americas," *Science*, vol. 319, no. 5869, pp. 1497–1502.

Goldin, D. S., 1996: "Statement from Daniel S. Goldin, NASA Administrator," press release from NASA Headquarters, Washington, DC, August 6, 1996, obtained May 19, 2008, from http://www.nasa.gov/home/hqnews/1996/96-159.txt.

Gordon, J. S., 2002: *A Thread across the Ocean: The Heroic Story of the Transatlantic Cable*. New York: Walker & Company, 240 pp.

———, 2004: *An Empire of Wealth: The Epic History of American Economic Power*. New York: HarperCollins, 460 pp.

Gore, A., 2006: *An Inconvenient Truth: The Planetary Emergency of Global Warming and What We Can Do About It*. Emmaus, PA: Rodale, 328 pp.

Gosnell, M., 2005: *Ice: The Nature, the History, and the Uses of an Astonishing Substance*. New York: Alfred A. Knopf, 563 pp.

Govindasamy, B., and K. Caldeira, 2000: "Geoengineering Earth's radiation balance to mitigate CO_2-induced climate change," *Geophysical Research Letters*, vol. 27, no. 14, pp. 2141–2144.

Gray, W. M., 2003: "Twentieth century challenges and milestones." In: Simpson, R., editor, *Hurricane! Coping with Disaster*. Washington, DC: American Geophysical Union, pp. 3–37.

Gribbin, J., 1977: *Forecasts, Famines, and Freezes: The World's Climate and the Future of Mankind*. New York: Pocket Books, 208 pp.

Grímsson, O. R., 2008: "Moving toward climate stabilization: Iceland's example." In: MacCracken, M. C., F. Moore, and J. C. Topping Jr., editors, *Sudden and Disruptive Climate Change: Exploring the Real Risks and How We Can Avoid Them*. London: Earthscan, pp. 231–236.

Grove, J. M., 1988: *The Little Ice Age*. London: Methuen, 498 pp.

Guilyardi, E., A. Wittenberg, A. Fedorov, M. Collins, C. Wang, A. Capotondi, G. J. van Oldenborgh, and T. Stockdale, 2009: "Understanding El Niño in ocean-atmosphere general circulation models: Progress and challenges," *Bulletin of the American Meteorological Society*, vol. 90, no. 3, pp. 325–340.

Gutro, R., and S. Cole, 2006: "NASA helps weed our national garden," *The Earth Observer*, vol. 18, no. 2, pp. 24–25.

Hall, A., 2004: "The role of surface albedo feedback in climate," *Journal of Climate*, vol. 17, no. 7, pp. 1550–1568.

Hallam, A., 1983: *Great Geological Controversies*. Oxford: Oxford University Press, 182 pp.

Hamill, T. M., 2007: "Toward making the AMS carbon neutral: Offsetting the impacts of flying to conferences," *Bulletin of the American Meteorological Society*, vol. 88, no. 11, pp. 1816–1819.

Hansen, J., A. Lacis, D. Rind, G. Russell, P. Stone, I. Fung, R. Ruedy, and J. Lerner, 1984: "Climate sensitivity: Analysis of feedback mechanisms." In: Hansen, J. E., and T. Takahashi, editors, *Climate Processes and Climate Sensitivity*. Washington, DC: American Geophysical Union, Maurice Ewing Series, vol. 5, pp. 130–163.

Hansen, J., A. Lacis, R. Ruedy, and M. Sato, 1992: "Potential climate impact of Mount Pinatubo eruption," *Geophysical Research Letters*, vol. 19, no. 2, pp. 215–218.

Hansen, J., and S. Lebedeff, 1987: "Global trends of measured surface air temperature," *Journal of Geophysical Research*, vol. 92, no. D11, pp. 13345–13372.

Hansen, J., R. Ruedy, J. Glascoe, and M. Sato, 1999: "GISS analysis of surface temperature change," *Journal of Geophysical Research*, vol. 104, no. D24, pp. 30997–31022.

Hansen, J., R. Ruedy, and M. Sato, 1996: "Global surface air temperature in 1995: Return to pre-Pinatubo level," *Geophysical Research Letters*, vol. 23, no. 13, pp. 1665–1668.

Hansen, J., M. Sato, P. Kharecha, G. Russell, D. W. Lea, and M. Siddall, 2007: "Climate change and trace gases," *Philosophical Transactions of the Royal Society A*, vol. 365, pp. 1925–1954.

Hansen, J., M. Sato, R. Ruedy, K. Lo, D. W. Lea, and M. Medina-Elizade, 2006: "Global temperature change," *Proceedings of the National Academy of Sciences of the United States of America*, vol. 103, no. 39, pp. 14288–14293.

Harland, W. B., and M. J. S. Rudwick, 1964: "The great infra-Cambrian ice age," *Scientific American*, vol. 211, no. 2, pp. 28–36.

Hastenrath, S., and P. D. Kruss, 1992: "The dramatic retreat of Mount Kenya's glaciers between 1963 and 1987: Greenhouse forcing," *Annals of Glaciology*, vol. 16, pp. 127–133.

Haszeldine, R. S., 2009: "Carbon capture and storage: How green can black be?" *Science*, vol. 325, no. 5948, pp. 1647–1652.

Hayes, R., and D. Grossman, 2006: *A Scientist's Guide to Talking with the Media: Practical Advice from the Union of Concerned Scientists*. New Brunswick, NJ: Rutgers University Press, 201 pp.

Hays, J. D., J. Imbrie, and N. J. Shackleton, 1976: "Variations in the Earth's orbit: Pacemaker of the ice ages," *Science*, vol. 194, no. 4270, pp. 1121–1132.

Heinrich, H., 1988: "Origin and consequences of cyclic ice rafting in the northeast Atlantic Ocean during the past 130,000 years," *Quaternary Research*, vol. 29, no. 2, pp. 142–152.

Helsen, M. M., M. R. van den Broeke, R. S. W. van de Wal, W. J. van de Berg, E. van Meijgaard, C. H. Davis, Y. Li, and I. Goodwin, 2008: "Elevation changes in Antarctica mainly determined by accumulation variability," *Science*, vol. 320, no. 5883, pp. 1626–1629.

Hemming, S. R., 2004: "Heinrich events: Massive late Pleistocene detritus layers of the North Atlantic and their global climate imprint," *Reviews of Geophysics*, vol. 42, RG1005, doi:10.1029/2003RG000128, 43 pp.

Hildebrand, A. R., G. T. Penfield, D. A. Kring, M. Pilkington, A. Camargo, S. B. Jacobsen, and W. V. Boynton, 1991: "Chicxulub crater: A possible Cretaceous/Tertiary boundary impact crater on the Yucatán Peninsula, Mexico," *Geology*, vol. 19, no. 9, pp. 867–871.

Hindmarsh, R. C. A., and E. LeMeur, 2001: "Dynamical processes involved in the retreat of marine ice sheets," *Journal of Glaciology*, vol. 47, no. 157, pp. 271–282.

Hoegh-Guldberg, O., 2005: "Low coral cover in a high-CO_2 world," *Journal of Geophysical Research*, vol. 110, C09S06, doi:10.1029/2004JC002528, 11 pp.

Hoffman, P. F., A. J. Kaufman, G. P. Halverson, and D. P. Schrag, 1998: "A Neoproterozoic snowball Earth," *Science*, vol. 281, no. 5381, pp. 1342–1346.

Hoffman, P. F., and D. P. Schrag, 2000: "Snowball Earth," *Scientific American*, vol. 282, no. 1, pp. 68–75.

Holman, J. A., 1995: *Pleistocene Amphibians and Reptiles in North America*. New York: Oxford University Press, 243 pp.

Hönisch, B., N. G. Hemming, D. Archer, M. Siddall, and J. F. McManus, 2009: "Atmospheric carbon dioxide concentration across the mid-Pleistocene transition," *Science*, vol. 324, no. 5934, pp. 1551–1554.

Hoyos, C. D., P. A. Agudelo, P. J. Webster, and J. A. Curry, 2006: "Deconvolution of the factors contributing to the increase in global hurricane intensity," *Science*, vol. 312, no. 5770, pp. 94–97.

Hughes, T., 1973: "Is the West Antarctic ice sheet disintegrating?" *Journal of Geophysical Research*, vol. 78, no. 33, pp. 7884–7910.

———, 1977: "West Antarctic ice streams," *Reviews in Geophysics and Space Physics*, vol. 15, no. 1, pp. 1–46.

———, 1981: "The weak underbelly of the West Antarctic ice sheet," *Journal of Glaciology*, vol. 27, no. 97, pp. 518–525.

Hulbe, C. L., 1997: "An ice shelf mechanism for Heinrich layer production," *Paleoceanography*, vol. 12, no. 5, pp. 711–717.

Hulbe, C. L., D. R. MacAyeal, G. H. Denton, J. Kleman, and T. V. Lowell, 2004: "Catastrophic ice shelf breakup as the source of Heinrich event icebergs," *Paleoceanography*, vol. 19, PA1004, doi:10.1029/2003PA000890, 15 pp.

Huybers, P., 2006: "Early Pleistocene glacial cycles and the integrated summer insolation forcing," *Science*, vol. 313, no. 5786, pp. 508–511.

Imbrie, J., and K. P. Imbrie, 1979: *Ice Ages: Solving the Mystery*. Hillside, NJ: Enslow Publishers, 224 pp.

Ingermann, R. L., M. L. Robinson, and J. G. Cloud, 2003: "Respiration of steelhead trout sperm: Sensitivity to pH and carbon dioxide," *Journal of Fish Biology*, vol. 62, no. 1, pp. 13–23.

Intergovernmental Panel on Climate Change (IPCC), 2007: "Summary for Policymakers." In: Solomon, S., D. Qin, M. Manning, Z. Chen, M. Marquis, K. B. Averyt, M. Tignor, and H. L. Miller, editors, *Climate Change 2007: The Physical Science Basis. Contribution of Working Group I to the Fourth Assessment Report of the Intergovernmental Panel on Climate Change*. Cambridge: Cambridge University Press, pp. 1–18.

Jansen, E., J. Overpeck, K. R. Briffa, J.-C. Duplessy, F. Joos, V. Masson-Delmotte, D. Olago, B. Otto-Bliesner, W. R. Peltier, S. Rahmstorf, R. Ramesh, D. Raynaud, D. Rind, O. Solomina, R. Villalba, and D. Zhang, 2007: "Palaeoclimate." In: Solomon, S., D. Qin, M. Manning, Z. Chen, M. Marquis, K. B. Averyt, M. Tignor, and H. L. Miller, editors, *Climate Change 2007: The Physical Science Basis. Contribution of Working Group I to the Fourth Assessment Report of the Intergovernmental Panel on Climate Change*. Cambridge: Cambridge University Press, pp. 433–497.

Jeffries, M. O., 1992: "Arctic ice shelves and ice islands: Origin, growth and disintegration, physical characteristics, structural-stratigraphic variability, and dynamics," *Reviews of Geophysics*, vol. 30, no. 3, pp. 245–267.

Jerch, K., 2008: "Capitalizing on carbon," *Bulletin of the Atomic Scientists*, vol. 64, no. 2, p. 16.

Johnsen, S. J., H. B. Clausen, W. Dansgaard, K. Fuhrer, N. Gundestrup, C. U. Hammer, P. Iversen, J. Jouzel, B. Stauffer, and J. P. Steffensen, 1992: "Irregular glacial interstadials recorded in a new Greenland ice core," *Nature*, vol. 359, no. 6393, pp. 311–313.

Johnson, R. G., 1997: "Climate control requires a dam at the Strait of Gibraltar," *Eos, Transactions, American Geophysical Union*, vol. 78, no. 27, pp. 277 and 280–281.

Johnson, R. G., and S.-E. Lauritzen, 1995: "Hudson Bay-Hudson Strait jökulhlaups and Heinrich events: A hypothesis," *Palaeogeography, Palaeoclimatology, Palaeoecology*, vol. 117, no. 1, pp. 123–137.

Jones, M. L., 2007: "The productivity of prediction and explanation," *Science*, vol. 317, no. 5843, p. 1327.

Jones, P. D., M. New, D. E. Parker, S. Martin, and I. G. Rigor, 1999: "Surface air temperature and its changes over the past 150 years," *Reviews of Geophysics*, vol. 37, no. 2, pp. 173–199.

Jones, P. D., T. M. L. Wigley, and P. B. Wright, 1986: "Global temperature variations between 1861 and 1984," *Nature*, vol. 322, no. 6078, pp. 430–434.

Joughin, I., S. B. Das, M. A. King, B. E. Smith, I. M. Howat, and T. Moon, 2008: "Seasonal speedup along the western flank of the Greenland ice sheet," *Science*, vol. 320, no. 5877, pp. 781–783.

Joughin, I., and S. Tulaczyk, 2002: "Positive mass balance of the Ross ice streams, West Antarctica," *Science*, vol. 295, no. 5554, pp. 476–480.

Kaiser, J., editor, 2005: "Random samples: Lifesaving pitchers," *Science*, vol. 308, no. 5719, p. 195.

Kaser, G., D. R. Hardy, T. Mölg, R. S. Bradley, and T. M. Hyera, 2004: "Modern glacier retreat on Kilimanjaro as evidence of climate change: Observations and facts," *International Journal of Climatology*, vol. 24, pp. 329–339.

Keeling, P. J., 2007: "Deep questions in the tree of life," *Science*, vol. 317, no. 5846, pp. 1875–1876.

Keeling, R. F., S. C. Piper, A. F. Bollenbacher, and S. J. Walker, 2009: "Monthly atmospheric CO_2 concentrations (ppm) derived from flask air samples, Mauna Loa, Hawaii 156W, 20N," downloaded August 16, 2009, from http://scrippsco2.ucsd.edu.

Keith, D. W., 2000: "Geoengineering the climate: History and prospect," *Annual Review of Energy and the Environment*, vol. 25, pp. 245–284.

———, 2009: "Why capture CO_2 from the atmosphere?" *Science*, vol. 325, no. 5948, pp. 1654–1655.

Keith, D. W., and H. Dowlatabadi, 1992: "A serious look at geoengineering," *Eos, Transactions, American Geophysical Union*, vol. 73, no. 27, pp. 289 and 292–293.

Kellogg, W. W., and S. H. Schneider, 1974: "Climate stabilization: For better or for worse?" *Science*, vol. 186, no. 4170, pp. 1163–1172.

Kennett, D. J., J. P. Kennett, A. West, C. Mercer, S. S. Que Hee, L. Bement, T. E. Bunch, M. Sellers, and W. S. Wolbach, 2009: "Nanodiamonds in the Younger Dryas boundary sediment layer," *Science*, vol. 323, no. 5910, p. 94.

Kennett, J. P., 1977: "Cenozoic evolution of Antarctic glaciation, the circum-Antarctic Ocean, and their impact on global paleoceanography," *Journal of Geophysical Research*, vol. 82, no. 27, pp. 3843–3860.

Keppler, F., J. T. G. Hamilton, M. Braß, and T. Röckmann, 2006: "Methane emissions from terrestrial plants under aerobic conditions," *Nature*, vol. 439, no. 7073, pp. 187–191.

Kerr, R. A., 1989: "Hansen vs. the world on the greenhouse threat," *Science*, vol. 244, no. 4908, pp. 1041–1043.

———, 2006: "False alarm: Atlantic Conveyor Belt hasn't slowed down after all," *Science*, vol. 314, no. 5802, p. 1064.

———, 2008: "A time war over the period we live in," *Science*, vol. 319, no. 5862, pp. 402–403.

Kintisch, E., 2007: "Should oceanographers pump iron?" *Science*, vol. 318, no. 5855, pp. 1368–1370.

Kirschvink, J. L., E. J. Gaidos, L. E. Bertani, N. J. Beukes, J. Gutzmer, L. N. Maepa, and R. E. Steinberger, 2000: "Paleoproterozoic snowball Earth: Extreme climatic and geochemical global change and its biological consequences," *Proceedings of the National Academy of Sciences of the United States of America*, vol. 97, no. 4, pp. 1400–1405.

Kitchell, J. F., D. E. Schindler, R. Ogutu-Ohwayo, and P. N. Reinthal, 1997: "The Nile perch in Lake Victoria: Interactions between predation and fisheries," *Ecological Applications*, vol. 7, no. 2, pp. 653–664.

Klebnikov, P., 2006: "First in nation, California caps global warming pollution," *Solutions* (newsletter of Environmental Defense), vol. 37, no. 5, pp. 1–2.

Klein Goldewijk, K., 2001: "Estimating global land use change over the past 300 years: The HYDE database," *Global Biogeochemical Cycles*, vol. 15, no. 2, pp. 417–433.

Kleiven, H. F., C. Kissel, C. Laj, U. S. Ninnemann, T. O. Richter, and E. Cortijo, 2008: "Reduced North Atlantic deep water coeval with the glacial Lake Agassiz freshwater outburst," *Science*, vol. 319, no. 5859, pp. 60–64.

Knoll, A. H., M. R. Walter, G. M. Narbonne, and N. Christie-Blick, 2006: "The Ediacaran Period: A new addition to the geologic time scale," *Lethaia*, vol. 39, pp. 13–30.

Knutson, T. R., J. J. Sirutis, S. T. Garner, G. A. Vecchi, and I. M. Held, 2008: "Simulated reduction in Atlantic hurricane frequency under twenty-first-century warming conditions," *Nature Geoscience*, vol. 1, no. 6, pp. 359–364.

Kolbert, E., 2006: *Field Notes from a Catastrophe*. New York: Bloomsbury Publishing, 213 pp.

Korycansky, D. G., G. Laughlin, and F. C. Adams, 2001: "Astronomical engineering: A strategy for modifying planetary orbits," *Astrophysics and Space Science*, vol. 275, no. 4, pp. 349–366.

Kuhn, T. S., 1970: *The Structure of Scientific Revolutions*, second edition. Chicago: University of Chicago Press, 210 pp.

Kukla, G., and J. Gavin, 1981: "Summer ice and carbon dioxide," *Science*, vol. 214, no. 4520, pp. 497–503.

Leatherman, S. P., and J. Williams, 2008: *Hurricanes: Causes, Effects, and the Future*. Minneapolis: Voyageur Press, 72 pp.

LeGrande, A. N., G. A. Schmidt, D. T. Shindell, C. V. Field, R. L. Miller, D. M. Koch, G. Faluvegi, and G. Hoffmann, 2006: "Consistent simulations of multiple proxy responses to an abrupt climate change event," *Proceedings of the National Academy of Sciences of the United States of America*, vol. 103, no. 4, pp. 837–842.

Lelieveld, J., 2006: "A nasty surprise in the greenhouse," *Nature*, vol. 443, no. 7110, pp. 405–406.

Lemke, P., J. Ren, R. B. Alley, I. Allison, J. Carrasco, G. Flato, Y. Fujii, G. Kaser, P. Mote, R. H. Thomas, and T. Zhang, 2007: "Observations: Changes in snow, ice and frozen ground." In: Solomon, S., D. Qin, M. Manning, Z. Chen, M. Marquis, K. B. Averyt, M. Tignor, and H. L. Miller, editors, *Climate Change 2007: The Physical Science Basis. Contribution of*

Working Group I to the Fourth Assessment Report of the Intergovernmental Panel on Climate Change. Cambridge: Cambridge University Press, pp. 337–383.

Leslie, S. W., 1980: "Thomas Midgley and the politics of industrial research," *Business History Review,* vol. 54, no. 4, pp. 480–503.

Lester, B., 2007: "Greening the meeting," *Science,* vol. 318, no. 5847, pp. 36–38.

Lin, B., B. A. Wielicki, L. H. Chambers, Y. Hu, and K.-M. Xu, 2002: "The Iris Hypothesis: A negative or positive cloud feedback?" *Journal of Climate,* vol. 15, no. 1, pp. 3–7.

Linden, E., 2006: *The Winds of Change: Climate, Weather, and the Destruction of Civilizations.* New York: Simon & Schuster, 302 pp.

Lindgren, E., and R. Gustafson, 2001: "Tick-borne encephalitis in Sweden and climate change," *Lancet,* vol. 358, no. 9275, pp. 16–18.

Lindsay, R. W., and J. Zhang, 2005: "The thinning of Arctic sea ice, 1988–2003: Have we passed a tipping point?" *Journal of Climate,* vol. 18, pp. 4879–4894.

Lindzen, R., 2006: "Climate of fear: Global-warming alarmists intimidate dissenting scientists into silence," *Wall Street Journal,* editorial page, April 12, 2006, p. A14.

Lindzen, R. S., M.-D. Chou, and A. Y. Hou, 2001: "Does the Earth have an adaptive infrared iris?" *Bulletin of the American Meteorological Society,* vol. 82, no. 3, pp. 417–432.

Loeb, N. G., and T. Wong, 2007: "Clouds and the Earth's radiation budget." In: King, M. D., C. L. Parkinson, K. C. Partington, and R. G. Williams, editors, *Our Changing Planet: The View from Space.* Cambridge: Cambridge University Press, pp. 21–25.

Lomborg, B., 2007: *Cool It: The Skeptical Environmentalist's Guide to Global Warming.* New York: Alfred A. Knopf, 253 pp.

Lovelock, J., 1988: *The Ages of Gaia: A Biography of Our Living Earth.* New York: W. W. Norton, 252 pp.

Lowell, T. V., T. G. Fisher, G. C. Comer, I. Haidas, N. Waterson, K. Glover, H. M. Loope, J. M. Schaefer, V. Rinterknecht, W. Broecker, G. Denton, and J. T. Teller, 2005: "Testing the Lake Agassiz meltwater trigger for the Younger Dryas," *Eos, Transactions, American Geophysical Union,* vol. 86, no. 40, pp. 365 and 372.

Lyon, B. E., A. S. Chaine, and D. W. Winkler, 2008: "A matter of timing," *Science,* vol. 321, no. 5892, pp. 1051–1052.

MacAyeal, D. R., 1993: "Binge/purge oscillations of the Laurentide ice sheet as a cause of the North Atlantic's Heinrich events," *Paleoceanography,* vol. 8, no. 6, pp. 775–784.

MacCracken, M. C., 2006: "Geoengineering: Worthy of cautious evaluation?" *Climatic Change,* vol. 77, pp. 235–243.

Macdougall, D., 2004: *Frozen Earth: The Once and Future Story of Ice Ages.* Berkeley: University of California Press, 256 pp.

MacKay, D. J. C., 2009: *Sustainable Energy—Without the Hot Air.* Cambridge: UIT Cambridge, 372 pp. (available free at http://www.withouthotair.com).

Manabe, S., and R. J. Stouffer, 1980: "Sensitivity of a global climate model to an increase of CO_2 concentration in the atmosphere," *Journal of Geophysical Research,* vol. 85, no. C10, pp. 5529–5554.

————, 1994: "Multiple-century response of a coupled ocean-atmosphere model to an increase of atmospheric carbon dioxide," *Journal of Climate*, vol. 7, no. 1, pp. 5–23.

Manabe, S., and R. T. Wetherald, 1967: "Thermal equilibrium of the atmosphere with a given distribution of relative humidity," *Journal of the Atmospheric Sciences*, vol. 24, no. 3, pp. 241–259.

————, 1975: "The effects of doubling the CO_2 concentration on the climate of a general circulation model," *Journal of the Atmospheric Sciences*, vol. 32, no. 1, pp. 3-15.

Mann, M. E., R. S. Bradley, and M. K. Hughes, 1999: "Northern Hemisphere temperatures during the past millennium: Inferences, uncertainties, and limitations," *Geophysical Research Letters*, vol. 26, no. 6, pp. 759–762.

Marchetti, C., 1977: "On geoengineering and the CO_2 problem," *Climatic Change*, vol. 1, pp. 59–68.

Margulis, L., 1998: *Symbiotic Planet: A New Look at Evolution*. New York: Basic Books, 147 pp.

Marshall, C. R., 2006: "Explaining the Cambrian 'explosion' of animals," *Annual Review of Earth and Planetary Sciences*, vol. 34, pp. 355–384, doi:10.1146/annurev. earth.33.031504.103001.

Martin, J. H., 1990: "Glacial-interglacial CO_2 change: The iron hypothesis," *Paleoceanography*, vol. 5, no. 1, pp. 1–13.

Martin, J. H., S. E. Fitzwater, and R. M. Gordon, 1990: "Iron deficiency limits phytoplankton growth in Antarctic waters," *Global Biogeochemical Cycles*, vol. 4, no. 1, pp. 5–12.

Martin, J. H., and R. M. Gordon, 1988: "Northeast Pacific iron distributions in relation to phytoplankton productivity," *Deep-Sea Research*, vol. 35, no. 2, pp. 177–196.

Martin, J. H., R. M. Gordon, S. Fitzwater, and W. W. Broenkow, 1989: "VERTEX: Phytoplankton/iron studies in the Gulf of Alaska," *Deep-Sea Research*, vol. 36, no. 5, pp. 649–680.

Martinson, D. G., P. D. Killworth, and A. L. Gordon, 1981: "A convective model for the Weddell polynya," *Journal of Physical Oceanography*, vol. 11, pp. 466–488.

Mayewski, P. A., and F. White, 2002: *The Ice Chronicles: The Quest to Understand Global Climate Change*. Hanover, NH: University Press of New England, 233 pp.

Maykut, G. A., and N. Untersteiner, 1971: "Some results from a time-dependent thermodynamic model of sea ice," *Journal of Geophysical Research*, vol. 76, no. 6, pp. 1550–1575.

McCormick, M. P., L. W. Thomason, and C. R. Trepte, 1995: "Atmospheric effects of the Mt Pinatubo eruption," *Nature*, vol. 373, no. 6513, pp. 399–404.

McDonald, K. C., B. Chapman, J. S. Kimball, and R. Zimmermann, 2007: "The tropical rain forest: Threatened powerhouse of the biosphere." In: King, M. D., C. L. Parkinson, K. C. Partington, and R. G. Williams, editors, *Our Changing Planet: The View from Space*. Cambridge: Cambridge University Press, pp. 274–279.

McIntyre, S., and R. McKitrick, 2003: "Corrections to the Mann et al. (1998) proxy data base and Northern Hemispheric average temperature series," *Energy and Environment*, vol. 14, no. 6, pp. 751–771.

————, 2005: "Hockey sticks, principal components, and spurious significance," *Geophysical Research Letters*, vol. 32, L03710, doi:10.1029/2004GL021750, 5 pp.

Meehl, G. A., T. F. Stocker, W. D. Collins, P. Friedlingstein, A. T. Gaye, J. M. Gregory, A. Kitoh, R. Knutti, J. M. Murphy, A. Noda, S. C. B. Raper, I. G. Watterson, A. J. Weaver, and Z.-C. Zhao, 2007: "Global climate projections." In: Solomon, S., D. Qin, M. Manning, Z. Chen, M. Marquis, K. B. Averyt, M. Tignor, and H. L. Miller, editors, *Climate Change 2007: The Physical Science Basis. Contribution of Working Group I to the Fourth Assessment Report of the Intergovernmental Panel on Climate Change.* Cambridge: Cambridge University Press, pp. 747–845.

Mercer, J. H., 1978: "West Antarctic ice sheet and CO_2 greenhouse effect: A threat of disaster," *Nature*, vol. 271, no. 5643, pp. 321–325.

Merlis, G., 2003: *How to Make the Most of Every Media Appearance.* New York: McGraw-Hill, 200 pp.

Metz, B., O. Davidson, P. Bosch, R. Dave, and L. Meyer, editors, 2007: *Climate Change 2007: Mitigation of Climate Change. Contribution of Working Group III to the Fourth Assessment Report of the Intergovernmental Panel on Climate Change.* Cambridge: Cambridge University Press, 851 pp.

Meyer, W. B., 1996: *Human Impact on the Earth.* Cambridge: Cambridge University Press, 253 pp.

Micklin, P. P., 1981: "A preliminary systems analysis of impacts of proposed Soviet river diversions on Arctic sea ice," *Eos, Transactions, American Geophysical Union*, vol. 62, no. 19, pp. 489–493.

————, 1986: "Water diversion in the Soviet Union," *Science*, vol. 234, no. 4775, p. 411.

————, 2007: "Destruction of the Aral Sea." In: King, M. D., C. L. Parkinson, K. C. Partington, and R. G. Williams, editors, *Our Changing Planet: The View from Space.* Cambridge: Cambridge University Press, pp. 320–325.

Midgley, T., Jr., 1937: "From the periodic table to production," *Industrial and Engineering Chemistry*, vol. 29, no. 2, pp. 241–244.

Midgley, T., Jr., and A. L. Henne, 1930: "Organic fluorides as refrigerants," *Industrial and Engineering Chemistry*, vol. 22, no. 5, pp. 542–545.

Mielke, P. W., Jr., G. W. Brier, L. O. Grant, G. J. Mulvey, and P. N. Rosenzweig, 1981: "A statistical reanalysis of the replicated Climax I and II wintertime orographic cloud seeding experiments," *Journal of Applied Meteorology*, vol. 20, no. 6, pp. 643–659.

Milankovitch, M., 1920: *Théorie Mathématique des Phénomènes Thermiques Produits par la Radiation Solaire.* Paris: Gauthier-Villars, 334 pp.

————, 1941: *Canon of Insolation and the Ice-Age Problem.* Jerusalem: Israel Program of Scientific Translations, 1969, 484 pp.

Mitchell, J. M., Jr., 1961: "Recent secular changes of global temperature," *Annals of the New York Academy of Sciences*, vol. 95, pp. 235–250.

Molina, M. J., and F. S. Rowland, 1974: "Stratospheric sink for chlorofluoromethanes: Chlorine atom-catalysed destruction of ozone," *Nature*, vol. 249, no. 5460, pp. 810–812.

Monbiot, G., 2007: *Heat: How to Stop the Planet from Burning.* Cambridge, MA: South End Press, 278 pp.

Moran, K., J. Backman, H. Brinkhuis, S. C. Clemens, T. Cronin, G. R. Dickens, F. Eynaud, J. Gattacceca, M. Jakobsson, R. W. Jordan, M. Kaminski, J. King, N. Koc, A. Krylov, N. Martinez, J. Matthiessen, D. McInroy, T. C. Moore, J. Onodera, M. O'Regan, H. Pälike, B. Rea, D. Rio, T. Sakamoto, D. C. Smith, R. Stein, K. St John, I. Suto, N. Suzuki, K. Takahashi, M. Watanabe, M. Yamamoto, J. Farrell, M. Frank, P. Kubik, W. Jokat, and Y. Kristoffersen, 2006: "The Cenozoic palaeoenvironment of the Arctic Ocean," *Nature,* vol. 441, no. 7093, pp. 601–605.

Morison, J., J. Wahr, R. Kwok, and C. Peralta-Ferriz, 2007: "Recent trends in Arctic Ocean mass distribution revealed by GRACE," *Geophysical Research Letters,* vol. 34, no. 7, L07602, doi:10.1029/2006GL029016, 6 pp.

Morrison, H. G., A. G. McArthur, F. D. Gillin, S. B. Aley, R. D. Adam, G. J. Olsen, A. A. Best, W. Z. Cande, F. Chen, M. J. Cipriano, B. J. Davids, S. C. Dawson, H. G. Elmendorf, A. B. Hehl, M. E. Holder, S. M. Huse, U. U. Kim, E. Lasek-Nesselquist, G. Manning, A. Nigam, J. E. J. Nixon, D. Palm, N. E. Passamaneck, A. Prabhu, C. I. Reich, D. S. Reiner, J. Samuelson, S. G. Svard, and M. L. Sogin, 2007: "Genomic minimalism in the early diverging intestinal parasite *Giardia lamblia,*" *Science,* vol. 317, no. 5846, pp. 1921–1926.

Murdoch, R., 2007: "Energy Initiative," speech given on May 9, 2007, at the Hudson Theatre, New York City, and available at http://www.newscorp.com/energy/full_speech.html (downloaded December 21, 2008).

Murphy, D. M., 2009: "Effect of stratospheric aerosols on direct sunlight and implications for concentrating solar power," *Environmental Science and Technology,* vol. 43, no. 8, pp. 2784–2786.

Murphy, J. M., and J. F. B. Mitchell, 1995: "Transient response of the Hadley Centre coupled ocean-atmosphere model to increasing carbon dioxide. Part II: Spatial and temporal structure of response," *Journal of Climate,* vol. 8, no. 1, pp. 57–80.

Murphy, J. M., D. M. H. Sexton, D. N. Barnett, G. S. Jones, M. J. Webb, M. Collins, and D. A. Stainforth, 2004: "Quantification of modelling uncertainties in a large ensemble of climate change simulations," *Nature,* vol. 430, no. 7001, pp. 768–772.

Narbonne, G. M., 2005: "The Ediacara biota: Neoproterozoic origin of animals and their ecosystems," *Annual Review of Earth and Planetary Sciences,* vol. 33, pp. 421–442, doi:10.1146/annurev.earth.33.092203.122519.

National Academy of Sciences (NAS), 1992: *Policy Implications of Greenhouse Warming: Mitigation, Adaptation, and the Science Base.* Washington, DC: National Academy Press, 918 pp.

Nel, A., 2005: "Air pollution-related illness: Effects of particles," *Science,* vol. 308, no. 5723, pp. 804–806.

Newman, P. A., 2008: "Ozone depletion," *Bulletin of the American Meteorological Society, Special Supplement on State of the Climate in 2007,* vol. 89, no. 7, pp. S104–S105.

Newman, P. A., S. R. Kawa, and E. R. Nash, 2004: "On the size of the Antarctic ozone hole," *Geophysical Research Letters,* vol. 31, L21104, doi:10.1029/2004GL020596, 4 pp.

Newman, P. A., E. R. Nash, S. R. Kawa, S. A. Montzka, and S. M. Schauffler, 2006: "When will the Antarctic ozone hole recover?" *Geophysical Research Letters*, vol. 33, L12814, doi:10.1029/2005GL025232, 5 pp.

NGRIP (North Greenland Ice Core Project) Members, 2004: "High-resolution record of Northern Hemisphere climate extending into the last interglacial period," *Nature*, vol. 431, no. 7005, pp. 147–151.

Normile, D., 2007: "Getting at the roots of killer dust storms," *Science*, vol. 317, no. 5836, pp. 314–316.

Nussey, D. H., E. Postma, P. Gienapp, and M. E. Visser, 2005: "Selection on heritable phenotypic plasticity in a wild bird population," *Science*, vol. 310, no. 5746, pp. 304–306.

Obasi, G. O. P., and M. K. Tolba, 1990: "Preface." In: Houghton, J. T., G. J. Jenkins, and J. J. Ephraums, editors, *Climate Change: The IPCC Scientific Assessment*. Cambridge: Cambridge University Press, p. iii.

Oeschger, H., J. Beer, U. Siegenthaler, B. Stauffer, W. Dansgaard, and C. C. Langway, 1984: "Late glacial climate history from ice cores." In: Hansen, J. E., and T. Takahashi, editors, *Climate Processes and Climate Sensitivity*. Washington, DC: American Geophysical Union, Maurice Ewing Series, vol. 5, pp. 299–306.

Officer, C., and J. Page, 1993: *Tales of the Earth: Paroxysms and Perturbations of the Blue Planet*. New York: Oxford University Press, 226 pp.

Ohfuchi, W., H. Sasaki, Y. Masumoto, and H. Nakamura, 2007: "'Virtual' atmospheric and oceanic circulation in the Earth Simulator," *Bulletin of the American Meteorological Society*, vol. 88, no. 6, pp. 861–866.

Ohmura, A., 2006: "Observed long-term variations of solar irradiance at the Earth's surface," *Space Science Reviews*, vol. 125, pp. 111–128.

Orr, F. M., Jr., 2009: "Onshore geologic storage of CO_2," *Science*, vol. 325, no. 5948, pp. 1656–1658.

Orr, J. C., V. J. Fabry, O. Aumont, L. Bopp, S. C. Doney, R. A. Feely, A. Gnanadesikan, N. Gruber, A. Ishida, F. Joos, R. M. Key, K. Lindsay, E. Maier-Reimer, R. Matear, P. Monfray, A. Mouchet, R. G. Najjar, G.-K. Plattner, K. B. Rodgers, C. L. Sabine, J. L. Sarmiento, R. Schlitzer, R. D. Slater, I. J. Totterdell, M.-F. Weirig, Y. Yamanaka, and A. Yool, 2005: "Anthropogenic ocean acidification over the twenty-first century and its impact on calcifying organisms," *Nature*, vol. 437, no. 7059, pp. 681–686.

Overpeck, J. T., and J. E. Cole, 2006: "Abrupt change in Earth's climate system," *Annual Review of Environment and Resources*, vol. 31, pp. 1–31.

Pacala, S., and R. Socolow, 2004: "Stabilization wedges: Solving the climate problem for the next 50 years with current technologies," *Science*, vol. 305, no. 5686, pp. 968–972.

Page, S. E., F. Siegert, J. O. Rieley, H.-D. V. Boehm, A. Jaya, and S. Limin, 2002: "The amount of carbon released from peat and forest fires in Indonesia during 1997," *Nature*, vol. 420, no. 6911, pp. 61–65.

Paillard, D., 2006: "What drives the ice age cycle?" *Science*, vol. 313, no. 5786, pp. 455–456.

Pala, C., 2005: "To save a vanishing sea," *Science*, vol. 307, no. 5712, pp. 1032–1034.

———, 2006: "Once a terminal case, the North Aral Sea shows new signs of life," *Science*, vol. 312, no. 5771, p. 183.

Pälike, H., R. D. Norris, J. O. Herrle, P. A. Wilson, H. K. Coxall, C. H. Lear, N. J. Shackleton, A. K. Tripati, and B. S. Wade, 2006: "The heartbeat of the Oligocene climate system," *Science*, vol. 314, no. 5807, pp. 1894–1898.

Panofsky, H. A., 1956: "Theories of climate change," *Weatherwise*, vol. 9, no. 6, pp. 183–187 and 204.

Parizek, B. R., and R. B. Alley, 2004: "Implications of increased Greenland surface melt under global-warming scenarios: Ice-sheet simulations," *Quaternary Science Reviews*, vol. 23, no. 9, pp. 1013–1027.

Parkinson, C. L., 1983: "On the development and cause of the Weddell polynya in a sea ice simulation," *Journal of Physical Oceanography*, vol. 13, no. 3, pp. 501–511.

———, 1985: *Breakthroughs: A Chronology of Great Achievements in Science and Mathematics 1200–1930.* Boston: G. K. Hall and Company, 576 pp.

———, 1996: "James Croll." In: Schneider, S. H., editor, *Encyclopedia of Climate and Weather.* New York: Oxford University Press, vol. 1, pp. 207–208.

———, 1997: *Earth from Above: Using Color-Coded Satellite Images to Examine the Global Environment.* Sausalito, CA: University Science Books, 176 pp.

———, 2004: "Southern Ocean sea ice and its wider linkages: Insights revealed from models and observations," *Antarctic Science*, vol. 16, no. 4, pp. 387–400.

———, 2006: "Earth's cryosphere: Current state and recent changes," *Annual Review of Environment and Resources*, vol. 31, pp. 33–60.

Parkinson, C. L., and R. A. Bindschadler, 1984: "Response of Antarctic sea ice to uniform atmospheric temperature increases." In: Hansen, J. E., and T. Takahashi, editors, *Climate Processes and Climate Sensitivity.* Washington, DC: American Geophysical Union, Maurice Ewing Series, vol. 5, pp. 254–264.

Parkinson, C. L., and D. J. Cavalieri, 2008: "Arctic sea ice variability and trends, 1979–2006," *Journal of Geophysical Research—Oceans*, vol. 113, C07003, doi:10.1029/2007JC004558, 28 pp.

Parkinson, C. L., J. C. Comiso, H. J. Zwally, D. J. Cavalieri, P. Gloersen, and W. J. Campbell, 1987: *Arctic Sea Ice, 1973–1976: Satellite Passive-Microwave Observations*, NASA SP-489. Washington, DC: National Aeronautics and Space Administration, 296 pp.

Parkinson, C. L., and W. W. Kellogg, 1979: "Arctic sea ice decay simulated for a CO_2-induced temperature rise," *Climatic Change*, vol. 2, pp. 149–162.

Parmesan, C., 2006: "Ecological and evolutionary responses to recent climate change," *Annual Review of Ecology, Evolution, and Systematics*, vol. 37, pp. 637–669.

Parry, M., O. Canziani, J. Palutikof, P. van der Linden, and C. Hanson, editors, 2007: *Climate Change 2007: Impacts, Adaptation and Vulnerability. Contribution of Working Group II to the Fourth Assessment Report of the Intergovernmental Panel on Climate Change.* Cambridge: Cambridge University Press, 976 pp.

Pearce, F., 2007: *With Speed and Violence: Why Scientists Fear Tipping Points in Climate Change.* Boston: Beacon Press, 278 pp.

Percival, O. V., 1890: "Runaway pond," *The Bizarre Notes and Queries: A Monthly Magazine of History, Folk-lore, Mathematics, Mysticism, Art, Science, Etc.*, vol. 7, pp. 130–132.

Peterson, T. C., W. M. Connolley, and J. Fleck, 2008: "The myth of the 1970s global cooling scientific consensus," *Bulletin of the American Meteorological Society*, vol. 89, no. 9, pp. 1325–1337.

Petit, J. R., J. Jouzel, D. Raynaud, N. I. Barkov, J.-M. Barnola, I. Basile, M. Bender, J. Chappellaz, M. Davis, G. Delaygue, M. Delmotte, V. M. Kotlyakov, M. Legrand, V. Y. Lipenkov, C. Lorius, L. Pépin, C. Ritz, E. Saltzman, and M. Stievenard, 1999: "Climate and atmospheric history of the past 420,000 years from the Vostok ice core, Antarctica," *Nature*, vol. 399, no. 6735, pp. 429–436.

Pfeffer, W. T., J. T. Harper, and S. O'Neel, 2008: "Kinematic constraints on glacier contributions to 21st-century sea-level rise," *Science*, vol. 321, no. 5894, pp. 1340–1343.

Pielke, R. A., Jr., C. Landsea, M. Mayfield, J. Laver, and R. Pasch, 2005: "Hurricanes and global warming," *Bulletin of the American Meteorological Society*, vol. 86, no. 11, pp. 1571–1575.

Pielke, R., Sr., J. Nielsen-Gammon, C. Davey, J. Angel, O. Bliss, N. Doesken, M. Cai, S. Fall, D. Niyogi, K. Gallo, R. Hale, K. G. Hubbard, X. Lin, H. Li, and S. Raman, 2007: "Documentation of uncertainties and biases associated with surface temperature measurement sites for climate change assessment," *Bulletin of the American Meteorological Society*, vol. 88, no. 6, pp. 913–928.

Pittock, A. B., 2008: "Ten reasons why climate change may be more severe than projected." In: MacCracken, M. C., F. Moore, and J. C. Topping Jr., editors, *Sudden and Disruptive Climate Change: Exploring the Real Risks and How We Can Avoid Them*. London: Earthscan, pp. 11–27.

Pollard, D., and S. L. Thompson, 1994: "Sea-ice dynamics and CO_2 sensitivity in a global climate model," *Atmosphere-Ocean*, vol. 32, no. 2, pp. 449–467.

Pope, K. O., A. C. Ocampo, and C. E. Duller, 1993: "Surficial geology of the Chicxulub impact crater, Yucatan, Mexico," *Earth, Moon, and Planets*, vol. 63, no. 2, pp. 93–104.

Pounds, J. A., M. R. Bustamante, L. A. Coloma, J. A. Consuegra, M. P. L. Fogden, P. N. Foster, E. La Marca, K. L. Masters, A. Merino-Viteri, R. Puschendorf, S. R. Ron, G. A. Sánchez-Azofeifa, C. J. Still, and B. E. Young, 2006: "Widespread amphibian extinctions from epidemic disease driven by global warming," *Nature*, vol. 439, no. 7073, pp. 161–167.

Pounds, J. A., M. P. L. Fogden, and J. H. Campbell, 1999: "Biological response to climate change on a tropical mountain," *Nature*, vol. 398, no. 6728, pp. 611–615.

Prentice, I. C., G. D. Farquhar, M. J. R. Fasham, M. L. Goulden, M. Heimann, V. J. Jaramillo, H. S. Kheshgi, C. LeQuere, R. J. Scholes, and D. W. R. Wallace, 2001: "The carbon cycle and atmospheric carbon dioxide." In: Houghton, J. T., Y. Ding, D. J. Griggs, M. Noguer, P. J. van der Linden, X. Dai, K. Maskell, and C. A. Johnson, editors, *Climate Change 2001: The Scientific Basis. Contribution of Working Group I to the Third Assessment Report of the Intergovernmental Panel on Climate Change*. Cambridge: Cambridge University Press, pp. 183–237.

Prestrud, P., and I. Stirling, 1994: "The International Polar Bear Agreement and the current status of polar bear conservation," *Aquatic Mammals*, vol. 20, no. 3, pp. 113–124.

Rahmstorf, S., 2002: "Ocean circulation and climate during the past 120,000 years," *Nature*, vol. 419, no. 6903, pp. 207–214.

———, 2007: "A semi-empirical approach to projecting future sea-level rise," *Science*, vol. 315, no. 5810, pp. 368–370.

Rainforest Alliance, 2007: *The Canopy*, vol. 20, no. 4, 8 pp.

Randall, D. A., R. A. Wood, S. Bony, R. Colman, T. Fichefet, J. Fyfe, V. Kattsov, A. Pitman, J. Shukla, J. Srinivasan, R. J. Stouffer, A. Sumi, and K. E. Taylor, 2007: "Climate models and their evaluation." In: Solomon, S., D. Qin, M. Manning, Z. Chen, M. Marquis, K. B. Averyt, M. Tignor, and H. L. Miller, editors, *Climate Change 2007: The Physical Science Basis. Contribution of Working Group I to the Fourth Assessment Report of the Intergovernmental Panel on Climate Change.* Cambridge: Cambridge University Press, pp. 589–662.

Rangno, A. L., and P. V. Hobbs, 1993: "Further analyses of the Climax cloud-seeding experiments," *Journal of Applied Meteorology*, vol. 32, no. 12, pp. 1837–1847.

———, 1995: "A new look at the Israeli cloud seeding experiments," *Journal of Applied Meteorology*, vol. 34, no. 5, pp. 1169–1193.

Rasch, P. J., P. J. Crutzen, and D. B. Coleman, 2008: "Exploring the geoengineering of climate using stratospheric sulfate aerosols: The role of particle size," *Geophysical Research Letters*, vol. 35, L02809, doi:10.1029/2007GL032179, 6 pp.

Rasmussen, S. O., K. K. Andersen, A. M. Svensson, J. P. Steffensen, B. M. Vinther, H. B. Clausen, M.-L. Siggaard-Andersen, S. J. Johnsen, L. B. Larsen, D. Dahl-Jensen, M. Bigler, R. Röthlisberger, H. Fischer, K. Goto-Azuma, M. E. Hansson, and U. Ruth, 2006: "A new Greenland ice core chronology for the last glacial termination," *Journal of Geophysical Research*, vol. 111, D06102, doi:10.1029/2005JD006079, 16 pp.

Rasool, S. I., and S. H. Schneider, 1971: "Atmospheric carbon dioxide and aerosols: Effects of large increases on global climate," *Science*, vol. 173, no. 3992, pp. 138–141.

Raymo, M. E., L. E. Lisiecki, and K. H. Nisancioglu, 2006: "Plio-Pleistocene ice volume, Antarctic climate, and the global $\delta^{18}O$ record," *Science*, vol. 313, no. 5786, pp. 492–495.

Rayner, N. A., D. E. Parker, E. B. Horton, C. K. Folland, L. V. Alexander, D. P. Rowell, E. C. Kent, and A. Kaplan, 2003: "Global analysis of sea surface temperature, sea ice, and night marine air temperature since the late nineteenth century," *Journal of Geophysical Research*, vol. 108, no. D14, 4407, doi:10.1029/2002JD002670, 29 pp.

Reichler, T., and J. Kim, 2008: "How well do coupled models simulate today's climate?" *Bulletin of the American Meteorological Society*, vol. 89, no. 3, pp. 303–311.

Renne, P. R., and A. R. Basu, 1991: "Rapid eruption of the Siberian Traps flood basalts at the Permo-Triassic boundary," *Science*, vol. 253, no. 5016, pp. 176–179.

Renne, P. R., Z. Zichao, M. A. Richards, M. T. Black, and A. R. Basu, 1995: "Synchrony and causal relations between Permian-Triassic boundary crises and Siberian flood volcanism," *Science*, vol. 269, no. 5229, pp. 1413–1416.

Revelle, R., and H. E. Suess, 1957: "Carbon dioxide exchange between atmosphere and ocean and the question of an increase of atmospheric CO_2 during the past decades," *Tellus*, vol. 9, no. 1, pp. 18–27.

Rignot, E. J., 1998: "Fast recession of a West Antarctic glacier," *Science*, vol. 281, pp. 549–551.

Rignot, E., G. Casassa, P. Gogineni, W. Krabill, A. Rivera, and R. Thomas, 2004: "Accelerated ice discharge from the Antarctic Peninsula following the collapse of Larsen B Ice Shelf," *Geophysical Research Letters*, vol. 31, L18401, doi:10.1029/2004GL020697.

Rind, D., R. Healy, C. Parkinson, and D. Martinson, 1995: "The role of sea ice in $2 \times CO_2$ climate model sensitivity. Part I: The total influence of sea ice thickness and extent," *Journal of Climate*, vol. 8, no. 3, pp. 449–463.

Ripple, W. J., and R. L. Beschta, 2007a: "Hardwood tree decline following large carnivore loss on the Great Plains, USA," *Frontiers in Ecology and the Environment*, vol. 5, no. 5, pp. 241–246.

———, 2007b: "Restoring Yellowstone's aspen with wolves," *Biological Conservation*, vol. 138, no. 4, pp. 514–519.

Robock, A., 2008a: "20 reasons why geoengineering may be a bad idea," *Bulletin of the Atomic Scientists*, vol. 64, no. 2, pp. 14–18 and 59.

———, 2008b: "Whither geoengineering," *Science*, vol. 320, no. 5880, pp. 1166–1167.

Rochelle, G. T., 2009: "Amine scrubbing for CO_2 capture," *Science*, vol. 325, no. 5948, pp. 1652–1654.

Roe, G. H., and M. B. Baker, 2007: "Why is climate sensitivity so unpredictable?" *Science*, vol. 318, no. 5850, pp. 629–632.

Root, T. L., J. T. Price, K. R. Hall, S. H. Schneider, C. Rosenzweig, and J. A. Pounds, 2003: "Fingerprints of global warming on wild animals and plants," *Nature*, vol. 421, no. 6918, pp. 57–60.

Rooth, C., 1982: "Hydrology and ocean circulation," *Progress in Oceanography*, vol. 11, no. 2, pp. 131–149.

Rosenlof, K. H., and G. C. Reid, 2008: "Trends in the temperature and water vapor content of the tropical lower stratosphere: Sea surface connection," *Journal of Geophysical Research*, vol. 113, D06107, doi:10.1029/2007JD009109, 15 pp.

Rosenthal, E., 2008: "Europe turns back to coal, raising climate fears," *New York Times*, April 23, 2008.

Rothrock, D. A., Y. Yu, and G. A. Maykut, 1999: "Thinning of the Arctic sea-ice cover," *Geophysical Research Letters*, vol. 26, no. 23, pp. 3469–3472.

Rott, H., W. Rack, P. Skvarca, and H. De Angelis, 2002: "Northern Larsen Ice Shelf, Antarctica: Further retreat after collapse," *Annals of Glaciology*, vol. 34, no. 1, pp. 277–282.

Rott, H., P. Skvarca, and T. Nagler, 1996: "Rapid collapse of Northern Larsen Ice Shelf, Antarctica," *Science*, vol. 271, no. 5250, pp. 788–792.

Rowlands, I. H., 1993: "The fourth meeting of the parties to the Montreal Protocol: Report and reflection," *Environment*, vol. 35, no. 6, pp. 25–34.

Royal Society, 2005: *Ocean Acidification Due to Increasing Atmospheric Carbon Dioxide.* Cardiff: Clyvedon Press, 60 pp. (available at http://www.royalsoc.ac.uk).

Ruddiman, W. F., 2005: *Plows, Plagues, and Petroleum: How Humans Took Control of Climate.* Princeton, NJ: Princeton University Press, 202 pp.

Running, S. W., 2008: "Ecosystem disturbance, carbon, and climate," *Science*, vol. 321, no. 5889, pp. 652–653.

Ryan, B. F., and W. D. King, 1997: "A critical review of the Australian experience in cloud seeding," *Bulletin of the American Meteorological Society*, vol. 78, no. 2, pp. 239–254.

Sabine, C. L., R. A. Feely, N. Gruber, R. M. Key, K. Lee, J. L. Bullister, R. Wanninkhof, C. S. Wong, D. W. R. Wallace, B. Tilbrook, F. J. Millero, T.-H. Peng, A. Kozyr, T. Ono, and A. F. Rios, 2004: "The oceanic sink for anthropogenic CO_2," *Science*, vol. 305, no. 5682, pp. 367–371.

Samuelson, R. J., 2007: "Greenhouse simplicities," *Newsweek*, August 20, 2007, p. 47.

Santee, M. L., 2007: "The chlorine threat to Earth's ozone shield." In: King, M. D., C. L. Parkinson, K. C. Partington, and R. G. Williams, editors, *Our Changing Planet: The View from Space.* Cambridge: Cambridge University Press, pp. 82–87.

Scambos, T. A., J. A. Bohlander, C. A. Shuman, and P. Skvarca, 2004: "Glacier acceleration and thinning after ice shelf collapse in the Larsen B embayment, Antarctica," *Geophysical Research Letters*, vol. 31, L18402, doi:10.1029/2004GL020670.

Scambos, T., C. Hulbe, M. Fahnestock, and J. Bohlander, 2000: "The link between climate warming and break-up of ice shelves in the Antarctic Peninsula," *Journal of Glaciology*, vol. 46, pp. 516–530.

Schaefer, V. J., 1946: "The production of ice crystals in a cloud of supercooled water droplets," *Science*, vol. 104, no. 2707, pp. 457–459.

———, 1949: "The formation of ice crystals in the laboratory and the atmosphere," *Chemical Reviews*, vol. 44, no. 2, pp. 291–320.

Scheffel, R. L., and S. J. Wernert, editors, 1980: *Natural Wonders of the World.* Pleasantville, NY: Reader's Digest Association, 463 pp.

Scher, H. D., and E. E. Martin, 2006: "Timing and climatic consequences of the opening of Drake Passage," *Science*, vol. 312, no. 5772, pp. 428–430.

Schneider, S. H., 1989: *Global Warming: Are We Entering the Greenhouse Century?* San Francisco: Sierra Club Books, 317 pp.

———, 1997: *Laboratory Earth: The Planetary Gamble We Can't Afford to Lose.* New York: Basic Books, 174 pp.

———, 2001: "Earth systems engineering and management," *Nature*, vol. 409, no. 6818, pp. 417–421.

Schneider, S. H., with L. E. Mesirow, 1976: *The Genesis Strategy: Climate and Global Survival.* New York: Plenum Press, 419 pp.

Schrag, D. P., 2009: "Storage of carbon dioxide in offshore sediments," *Science*, vol. 325, no. 5948, pp. 1658–1659.

Segrè, G., 2002: *A Matter of Degrees: What Temperature Reveals about the Past and Future of Our Species, Planet, and Universe.* New York: Viking, 300 pp.

Seifritz, W., 1989: "Mirrors to halt global warming?" *Nature,* vol. 340, no. 6235, p. 603.

Seitter, K. L., 2007: "AMS meetings go more green," *Bulletin of the American Meteorological Society,* vol. 88, no. 11, pp. 1811–1812.

Sellers, W. D., 1969: "A global climatic model based on the energy balance of the Earth-atmosphere system," *Journal of Applied Meteorology,* vol. 8, no. 3, pp. 392–400.

Senior, C. A., and J. F. B. Mitchell, 1993: "Carbon dioxide and climate: The impact of cloud parameterization," *Journal of Climate,* vol. 6, no. 3, pp. 393–418.

Sepkoski, J. J., Jr., 1993: "Foundations: Life in the oceans." In: Gould, S. J., editor, *The Book of Life.* New York: W. W. Norton, pp. 36–63.

Serreze, M. C., and J. A. Francis, 2006: "The Arctic on the fast track of change," *Weather,* vol. 61, no. 3, pp. 65–69.

Service, R. F., 2009: "Hydrogen cars: Fad or the future?" *Science,* vol. 324, no. 5932, pp. 1257–1259.

Shabtaie, S., and C. R. Bentley, 1987: "West Antarctic ice streams draining into the Ross Ice Shelf: Configuration and mass balance," *Journal of Geophysical Research,* vol. 92, no. B2, pp. 1311–1336.

Sheets, R. C., 2003: "Hurricane surveillance by specially instrumented aircraft." In: Simpson, R., editor, *Hurricane! Coping with Disaster.* Washington, DC: American Geophysical Union, pp. 63–101.

Shen, B., L. Dong, S. Xiao, and M. Kowalewski, 2008: "The Avalon explosion: Evolution of Ediacara morphospace," *Science,* vol. 319, no. 5859, pp. 81–84.

Shepherd, A., and D. Wingham, 2007: "Recent sea-level contributions of the Antarctic and Greenland ice sheets," *Science,* vol. 315, no. 5818, pp. 1529–1532.

Shepherd, A., D. J. Wingham, J. A. D. Mansley, and H. F. J. Corr, 2001: "Inland thinning of Pine Island Glacier, West Antarctica," *Science,* vol. 291, no. 5505, pp. 862–864.

Shindell, D., D. Rind, N. Balachandran, J. Lean, and P. Lonergan, 1999: "Solar cycle variability, ozone, and climate," *Science,* vol. 284, no. 5412, pp. 305–308.

Shindell, D. T., G. A. Schmidt, M. E. Mann, D. Rind, and A. Waple, 2001: "Solar forcing of regional climate change during the Maunder Minimum," *Science,* vol. 294, no. 5549, pp. 2149–2152.

Silverman, B. A., 2001: "A critical assessment of glaciogenic seeding of convective clouds for rainfall enhancement," *Bulletin of the American Meteorological Society,* vol. 82, no. 5, pp. 903–923.

Simpson, J., G. W. Brier, and R. H. Simpson, 1967: "Stormfury cumulus seeding experiment 1965: Statistical analysis and main results," *Journal of the Atmospheric Sciences,* vol. 24, no. 5, pp. 508–521.

Simpson, R. H., and J. S. Malkus, 1964: "Experiments in hurricane modification," *Scientific American,* vol. 211, no. 6, pp. 27–37.

Smith, E., 1999: "Atlantic and east coast hurricanes 1900–98: A frequency and intensity study for the twenty-first century," *Bulletin of the American Meteorological Society*, vol. 80, no. 12, pp. 2717–2720.

Socolow, R., R. Hotinski, J. B. Greenblatt, and S. Pacala, 2004: "Solving the climate problem: Technologies available to curb CO_2 emissions," *Environment*, vol. 46, no. 10, pp. 8–19.

Soden, B. J., R. T. Wetherald, G. L. Stenchikov, and A. Robock, 2002: "Global cooling after the eruption of Mount Pinatubo: A test of climate feedback by water vapor," *Science*, vol. 296, no. 5568, pp. 727–730.

Solomon, S., 1990: "Progress towards a quantitative understanding of Antarctic ozone depletion," *Nature*, vol. 347, no. 6291, pp. 347–354.

Solomon, S., D. Qin, M. Manning, R. B. Alley, T. Berntsen, N. L. Bindoff, Z. Chen, A. Chidthaisong, J. M. Gregory, G. C. Hegerl, M. Heimann, B. Hewitson, B. J. Hoskins, F. Joos, J. Jouzel, V. Kattsov, U. Lohmann, T. Matsuno, M. Molina, N. Nicholls, J. Overpeck, G. Raga, V. Ramaswamy, J. Ren, M. Rusticucci, R. Somerville, T. F. Stocker, R. J. Stouffer, P. Whetton, R. A. Wood, and D. Wratt, 2007a: "Technical Summary." In: Solomon, S., D. Qin, M. Manning, Z. Chen, M. Marquis, K. B. Averyt, M. Tignor, and H. L. Miller, editors, *Climate Change 2007: The Physical Science Basis. Contribution of Working Group I to the Fourth Assessment Report of the Intergovernmental Panel on Climate Change.* Cambridge: Cambridge University Press, pp. 19–91.

Solomon, S., D. Qin, M. Manning, Z. Chen, M. Marquis, K. B. Averyt, M. Tignor, and H. L. Miller, editors, 2007b: *Climate Change 2007: The Physical Science Basis. Contribution of Working Group I to the Fourth Assessment Report of the Intergovernmental Panel on Climate Change.* Cambridge: Cambridge University Press, 996 pp.

Soon, W., S. Baliunas, C. Idso, S. Idso, and D. R. Legates, 2003: "Reconstructing climatic and environmental changes of the past 1000 years: A reappraisal," *Energy and Environment*, vol. 14, nos. 2 and 3, pp. 233–296.

Spencer, R. W., 2008: *Climate Confusion: How Global Warming Hysteria Leads to Bad Science, Pandering Politicians and Misguided Policies That Hurt the Poor.* New York: Encounter Books, 191 pp.

Spencer, R. W., W. D. Braswell, J. R. Christy, and J. Hnilo, 2007: "Cloud and radiation budget changes associated with tropical intraseasonal oscillations," *Geophysical Research Letters*, vol. 34, L15707, doi:10.1029/2007GL029698, 5 pp.

Spoor, F., M. G. Leakey, P. N. Gathogo, F. H. Brown, S. C. Antón, I. McDougall, C. Kiarie, F. K. Manthi, and L. N. Leakey, 2007: "Implications of new early *Homo* fossils from Ileret, east of Lake Turkana, Kenya," *Nature*, vol. 448, no. 7154, pp. 688–691.

Stainforth, D. A., T. Alna, C. Christensen, M. Collins, N. Faull, D. J. Frame, J. A. Kettleborough, S. Knight, A. Martin, J. M. Murphy, C. Piani, D. Sexton, L. A. Smith, R. A. Spicer, A. J. Thorpe, and M. R. Allen, 2005: "Uncertainty in predictions of the climate response to rising levels of greenhouse gases," *Nature*, vol. 433, no. 7024, pp. 403–406.

Standage, T., 1998: *The Victorian Internet: The Remarkable Story of the Telegraph and the Nineteenth Century's On-line Pioneers.* New York: Berkley Books, 228 pp.

Steffensen, J. P., K. K. Andersen, M. Bigler, H. B. Clausen, D. Dahl-Jensen, H. Fischer, K. Goto-Azuma, M. Hansson, S. J. Johnsen, J. Jouzel, V. Masson-Delmotte, T. Popp, S. O. Rasmussen, R. Röthlisberger, U. Ruth, B. Stauffer, M.-L. Siggaard-Andersen, Á. E. Sveinbjörnsdóttir, A. Svensson, and J. W. C. White, 2008: "High-resolution Greenland ice core data show abrupt climate change happens in few years," *Science*, vol. 321, no. 5889, pp. 680–684.

Stern, D. I., 2005: "Global sulfur emissions from 1850 to 2000," *Chemosphere*, vol. 58, no. 2, pp. 163–175.

Stern, N., 2007: *The Economics of Climate Change: The Stern Review.* Cambridge: Cambridge University Press, 692 pp.

Stirling, I., and C. L. Parkinson, 2006: "Possible effects of climate warming on selected populations of polar bears (*Ursus maritimus*) in the Canadian Arctic," *Arctic*, vol. 59, no. 3, pp. 261–275.

Stocker, T. F., and O. Marchal, 2000: "Abrupt climate change in the computer: Is it real?" *Proceedings of the National Academy of Sciences of the United States of America*, vol. 97, no. 4, pp. 1362–1365.

Stokstad, E., 2006: "Meeting briefs: Society of Vertebrate Paleontology, 18–21 October, Ottawa, Canada," *Science*, vol. 314, no. 5801, pp. 920–926.

Stolarski, R. S., 1988: "The Antarctic ozone hole," *Scientific American*, vol. 258, no. 1, pp. 30–36.

Stommel, H., 1961: "Thermohaline convection with two stable regimes of flow," *Tellus*, vol. 13, no. 2, pp. 224–230.

Stone, R., 2008: "Three Gorges Dam: Into the unknown," *Science*, vol. 321, no. 5889, pp. 628–632.

Stothers, R. B., 1984: "The great Tambora eruption in 1815 and its aftermath," *Science*, vol. 224, no. 4654, pp. 1191–1198.

Stott, L., A. Timmermann, and R. Thunell, 2007: "Southern Hemisphere and deep-sea warming led deglacial atmospheric CO_2 rise and tropical warming," *Science*, vol. 318, no. 5849, pp. 435–438.

Stroeve, J., M. M. Holland, W. Meier, T. Scambos, and M. Serreze, 2007: "Arctic sea ice decline: Faster than forecast," *Geophysical Research Letters*, vol. 34, L09501, doi:10.1029/2007GL029703, 5 pp.

Tarasov, L., and W. R. Peltier, 2005: "Arctic freshwater forcing of the Younger Dryas cold reversal," *Nature*, vol. 435, no. 7042, pp. 662–665.

Teller, J. T., D. W. Leverington, and J. D. Mann, 2002: "Freshwater outbursts to the oceans from glacial Lake Agassiz and their role in climate change during the last deglaciation," *Quaternary Science Reviews*, vol. 21, no. 8, pp. 879–887.

Textor, C., H.-F. Graf, C. Timmreck, and A. Robock, 2004: "Emissions from volcanoes." In: Granier, C., P. Artaxo, and C. E. Reeves, editors, *Emissions of Atmospheric Trace Compounds*. Dordrecht: Kluwer Academic Publishers, pp. 269–303.

Thomas, C. D., A. Cameron, R. E. Green, M. Bakkenes, L. J. Beaumont, Y. C. Collingham, B. F. N. Erasmus, M. Ferreira de Siqueira, A. Grainger, L. Hannah, L. Hughes, B. Huntley, A. S. van Jaarsveld, G. F. Midgley, L. Miles, M. A. Ortega-Huerta, A. T. Peterson, O. L. Phillips, and S. E. Williams, 2004: "Extinction risk from climate change," *Nature*, vol. 427, no. 6970, pp. 145–148.

Thompson, L. G., 2000: "Ice core evidence for climate change in the tropics: Implications for our future," *Quaternary Science Reviews*, vol. 19, no. 1, pp. 19–35.

Thompson, L. G., E. Mosley-Thompson, H. Brecher, M. Davis, B. Leon, D. Les, P.-N. Lin, T. Mashiotta, and K. Mountain, 2006: "Abrupt tropical climate change: Past and present," *Proceedings of the National Academy of Sciences of the United States of America*, vol. 103, no. 28, pp. 10536–10543.

Thompson, L. G., E. Mosley-Thompson, M. E. Davis, K. A. Henderson, H. H. Brecher, V. S. Zagorodnov, T. A. Mashiotta, P.-N. Lin, V. N. Mikhalenko, D. R. Hardy, and J. Beer, 2002: "Kilimanjaro ice core records: Evidence of Holocene climate change in tropical Africa," *Science*, vol. 298, no. 5593, pp. 589–593.

Thompson, L. G., E. Mosley-Thompson, M. E. Davis, P.-N. Lin, K. Henderson, and T. A. Mashiotta, 2003: "Tropical glacier and ice core evidence of climate change on annual to millennial time scales," *Climatic Change*, vol. 59, nos. 1–2, pp. 137–155.

Thompson, L. G., E. Mosley-Thompson, and K. A. Henderson, 2000: "Ice-core palaeoclimate records in tropical South America since the last glacial maximum," *Journal of Quaternary Science*, vol. 15, no. 4, pp. 377–394.

Thompson, L. G., T. Yao, M. E. Davis, K. A. Henderson, E. Mosley-Thompson, P.-N. Lin, J. Beer, H.-A. Synal, J. Cole-Dai, and J. F. Bolzan, 1997: "Tropical climate instability: The last glacial cycle from a Qinghai-Tibetan ice core," *Science*, vol. 276, no. 5320, pp. 1821–1825.

Tilmes, S., R. Müller, and R. Salawitch, 2008: "The sensitivity of polar ozone depletion to proposed geoengineering schemes," *Science*, vol. 320, no. 5880, pp. 1201–1204.

Trenberth, K. E., editor, 1992: *Climate System Modeling*. Cambridge: Cambridge University Press, 788 pp.

Trenberth, K. E., P. D. Jones, P. Ambenje, R. Bojariu, D. Easterling, A. Klein Tank, D. Parker, F. Rahimzadeh, J. A. Renwick, M. Rusticucci, B. Soden, and P. Zhai, 2007: "Observations: Surface and atmospheric climate change." In: Solomon, S., D. Qin, M. Manning, Z. Chen, M. Marquis, K. B. Averyt, M. Tignor, and H. L. Miller, editors, *Climate Change 2007: The Physical Science Basis. Contribution of Working Group I to the Fourth Assessment Report of the Intergovernmental Panel on Climate Change*. Cambridge: Cambridge University Press, pp. 235-336.

Tudge, C., 1997: *The Time before History: 5 Million Years of Human Impact*. New York: Touchstone, 366 pp.

Turner, J., J. C. Comiso, G. J. Marshall, T. A. Lachlan-Cope, T. Bracegirdle, T. Maksym, M. P. Meredith, Z. Wang, and A. Orr, 2009: "Non-annular atmospheric circulation change induced by stratospheric ozone depletion and its role in the recent increase of Antarctic sea ice extent," *Geophysical Research Letters*, vol. 36, L08502, doi:10.1029/2009GL037524, 5 pp.

Tyndall, J., 1861: "On the absorption and radiation of heat by gases and vapours, and on the physical connexion of radiation, absorption, and conduction," *The London, Edinburgh and Dublin Philosophical Magazine and Journal of Science*, series 4, vol. 22, no. 146, pp. 169–194 and 273–285.

Vaughan, D. G., and C. S. M. Doake, 1996: "Recent atmospheric warming and retreat of ice shelves on the Antarctic Peninsula," *Nature*, vol. 379, no. 6563, pp. 328–331.

Victor, D. G., 2008: "On the regulation of geoengineering," *Oxford Review of Economic Policy*, vol. 24, no. 2, pp. 322–336.

Vimeux, F., K. M. Cuffey, and J. Jouzel, 2002: "New insights into Southern Hemisphere temperature changes from Vostok ice cores using deuterium excess correction," *Earth and Planetary Science Letters*, vol. 203, no. 3–4, pp. 829–843.

Vincent, W. F., J. A. E. Gibson, and M. O. Jeffries, 2001: "Ice-shelf collapse, climate change, and habitat loss in the Canadian high Arctic," *Polar Record*, vol. 37, no. 201, pp. 133–142.

Vonnegut, B., 1947: "The nucleation of ice formation by silver iodide," *Journal of Applied Physics*, vol. 18, no. 7, pp. 593–595.

Walker, G., 2006: "The tipping point of the iceberg," *Nature*, vol. 441, no. 7095, pp. 802–805.

Walsh, J. E., 1991: "The Arctic as a bellwether," *Nature*, vol. 352, no. 6330, pp. 19–20.

Walsh, J. E., O. Anisimov, J. O. M. Hagen, T. Jakobsson, J. Oerlemans, T. D. Prowse, V. Romanovsky, N. Savelieva, M. Serreze, A. Shiklomanov, I. Shiklomanov, and S. Solomon, 2005: "Cryosphere and hydrology." In: ACIA, *Arctic Climate Impact Assessment*. Cambridge: Cambridge University Press, pp. 183–242.

Walsh, J. E., and W. L. Chapman, 2001: "20th-century sea-ice variations from observational data," *Annals of Glaciology*, vol. 33, pp. 444–448.

Walsh, J. E., V. M. Kattsov, W. L. Chapman, V. Govorkova, and T. Pavlova, 2002: "Comparison of Arctic climate simulations by uncoupled and coupled global models," *Journal of Climate*, vol. 15, no. 12, pp. 1429–1446.

Walter, K. M., M. E. Edwards, G. Grosse, S. A. Zimov, and F. S. Chapin III, 2007a: "Thermokarst lakes as a source of atmospheric CH_4 during the last deglaciation," *Science*, vol. 318, no. 5850, pp. 633–636.

Walter, K. M., L. C. Smith, and F. S. Chapin III, 2007b: "Methane bubbling from northern lakes: Present and future contributions to the global methane budget," *Philosophical Transactions of the Royal Society A*, vol. 365, pp. 1657–1676.

Walter, K. M., S. A. Zimov, J. P. Chanton, D. Verbyla, and F. S. Chapin III, 2006: "Methane bubbling from Siberian thaw lakes as a positive feedback to climate warming," *Nature*, vol. 443, no. 7107, pp. 71–74.

Wan, Z., 2007: "Temperature of the land surface." In: King, M. D., C. L. Parkinson, K. C. Partington, and R. G. Williams, editors, *Our Changing Planet: The View from Space*. Cambridge: Cambridge University Press, pp. 136–141.

Ward, K., editor, 1989: *Great Disasters: Dramatic True Stories of Nature's Awesome Powers*. Pleasantville, NY: Reader's Digest Association, 320 pp.

Washington, W. M., and G. A. Meehl, 1989: "Climate sensitivity due to increased CO_2: Experiments with a coupled atmosphere and ocean general circulation model," *Climate Dynamics*, vol. 4, no. 1, pp. 1–38.

Washington, W. M., and C. L. Parkinson, 2005: *An Introduction to Three-Dimensional Climate Modeling*, second edition. Sausalito, CA: University Science Books, 354 pp.

Webster, P. J., G. J. Holland, J. A. Curry, and H.-R. Chang, 2005: "Changes in tropical cyclone number, duration, and intensity in a warming environment," *Science*, vol. 309, no. 5742, pp. 1844–1846.

Wentz, F. J., L. Ricciardulli, K. Hilburn, and C. Mears, 2007: "How much more rain will global warming bring?" *Science*, vol. 317, no. 5835, pp. 233–235.

Wexler, H., 1958: "Modifying weather on a large scale," *Science*, vol. 128, no. 3331, pp. 1059–1063.

Whaley, J., 2007: "The *Azolla* story: Climate change and Arctic hydrocarbons," *GEO ExPro*, September 2007, pp. 66–72.

Wielicki, B. A., T. Wong, R. P. Allan, A. Slingo, J. T. Kiehl, B. J. Soden, C. T. Gordon, A. J. Miller, S.-K. Yang, D. A. Randall, F. Robertson, J. Susskind, and H. Jacobowitz, 2002: "Evidence for large decadal variability in the tropical mean radiative energy budget," *Science*, vol. 295, no. 5556, pp. 841–844.

Wigley, T. M. L., 2006: "A combined mitigation/geoengineering approach to climate stabilization," *Science*, vol. 314, no. 5798, pp. 452–454.

Wilbanks, T., P. Romero Lankao, M. Bao, F. Berkhout, S. Cairncross, J.-P. Ceron, M. Kapshe, R. Muir-Wood, and R. Zapata-Marti, 2007: "Industry, settlement and society." In: Parry, M., O. Canziani, J. Palutikof, P. van der Linden, and C. Hanson, editors, *Climate Change 2007: Impacts, Adaptation and Vulnerability. Contribution of Working Group II to the Fourth Assessment Report of the Intergovernmental Panel on Climate Change*. Cambridge: Cambridge University Press, pp. 357–390.

Wilcove, D. S., 2008: *No Way Home: The Decline of the World's Great Animal Migrations*. Washington, DC: Island Press, 245 pp.

Wilford, J. N., 2000a: "Ages-old icecap at North Pole is now liquid, scientists find," *New York Times*, August 19, 2000, p. A1.

———, 2000b: "Open water at Pole is not surprising, experts say," *New York Times*, August 29, 2000, p. F3.

Williams, R. S., Jr., and J. G. Ferrigno, editors, 1993: *Satellite Image Atlas of Glaciers of the World: Glaciers of Europe*. Washington, DC: U.S. Geological Survey, 164 pp.

———, editors, 1998: *Satellite Image Atlas of Glaciers of the World: Glaciers of South America.* Washington, DC: U.S. Geological Survey, 206 pp.

———, editors, 2002: *Satellite Image Atlas of Glaciers of the World: Glaciers of North America.* Washington, DC: U.S. Geological Survey, 405 pp.

Wilson, E. O., 1992: *The Diversity of Life.* Cambridge, MA: The Belknap Press of Harvard University Press, 424 pp.

Winckler, G., R. F. Anderson, M. Q. Fleisher, D. McGee, and N. Mahowald, 2008: "Covariant glacial-interglacial dust fluxes in the equatorial Pacific and Antarctica," *Science,* vol. 320, no. 5872, pp. 93–96.

Wood, B., and M. Collard, 1999: "The human genus," *Science,* vol. 284, no. 5411, pp. 65–71.

Woodley, W. L., and D. Rosenfeld, 2004: "The development and testing of a new method to evaluate the operational cloud-seeding programs in Texas," *Journal of Applied Meteorology,* vol. 43, no. 2, pp. 249–263.

Wray, G. A., J. S. Levinton, and L. H. Shapiro, 1996: "Molecular evidence for deep Precambrian divergences among metazoan phyla," *Science,* vol. 274, no. 5287, pp. 568–573.

Xie, S.-P., 2004: "Satellite observations of cool ocean-atmosphere interactions," *Bulletin of the American Meteorological Society,* vol. 85, no. 2, pp. 195–208.

Xu, D.-Y., S.-L. Ma, Z.-F. Chai, X.-Y. Mao, Y.-Y. Sun, Q.-W. Zhang, and Z.-Z. Yang, 1985: "Abundance variation of iridium and trace elements at the Permian/Triassic boundary at Shangsi in China," *Nature,* vol. 314, no. 6007, pp. 154–156.

Yu, L., and R. A. Weller, 2007: "Objectively analyzed air-sea heat fluxes for the global ice-free oceans (1981–2005)," *Bulletin of the American Meteorological Society,* vol. 88, no. 4, pp. 527–539.

Yu, Y., G. A. Maykut, and D. A. Rothrock, 2004: "Changes in the thickness distribution of Arctic sea ice between 1958–1970 and 1993–1997," *Journal of Geophysical Research,* vol. 109, C08004, doi:10.1029/2003JC001982, 13 pp.

Zhang, X., F. W. Zwiers, G. C. Hegerl, F. H. Lambert, N. P. Gillett, S. Solomon, P. A. Stott, and T. Nozawa, 2007: "Detection of human influence on twentieth-century precipitation trends," *Nature,* vol. 448, no. 7152, pp. 461–465.

Zimov, S. A., E. A. G. Schuur, and F. S. Chapin III, 2006: "Permafrost and the global carbon budget," *Science,* vol. 312, no. 5780, pp. 1612–1613.

Zwally, H. J., 1989: "Growth of Greenland ice sheet: Interpretation," *Science,* vol. 246, no. 4937, pp. 1589–1591.

Zwally, H. J., W. Abdalati, T. Herring, K. Larson, J. Saba, and K. Steffen, 2002: "Surface melt-induced acceleration of Greenland ice-sheet flow," *Science,* vol. 297, no. 5579, pp. 218–222.

Zwally, H. J., A. C. Brenner, J. A. Major, R. A. Bindschadler, and J. G. Marsh, 1989: "Growth of Greenland ice sheet: Measurement," *Science,* vol. 246, no. 4937, pp. 1587–1589.

Zwally, H. J., J. C. Comiso, C. L. Parkinson, W. J. Campbell, F. D. Carsey, and P. Gloersen, 1983a: *Antarctic Sea Ice, 1973-1976: Satellite Passive-Microwave Observations,* NASA SP-459. Washington, DC: National Aeronautics and Space Administration, 206 pp.

Zwally, H. J., M. B. Giovinetto, J. Li, H. G. Cornejo, M. A. Beckley, A. C. Brenner, J. L. Saba, and D. Yi, 2005: "Mass changes of the Greenland and Antarctic ice sheets and shelves and contributions to sea-level rise: 1992–2002," *Journal of Glaciology*, vol. 51, no. 175, pp. 509–527.

Zwally, H. J., C. L. Parkinson, and J. C. Comiso, 1983b: "Variability of Antarctic sea ice and changes in carbon dioxide," *Science*, vol. 220, no. 4601, pp. 1005–1012.

Index

About the Author

Claire L. Parkinson is a climatologist with a specialty in polar sea ice and an interest in the history and philosophy of science. She received a bachelor's degree in mathematics from Wellesley College and master's and PhD degrees in geography/climatology from the Ohio State University. For most of her career, she has been employed at NASA's Goddard Space Flight Center in Greenbelt, Maryland, where her research has centered largely on polar sea ice and its connections to the broader climate system and where she is also the project scientist for the Earth-observing Aqua satellite mission. She has written books on satellite imagery and the history of science, has coauthored with Warren Washington a book on climate modeling, and has coauthored with several colleagues books on Arctic and Antarctic sea ice.

Dr. Parkinson has received a NASA Exceptional Achievement Medal for her efforts in public outreach, a NASA Outstanding Leadership Medal for her role as the Aqua Project Scientist, and the Richard P. Goldthwait Medal from the Byrd Polar Research Center for her sea ice research. She is a senior fellow at NASA Goddard Space Flight Center, a member of the National Academy of Engineering, a fellow of the American Meteorological Society and of Phi Beta Kappa, and a member of the Council of the American Association for the Advancement of Science.